a shorter model theory

wilfrid
hodges

Professor of Mathematics
School of Mathematical Sciences
Queen Mary and Westfield College
University of London

a shorter model theory

 **CAMBRIDGE**
UNIVERSITY PRESS

PUBLISHED BY THE PRESS SYNDICATE OF THE UNIVERSITY OF CAMBRIDGE
The Pitt Building, Trumpington Street, Cambridge, United Kingdom

CAMBRIDGE UNIVERSITY PRESS
The Edinburgh Building, Cambridge CB2 2RU, UK
40 West 20th Street, New York, NY 10011–4211, USA
477 Williamstown Road, Port Melbourne VIC 3207, Australia
Ruiz de Alarcón 13, 28014 Madrid, Spain
Dock House, The Waterfront, Cape Town

First published 1997
Reprinted 1999, 2002

Typeset in 10/12pt Times

A catalogue record for this book is available from the British Library

Library of Congress Cataloguing in Publication data
Hodges, Wilfrid.
    A shorter model theory / Wilfrid Hodges.
        p.   cm.
    Includes index.
    ISBN 0-521-58713-1
    1. Model theory.   I. Title.
    QA9.7.H65   1997
    511'.8–dc20   96-36603 CIP

ISBN 0 521 58713 1 paperback

Transferred to digital printing 2003

CONTENTS

INTRODUCTION

Shorter than what?

Shorter than another book I once wrote, called *Model theory* and published in 1993 by Cambridge University Press. It was large (772 pages), so Cambridge University Press asked me for a shorter version to be published in paperback as a textbook. This is that shorter version.

And what is model theory?

Model theory is about the classification of mathematical structures, maps and sets by means of logical formulas. One can classify structures according to what logical sentences are true in them; in fact the term 'model' comes from the usage 'structure A is a model of sentence ϕ', meaning that ϕ is true in A. Model theorists look for ways of constructing models of given sentences; and so a large part of model theory is directly about constructions and only indirectly about classification. This book describes some very pretty constructions, including ultraproducts, Ehrenfeucht–Mostowski models and existentially closed models.

In 1973 C. C. Chang and Jerry Keisler characterised model theory as

universal algebra plus logic.

They meant the universal algebra to stand for structures and the logic to stand for logical formulas. This is neat, but it might suggest that model theorists and universal algebraists have closely related interests, which is debatable. Also it leaves out the fact that model theorists study the sets definable in a single structure by a logical formula. In this respect model theorists are much closer to algebraic geometers, who study the sets of points definable by equations over a field. A more up-to-date slogan might be that model theory is

algebraic geometry minus fields.

In fact some of the most striking successes of model theory have been theorems about the existence of solutions of equations over fields. Examples are the work of Ax, Kochen and Ershov on Artin's conjecture in 1965, and the proof of the Mordell–Lang conjecture for function fields by Hrushovski in 1993. These examples both lie beyond the scope of this book.

Which of the items in the longer book did you most regret leaving out here?

Probably the Hrushovski construction for extracting a group from a Zil'ber configuration in a strongly minimal set. The construction is tough and exciting, but still reasonably elementary. It is also a foretaste of important things in geometric stability theory. This gives me an excuse to correct the most embarrassing slip in the longer book (pointed out to me by Byunghan Kim): Lemma 4.7.11(c) is clearly wrong, but there is no problem in doing without it.

The three other main omissions are direct products (together with Horn sentences and the Feferman-Vaught theorem); constructions of many non-iso-morphic models; and an appendix surveying the model theory of modules, abelian groups, groups, fields and linear orderings.

Did you add anything?

Yes. There is a new chapter proving Morley's theorem and developing the machinery needed for the proof. It replaces a chapter which gave a good deal of information about categoricity (for example a finitely axiomatised uncount-ably categorical complete theory), but few proofs. This new chapter was written with an eye to introductory graduate courses in model theory.

I also added to each chapter a set of suggested readings. These range from philosophical or historical background to examples of work at the present frontiers of the subject. They replace the forty-odd pages of references in the longer book; readers who want detailed information on the sources of results can find it there but not here.

Were there things in the longer book that you were particularly glad to be able to improve here?

Mercifully no; there are few changes of substance. But I welcome the chance to thank—in addition to the people I thanked in the longer volume—several colleagues who generously sent me corrections to the longer book, notably Peter Cameron, Matthias Clasen, Solomon Feferman, Ed Hamada, Ingo Kraus, Byunghan Kim, Don Monk, Rahim Moosa, Philipp Rothmaler, Gabriel Sabbagh, Malcolm Schonfield and Ann-Marie Westgate. Lists of corrections for the shorter and longer versions are at:

`ftp.maths.qmw.ac.uk/pub/preprints/hodges/smtcorrig.tex`

and

`ftp.maths.qmul.ac.uk/pub/preprints/hodges/mtcorrig.tex`

respectively.

NOTE ON NOTATION

I assume Zermelo–Fraenkel set theory, ZFC. In particular I always assume the axiom of choice (except where the axiom itself is under discussion). I never assume the continuum hypothesis, existence of uncountable inaccessibles etc., without being honest about it.

The notation $x \subseteq y$ means that x is a subset of y; $x \subset y$ means that x is a proper subset of y. I write dom(f), im(f) for the domain and image of a function f. 'Greater than' means greater than, never 'greater than or equal to'. $\mathcal{P}(x)$ is the power set of x.

Ordinals are von Neumann ordinals, i.e. the predecessors of an ordinal α are exactly the elements of α. I use symbols α, β, γ, δ, i, j etc. for ordinals; δ is usually a limit ordinal. A cardinal κ is the smallest ordinal of cardinality κ, and the infinite cardinals are listed as ω_0, ω_1 etc. I use symbols κ, λ, μ, ν for cardinals; they are not assumed to be infinite unless the context clearly requires it (though I have probably slipped on this point once or twice). Natural numbers m, n etc. are the same thing as finite cardinals.

'Countable' means of cardinality ω. An infinite cardinal λ is a **regular** cardinal if it can't be written as the sum of fewer than λ cardinals which are all smaller than λ; otherwise it is **singular**. Every infinite successor cardinal κ^+ is regular. The smallest singular cardinal is $\omega_\omega = \sum_{n<\omega} \omega_n$. The **cofinality** cf($\alpha$) of an ordinal α is the least ordinal β such that α has a cofinal subset of order-type β; it can be shown that this ordinal β is either finite or regular. If α and β are ordinals, $\alpha\beta$ is the ordinal product consisting of β copies of α laid end to end. If κ and λ are cardinals, $\kappa\lambda$ is the cardinal product. The context should always show which of these products is intended.

Sequences are well-ordered (except for indiscernible sequences in Chapter 9, and it is explicit there what is happening). I use the notation $\bar{x}$, $\bar{a}$ etc. for sequences $(x_0, x_1, \ldots)$, $(a_0, a_1, \ldots)$ etc., but loosely: the nth term of a sequence $\bar{x}$ may be x_n or $x(n)$ or something else, depending on the context, and some sequences start at x_1. Sequences of finite length are called **tuples**. The terms of a sequence are sometimes called its **items**, to avoid the ambiguity in the term 'term'. A sequence is said to be **non-repeating** if no item occurs twice or more in it. If $\bar{a}$ is a sequence $(a_0, a_1, \ldots)$ and f is a map, then $f\bar{a}$ is $(fa_0, fa_1, \ldots)$. The length of a sequence σ is written lh(σ). If σ is a sequence of length m and $n \leqslant m$, then $\sigma|n$ is the initial segment consisting of the first n terms of σ. The set of sequences of length γ whose items all come from the set X is written $^\gamma X$. Thus $^n 2$ is the set of ordered

n-tuples of 0's and 1's; $^{<\gamma}X$ is $\bigcup_{\alpha<\gamma} {}^{\alpha}X$. I write η, ζ, θ etc. for linear orderings; η^* is the ordering η run backwards.

I don't distinguish systematically between tuples and strings. If $\bar{a}$ and $\bar{b}$ are strings, $\bar{a}\,\hat{}\,\bar{b}$ is the concatenated string consisting of $\bar{a}$ followed by $\bar{b}$; but often for simplicity I write it $\bar{a}\bar{b}$. There is a clash between the usual notation of model theory and the usual notation of groups: in model theory xy is the string consisting of x followed by y, but in groups it is x times y. One has to live with this; but where there is any ambiguity I have used $x\,\hat{}\,y$ for the concatenated string and $x \cdot y$ for the group product.

Model-theoretic notation is defined as and when we need it. The most basic items appear in Chapter 1 and the first five sections of Chapter 2.

'I' means I, 'we' means we.

1

Naming of parts

Every person had in the beginning one only proper name, except the savages of Mount Atlas in Barbary, which were reported to be both nameless and dreamless.

William Camden

In this first chapter we meet the main subject-matter of model theory: structures.

Every mathematician handles structures of some kind – be they modules, groups, rings, fields, lattices, partial orderings, Banach algebras or whatever. This chapter will define basic notions like 'element', 'homomorphism', 'substructure', and the definitions are not meant to contain any surprises. The notion of a (Robinson) 'diagram' of a structure may look a little strange at first, but really it is nothing more than a generalisation of the multiplication table of a group.

Nevertheless there is something that the reader may find unsettling. Model theorists are forever talking about symbols, names and labels. A group theorist will happily write the same abelian group multiplicatively or additively, whichever is more convenient for the matter in hand. Not so the model theorist: for him or her the group with '·' is one structure and the group with '+' is a different structure. Change the name and you change the structure.

This must look like pedantry. Model theory is an offshoot of mathematical logic, and I can't deny that some distinguished logicians have been pedantic about symbols. Nevertheless there are several good reasons why model theorists take the view that they do. For the moment let me mention two.

In the first place, we shall often want to compare two structures and study the homomorphisms from one to the other. What is a homomorphism? In the particular case of groups, a homomorphism from group G to group H is a map that carries multiplication in G to multiplication in H. There is an obvious way to generalise this notion to arbitrary structures: a homomorphism from structure A to structure B is a map which carries each operation of A to *the operation with the same name in B*.

Secondly, we shall often set out to build a structure with certain properties. One of the maxims of model theory is this: *name the elements of your structure first, then decide how they should behave.* If the names are well chosen, they will serve both as a scaffolding for the construction, and as raw materials.

Aha – says the group theorist – I see you aren't really talking about *written* symbols at all. For the purposes you have described, you only need to have formal labels for some parts of your structures. It should be quite irrelevant what kinds of thing your labels are; you might even want to have uncountably many of them.

Quite right. In fact we shall follow the lead of A. I. Mal'tsev and put no restrictions at all on what can serve as a name. For example any ordinal can be a name, and any mathematical object can serve as a name of itself. The items called 'symbols' in this book need not be written down. They need not even be dreamed.

1.1 Structures

We begin with a definition of 'structure'. It would have been possible to set up the subject with a slicker definition – say by leaving out clauses (1.2) and (1.4) below. But a little extra generality at this stage will save us endless complications later on.

A **structure** A is an object with the following four ingredients.

(1.1)   A set called the **domain** of A, written $\mathrm{dom}(A)$ or $\mathrm{dom}\,A$ (some people call it the **universe** or **carrier** of A). The elements of $\mathrm{dom}(A)$ are called the **elements** of the structure A. The **cardinality** of A, in symbols $|A|$, is defined to be the cardinality $|\mathrm{dom}\,A|$ of $\mathrm{dom}(A)$.

(1.2)   A set of elements of A called **constant elements**, each of which is named by one or more **constants**. If c is a constant, we write c^A for the constant element named by c.

(1.3)   For each positive integer n, a set of n-ary relations on $\mathrm{dom}(A)$ (i.e. subsets of $(\mathrm{dom}\,A)^n$), each of which is named by one or more n-ary **relation symbols**. If R is a relation symbol, we write R^A for the relation named by R.

(1.4)   For each positive integer n, a set of n-ary operations on $\mathrm{dom}(A)$ (i.e maps from $(\mathrm{dom}\,A)^n$ to $\mathrm{dom}(A)$), each of which is named by one or more n-ary **function symbols**. If F is a function symbol, we write F^A for the function named by F.

Except where we say otherwise, any of the sets (1.1)–(1.4) may be empty. As mentioned in the chapter introduction, the constant, relation and function 'symbols' can be any mathematical objects, not necessarily written symbols;

but for peace of mind one normally assumes that, for instance, a 3-ary relation symbol doesn't also appear as a 3-ary function symbol or a 2-ary relation symbol. We shall use capital letters A, B, C, . . . for structures.

Sequences of elements of a structure are written $\bar{a}$, $\bar{b}$ etc. A **tuple in** A (or **from** A) is a finite sequence of elements of A; it is an n-**tuple** if it has length n. Usually we leave it to the context to determine the length of a sequence or tuple.

This concludes the definition of 'structure'.

Example 1: *Graphs.* A **graph** consists of a set V (the set of **vertices**) and a set E (the set of **edges**), where each edge is a set of two distinct vertices. An edge $\{v, w\}$ is said to **join** the two vertices v and w. We can picture a finite graph by putting dots for the vertices and joining two vertices v, w by a line when $\{v, w\}$ is an edge:

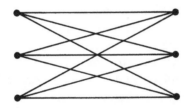

One natural way to make a graph G into a structure is as follows. The elements of G are the vertices. There is one binary relation R^G; the ordered pair (v, w) lies in R^G if and only if there is an edge joining v to w.

Example 2: *Linear orderings.* Suppose $\leqslant$ linearly orders a set X. Then we can make $(X, \leqslant)$ into a structure A as follows. The domain of A is the set X. There is one binary relation symbol R, and its interpretation R^A is the ordering $\leqslant$. (In practice we would usually write the relation symbol as $\leqslant$ rather than R.)

Example 3: *Groups.* We can think of a group as a structure G with one constant 1 naming the identity 1^G, one binary function symbol $\cdot$ naming the group product operation $\cdot^G$, and one unary function symbol $^{-1}$ naming the inverse operation $^{(-1)G}$. Another group H will have the same symbols 1, $\cdot$, $^{-1}$; then 1^H is the identity element of H, $\cdot^H$ is the product operation of H, and so on.

Example 4: *Vector spaces.* There are several ways to make a vector space into a structure, but here is the most convenient. Suppose V is a vector space over a field of scalars K. Take the domain of V to be the set of vectors of V. There is one constant element 0^V, the origin of the vector space. There is one binary operation, $+^V$, which is addition of vectors. There is a 1-ary operation $-^V$ for additive inverse; and for every scalar k there is a 1-ary operation k^V to represent multiplying a vector by k. Thus each scalar serves as a 1-ary function symbol. (In fact the symbol '$-$' is redundant, because $-^V$ is the same operation as $(-1)^V$.)

When we speak of vector spaces below, we shall assume that they are structures of this form (unless anything is said to the contrary). The same goes for modules, replacing the field K by a ring.

Two questions spring to mind. First, aren't these examples a little arbitrary? For example, why did we give the group structure a symbol for the multiplicative inverse $^{-1}$, but not a symbol for the commutator $[\,,\,]$? Why did we put into the linear ordering structure a symbol for the ordering $\leqslant$, but not one for the corresponding strict ordering $<$?

The answer is yes; these choices were arbitrary. But some choices are more sensible than others. We shall come back to this in the next section.

And second, *exactly* what is a structure? Our definition said nothing about the way in which the ingredients (1.1)–(1.4) are packed into a single entity.

True again. But this was a deliberate oversight – the packing arrangements will never matter to us. Some writers define A to be an ordered pair $\langle \mathrm{dom}(A), f \rangle$ where f is a function taking each symbol S to the corresponding item S^A. The important thing is to know what the symbols and the ingredients are, and this can be indicated in any reasonable way.

For example a model theorist may refer to the structure

$$\langle \mathbb{R}, +, -, \cdot, 0, 1, \leqslant \rangle.$$

With some common sense the reader can guess that this means the structure whose domain is the set of real numbers, with constants 0 and 1 naming the numbers 0 and 1, a 2-ary relation symbol $\leqslant$ naming the relation $\leqslant$, 2-ary function symbols $+$ and $\cdot$ naming addition and multiplication respectively, and a 1-ary function symbol naming minus.

Signatures

The **signature** of a structure A is specified by giving

(1.5)   the set of constants of A, and for each separate $n > 0$, the set of n-ary relation symbols and the set of n-ary function symbols of A.

We shall assume that the signature of a structure can be read off uniquely from the structure.

The symbol L will be used to stand for signatures. Later it will also stand for languages – think of the signature of A as a kind of rudimentary language for talking about A. If A has signature L, we say A is an L-**structure**.

A signature L with no constants or function symbols is called a **relational signature**, and an L-structure is then said to be a **relational structure**. A signature with no relation symbols is sometimes called an **algebraic signature**.

Exercises for section 1.1

1. According to Thomas Aquinas, God is a structure G with three elements '*pater*', '*filius*' and '*spiritus sanctus*', in a signature consisting of one asymmetric binary relation ('*relatio opposita*') R, read as '*relatio originis*'. Aquinas asserts also that the three elements can be uniquely identified in terms of R^G. Deduce – as Aquinas did – that if the pairs (*pater*, *filius*) and (*pater*, *spiritus sanctus*) lie in R^G, then exactly one of the pairs (*filius*, *spiritus sanctus*) and (*spiritus sanctus*, *filius*) lies in R^G.

2. Let X be a set and L a signature; write $\kappa(X, L)$ for the number of distinct L-structures which have domain X. Show that if X is a finite set then $\kappa(X, L)$ is either finite or at least 2^ω.

1.2 Homomorphisms and substructures

The following definition is meant to take in, with one grand sweep of the arm, virtually all the things that are called 'homomorphism' in any branch of algebra.

Let L be a signature and let A and B be L-structures. By a **homomorphism** f from A to B, in symbols $f: A \to B$, we shall mean a function f from dom(A) to dom(B) with the following three properties.

(2.1)   For each constant c of L, $f(c^A) = c^B$.

(2.2)   For each $n > 0$ and each n-ary relation symbol R of L and n-tuple $\bar{a}$ from A, if $\bar{a} \in R^A$ then $f\bar{a} \in R^B$.

(2.3)   For each $n > 0$ and each n-ary function symbol F of L and n-tuple $\bar{a}$ from A, $f(F^A(\bar{a})) = F^B(f\bar{a})$.

(If $\bar{a}$ is $(a_0, \ldots, a_{n-1})$ then $f\bar{a}$ means $(fa_0, \ldots, fa_{n-1})$; cf. Note on notation.) By an **embedding** of A into B we mean a homomorphism $f: A \to B$ which is injective and satisfies the following stronger version of (2.2).

(2.4)   For each $n > 0$, each n-ary relation symbol R of L and each n-tuple $\bar{a}$ from A, $\bar{a} \in R^A \Leftrightarrow f\bar{a} \in R^B$.

An **isomorphism** is a surjective embedding. Homomorphisms $f: A \to A$ are called **endomorphisms** of A. Isomorphisms $f: A \to A$ are called **automorphisms** of A.

For example if G and H are groups and $f: G \to H$ is a homomorphism, then (2.1) says that $f(1^G) = 1^H$, and (2.3) says that for all elements a, b of G, $f(a \cdot {}^G b) = f(a) \cdot {}^H f(b)$ and $f(a^{(-1)^G}) = f(a)^{(-1)^H}$. This is exactly the usual definition of homomorphism between groups. Clause (2.2) adds nothing in this case since there are no relation symbols in the signature. For the same reason (2.4) is vacuous for groups. So a homomorphism between groups is an embedding if and only if it is an injective homomorphism.

We sometimes write 1_A for the identity map on $\mathrm{dom}(A)$. Clearly it is a homomorphism from A to A, in fact an automorphism of A. We say that A is **isomorphic** to B, in symbols $A \cong B$, if there is an isomorphism from A to B.

The following facts are nearly all immediate from the definitions.

Theorem 1.2.1. *Let L be a signature.*

(a) *If A, B, C are L-structures and $f: A \to B$ and $g: B \to C$ are homomorphisms, then the composed map gf is a homomorphism from A to C. If moreover f and g are both embeddings then so is gf.*

(b) *If A, B are L-structures and $f: A \to B$ is a homomorphism then $1_B f = f = f 1_A$.*

(c) *Let A, B, C be L-structures. Then 1_A is an isomorphism. If $f: A \to B$ is an isomorphism then the inverse map $f^{-1}: \mathrm{dom}(B) \to \mathrm{dom}(A)$ exists and is an isomorphism from B to A. If $f: A \to B$ and $g: B \to C$ are isomorphisms then so is gf.*

(d) *The relation $\cong$ is an equivalence relation on the class of L-structures.*

(e) *If A, B are L-structures, $f: A \to B$ is a homomorphism and there exist homomorphisms $g: B \to A$ and $h: B \to A$ such that $gf = 1_A$ and $fh = 1_B$, then f is an isomorphism and $g = h = f^{-1}$.*

Proof. (e) $g = g 1_B = g f h = 1_A h = h$. Since $gf = 1_A$, f is an embedding. Since $fh = 1_B$, f is surjective. $\qquad\square$

Substructures

If A and B are L-structures with $\mathrm{dom}(A) \subseteq \mathrm{dom}(B)$ and the inclusion map $i: \mathrm{dom}(A) \to \mathrm{dom}(B)$ is an embedding, then we say that B is an **extension** of A, or that A is a **substructure** of B, in symbols $A \subseteq B$. Note that if i is the inclusion map from $\mathrm{dom}(A)$ to $\mathrm{dom}(B)$, then condition (2.1) above says that $c^A = c^B$ for each constant c, condition (2.2) says that $R^A = R^B \cap (\mathrm{dom}\, A)^n$ for each n-ary relation symbol R, and finally condition (2.3) says that $F^A = F^B|(\mathrm{dom}\, A)^n$ (the restriction of F^B to $(\mathrm{dom}\, A)^n$) for each n-ary function symbol F.

When does a set of elements of a structure form the domain of a substructure? The next lemma gives us a criterion.

Lemma 1.2.2. *Let B be an L-structure and X a subset of* dom (B). *Then the following are equivalent.*
(a) $X = $ dom (A) *for some $A \subseteq B$.*
(b) *For every constant c of L, $c^B \in X$; and for every $n > 0$, every n-ary function symbol F of L and every n-tuple $\bar{a}$ of elements of X, $F^B(\bar{a}) \in X$.*
If (a) *and* (b) *hold, then A is uniquely determined.*

Proof. Suppose (a) holds. Then for every constant c of L, $c^B = c^A$ by (2.1); but $c^A \in $ dom $(A) = X$, so $c^B \in X$. Similarly, for each n-ary function symbol F of L and each n-tuple $\bar{a}$ of elements of X, $\bar{a}$ is an n-tuple in A and so $F^B(\bar{a}) = F^A(\bar{a}) \in $ dom $(A) = X$. This proves (b).

Conversely, if (b) holds, then we can define A by putting dom $(A) = X$, $c^A = c^B$ for each constant c of L, $F^A = F^B | X^n$ for each n-ary function symbol F of L, and $R^A = R^B \cap X^n$ for each n-ary relation symbol R of L. Then $A \subseteq B$; moreover this is the only possible definition of A, given that $A \subseteq B$ and dom $(A) = X$. $\qquad\square$

Let B be an L-structure and Y a set of elements of B. Then it follows easily from Lemma 1.2.2 that there is a unique smallest substructure A of B whose domain includes Y; we call A the **substructure of B generated by Y,** or the **hull** of Y in B, in symbols $A = \langle Y \rangle_B$. We call Y a **set of generators** for A. A structure B is said to be **finitely generated** if B is of form $\langle Y \rangle_B$ for some finite set Y of elements.

When the context allows, we write just $\langle Y \rangle$ instead of $\langle Y \rangle_B$. Sometimes it's convenient to list the generators Y as a sequence $\bar{a}$; then we write $\langle \bar{a} \rangle_B$ for $\langle Y \rangle_B$.

For many purposes we shall need to know the cardinality of the structure $\langle Y \rangle_B$. This can be estimated as follows. We define the **cardinality** of L, in symbols $|L|$, to be the least infinite cardinal $\geqslant$ the number of symbols in L. (In fact we shall see in Exercise 2.1.7 below that $|L|$ is equal to the number of first-order formulas of L, up to choice of variables; this is one reason why $|L|$ is taken to be infinite even when L contains only finitely many symbols.)

Warning. Occasionally it's important to know that a signature L contains only finitely many symbols. In this case we say that L is **finite**, in spite of the definition just given for $|L|$. See Exercise 6.

Theorem 1.2.3. *Let B be an L-structure and Y a set of elements of B. Then $|\langle Y \rangle_B| \leqslant |Y| + |L|$.*

Proof. We shall construct $\langle Y \rangle_B$ explicitly, thus proving its existence and uniqueness at the same time. We define a set $Y_m \subseteq \mathrm{dom}\,(B)$ for each $m < \omega$, by induction on m:

$$Y_0 = Y \cup \{c^B\colon c \text{ a constant of } L\},$$
$$Y_{m+1} = Y_m \cup \{F^B(\bar{a})\colon \text{for some } n > 0,\ F \text{ is an } n\text{-ary function}$$
$$\text{symbol of } L \text{ and } \bar{a} \text{ is an } n\text{-tuple of elements of } Y_m\}.$$

Finally we put $X = \bigcup_{m<\omega} Y_m$. Clearly X satisfies condition (b) of Lemma 1.2.2, so by that lemma there is a unique substructure A of B with $X = \mathrm{dom}\,(A)$. If A' is any substructure of B with $Y \subseteq \mathrm{dom}\,(A')$, then by induction on m we see that each Y_m is included in $\mathrm{dom}\,(A')$ (by the implication (a) $\Rightarrow$ (b) in Lemma 1.2.2), and hence $X \subseteq \mathrm{dom}\,(A')$. Therefore A is the unique smallest substructure of W whose domain includes Y, in short $A = \langle Y \rangle_B$.

Now we estimate the cardinality of A. Put $\kappa = |Y| + |L|$. Clearly $|Y_0| \leqslant \kappa$. For each fixed n, if Z is a subset of $\mathrm{dom}\,(B)$ of cardinality $\leqslant \kappa$, then the set

$$\{F^B(\bar{a})\colon F \text{ is an } n\text{-ary function symbol of } L \text{ and } \bar{a} \in Z^n\}$$

has cardinality at most $\kappa \cdot \kappa^n = \kappa$, since κ is infinite. Hence if $|Y_m| \leqslant \kappa$, then $|Y_{m+1}| \leqslant \kappa + \kappa = \kappa$. Thus by induction on m, each $|Y_m| \leqslant \kappa$, and so $|X| \leqslant \omega \cdot \kappa = \kappa$. Since $|\langle Y \rangle_B| = |X|$ by definition, this proves the theorem. $\square$

Choice of signature

As we remarked earlier, one and the same mathematical object can be interpreted as a structure in several different ways. The same function or relation can be named by different symbols: for example the identity element in a group can be called e or 1. We also have some choice about which elements, functions or relations should be given names at all. A group should surely have a name $\cdot$ for the product operation; ought it also to have names $^{-1}$ and 1 for the inverse and identity? How about $[\,,]$ for the commutator, $[a, b] = a^{-1}b^{-1}ab$?

A healthy general principle is that, other things being equal, *signatures should be chosen so that the notions of homomorphism and substructure agree with the usual notions for the relevant branch of mathematics*.

For example in the case of groups, if the only named operation is the product $\cdot$, then the substructures of a group will be its subsemigroups, closed under $\cdot$ but not necessarily containing inverses or identity. If we name $\cdot$ and the identity 1, then the substructures will be the submonoids. To ensure that substructure equals subgroup, we also need to put in a symbol for $^{-1}$. Given that symbols for $\cdot$ and $^{-1}$ are included, we would get exactly the same substructures and homomorphisms if we also included a symbol for the commutator; most model theorists use Ockham's razor and leave it out.

Thus for some classes of objects there is a natural choice of signature. For groups the natural choice is to name $\cdot$, 1 (or e) and $^{-1}$; we call this the **signature of groups**. We shall always assume (unless otherwise stated) that rings have a 1. So the natural choice of signature for rings is to name $+$, $-$, $\cdot$, 0 and 1; we call this the **signature of rings**. The **signature of partial orderings** has just the symbol $\leqslant$. The **signature of lattices** has just $\wedge$ and $\vee$.

What if we want to add new symbols, for example to name a particular element? The next definitions take care of this situation.

Reduction and expansion

Suppose L^- and L^+ are signatures, and L^- is a subset of L^+. Then if A is an L^+-structure, we can turn A into an L^--structure by simply forgetting the symbols of L^+ which are not in L^-. (We don't remove any elements of A, though some constant elements in A may cease to be constant elements in the new structure.) The resulting L^--structure is called the **L^--reduct** of A or the **reduct of A to L^-**, in symbols $A|L^-$. If $f: A \to B$ is a homomorphism of L^+-structures, then the same map f is also a homomorphism $f: A|L^- \to B|L^-$ of L^--structures.

When A is an L^+-structure and C is its L^--reduct, we say that A is an **expansion** of C to L^+. In general C may have many different expansions to L^+. There is a useful but imprecise notation for expansions. Suppose the symbols which are in L^+ but not in L^- are constants c, d and a function symbol F; then c^A, d^A are respectively elements a, b of A, and F^A is some operation f on $\mathrm{dom}(A)$. We write

$$(2.5) \qquad\qquad A = (C, a, b, f)$$

to express that A is an expansion of C got by adding symbols to name a, b and f. The notation is imprecise because it doesn't say what symbols are used to name a, b and f respectively; but often the choice of symbols is unimportant or obvious from the context.

Exercises for section 1.2

1. Verify all parts of Theorem 1.2.1.

2. Let L be a signature and **K** a class of L-structures. Suppose that A and B are in **K**, and for every structure C in **K** there are unique homomorphisms $f: A \to C$ and $g: B \to C$. Show that there is a unique isomorphism from A to B.

3. Let A, B be L-structures, X a subset of $\mathrm{dom}(A)$ and $f: \langle X \rangle_A \to B$ and $g: \langle X \rangle_A \to B$ homomorphisms. Show that if $f|X = g|X$ then $f = g$.

4. Prove the following, in which all structures are assumed to be L-structures for some fixed signature L.

(a) Every homomorphism $f: A \to C$ can be factored as $f = hg$ for some surjective homomorphism $g: A \to B$ and extension $h: B \to C$:

(2.6)

$$A \xrightarrow{\quad f \quad} C$$
$$g \searrow \quad \nearrow h$$
$$B$$

The unique structure B is called the **image** *of g,* img *for short. More generally we say that an L-structure B is a* **homomorphic image** *of A if there exists a surjective homomorphism $g: A \to B$.*

(b) Every embedding $f: A \to C$ can be factored as $f = hg$ where g is an extension and h is an isomorphism. *This is a small piece of set theory which often gets swept under the carpet. It says that when we have embedded A into C, we can assume that A is a substructure of C. The embedding of the rationals in the reals is a familiar example.*

5. Let A and B be L-structures with A a substructure of B. A **retraction** from B to A is a homomorphism $f: B \to A$ such that $f(a) = a$ for every element a of A. Show (a) if $f: B \to A$ is a retraction, then $f^2 = f$, (b) if B is any L-structure and f an endomorphism of B such that $f^2 = f$, then f is a retraction from B to a substructure A of B.

6. Let L be a finite signature with no function symbols. (a) Show that every finitely generated L-structure is finite. (b) Show that for each $n < \omega$ there are up to isomorphism only finitely many L-structures of cardinality n.

7. Show that we can give fields a signature which makes substructure = subfield and homomorphism = field embedding, provided we are willing to let 0^{-1} be 0. *Most mathematicians seem to be unwilling, and so the normal custom is to give fields the same signature as rings.*

8. Many model theorists require the domain of any structure to be non-empty. How would this affect Lemma 1.2.2 and the definition of $\langle Y \rangle_B$?

9. Give an example of a structure of cardinality ω_2 which has a substructure of cardinality ω but no substructure of cardinality ω_1.

1.3 Terms and atomic formulas

In Chapter 2 we shall introduce a formal language for talking about L-structures. This language will be built up from the atomic formulas of L, which we must now define.

Each atomic formula will be a string of symbols including symbols of L. Since the symbols in L can be any kinds of object and not necessarily written expressions, the idea of a 'string of symbols' has to be taken with a pinch of set-theoretic coding.

Terms

Every language has a stock of **variables**. These are symbols written v, x, y, z, t, x_0, x_1 etc., and one of their purposes is to serve as temporary labels for elements of a structure. Any symbol can be used as a variable, provided it is not already being used for something else. The choice of variables is never important, and for theoretical purposes many model theorists restrict them to be the expressions v_0, v_1, v_2, ... with natural number or ordinal subscripts.

The **terms** of the signature L are strings of symbols defined as follows (where the symbols '(', ')' and ',' are assumed not to occur anywhere in L – henceforth points like this go without saying).

(3.1)   Every variable is a term of L.

(3.2)   Every constant of L is a term of L.

(3.3)   If $n > 0$, F is an n-ary function symbol of L and $t_1, \ldots, t_n$ are terms of L then the expression $F(t_1, \ldots, t_n)$ is a term of L.

(3.4)   Nothing else is a term of L.

A term is said to be **closed** (computer scientists say **ground**) if no variables occur in it. The **complexity** of a term is the number of symbols occurring in it, counting each occurrence separately. (The important point is that if t occurs as part of s then s has higher complexity than t.)

If we introduce a term t as $t(\bar{x})$, this will always mean that $\bar{x}$ is a sequence $(x_0, x_1, \ldots)$, possibly infinite, of distinct variables, and every variable which occurs in t is among the variables in $\bar{x}$. Later in the same context we may write $t(\bar{s})$, where $\bar{s}$ is a sequence of terms $(s_0, s_1, \ldots)$; then $t(\bar{s})$ means the term got from t by putting s_0 in place of x_0, s_1 in place of x_1, etc., throughout t. (For example if $t(x, y)$ is the term $y + x$, then $t(0, 2y)$ is the term $2y + 0$ and $t(t(x, y), y)$ is the term $y + (y + x)$.)

To make variables and terms stand for elements of a structure, we use the following convention. Let $t(\bar{x})$ be a term of L, where $\bar{x} = (x_0, x_1, \ldots)$. Let A be an L-structure and $\bar{a} = (a_0, a_1, \ldots)$ a sequence of elements of A; we assume that $\bar{a}$ is at least as long as $\bar{x}$. Then $t^A(\bar{a})$ (or $t^A[\bar{a}]$ when we need a more distinctive notation) is defined to be the element of A which is named by t when x_0 is interpreted as a name of a_0, and x_1 as a name of a_1, and so on. More precisely, using induction on the complexity of t,

(3.5)   if t is the variable x_i then $t^A[\bar{a}]$ is a_i,

(3.6)   if t is a constant c then $t^A[\bar{a}]$ is the element c^A,

(3.7)   if t is of the form $F(s_1, \ldots, s_n)$ where each s_i is a term $s_i(\bar{x})$,
         then $t^A[\bar{a}]$ is the element $F^A(s_1{}^A[\bar{a}], \ldots, s_n{}^A[\bar{a}])$.

(Cf. (1.2), (1.4) above.) If t is a closed term then $\bar{a}$ plays no role and we simply write t^A for $t^A[\bar{a}]$.

Atomic formulas

The **atomic formulas** of L are the strings of symbols given by (3.8) and (3.9) below.

(3.8)   If s and t are terms of L then the string $s = t$ is an atomic formula of L.

(3.9)   If $n > 0$, R is an n-ary relation symbol of L and $t_1, \ldots, t_n$ are terms of L then the expression $R(t_1, \ldots, t_n)$ is an atomic formula of L.

(Note that the symbol '=' is not assumed to be a relation symbol in the signature.) An **atomic sentence** of L is an atomic formula in which there are no variables.

Just as with terms, if we introduce an atomic formula ϕ as $\phi(\bar{x})$, then $\phi(\bar{s})$ means the atomic formula got from ϕ by putting terms from the sequence $\bar{s}$ in place of all occurrences of the corresponding variables from $\bar{x}$.

If the variables $\bar{x}$ in an atomic formula $\phi(\bar{x})$ are interpreted as names of elements $\bar{a}$ in a structure A, then ϕ makes a statement about A. The statement may be true or false. If it's true, we say ϕ is **true of** $\bar{a}$ **in** A, or that $\bar{a}$ **satisfies** ϕ **in** A, in symbols

$$A \vDash \phi[\bar{a}] \quad \text{or equivalently} \quad A \vDash \phi(\bar{a}).$$

We can give a formal definition of this relation $\vDash$. Let $\phi(\bar{x})$ be an atomic formula of L with $\bar{x} = (x_0, x_1, \ldots)$. Let A be an L-structure and $\bar{a}$ a sequence $(a_0, a_1, \ldots)$ of elements of A; we assume that $\bar{a}$ is at least as long as $\bar{x}$. Then

(3.10)   if ϕ is the formula $s = t$ where $s(\bar{x})$, $t(\bar{x})$ are terms, then $A \vDash \phi[\bar{a}]$ iff $s^A[\bar{a}] = t^A[\bar{a}]$,

(3.11)   if ϕ is the formula $R(s_1, \ldots, s_n)$ where $s_1(\bar{x}), \ldots, s_n(\bar{x})$ are terms, then $A \vDash \phi[\bar{a}]$ iff the ordered n-tuple $(s_1{}^A[\bar{a}], \ldots, s_n{}^A[\bar{a}])$ is in R^A.

(Cf. (1.3) above.) When ϕ is an atomic sentence, we can omit the sequence $\bar{a}$ and write simply $A \vDash \phi$ in place of $A \vDash \phi[\bar{a}]$.

We say that A is a **model** of ϕ, or that ϕ is **true in** A, if $A \vDash \phi$. When T is a set of atomic sentences, we say that A is a **model** of T (in symbols, $A \vDash T$) if A is a model of every atomic sentence in T.

Theorem 1.3.1. *Let A and B be L-structures and f a map from* $\mathrm{dom}(A)$ *to* $\mathrm{dom}(B)$.

(a) *If f is a homomorphism then, for every term $t(\bar{x})$ of L and tuple $\bar{a}$ from A, $f(t^A[\bar{a}]) = t^B[f\bar{a}]$.*

(b) *f is a homomorphism if and only if, for every atomic formula $\phi(\bar{x})$ of L and tuple $\bar{a}$ from A,*

$$(3.12) \qquad A \vDash \phi[\bar{a}] \Rightarrow B \vDash \phi[f\bar{a}].$$

(c) *f is an embedding if and only if, for every atomic formula $\phi(\bar{x})$ of L and tuple $\bar{a}$ from A,*

$$(3.13) \qquad A \vDash \phi[\bar{a}] \Leftrightarrow B \vDash \phi[f\bar{a}].$$

Proof. (a) is easily proved by induction on the complexity of t, using (3.5)–(3.7).

(b) Suppose first that f is a homomorphism. As a typical example, suppose $\phi(\bar{x})$ is $R(s, t)$, where $s(\bar{x})$ and $t(\bar{x})$ are terms. Assume $A \vDash \phi[\bar{a}]$. Then by (3.11) we have

$$(3.14) \qquad (s^A[\bar{a}], \, t^A[\bar{a}]) \in R^A.$$

Then by part (a) and the fact that f is a homomorphism (see (2.2) above),

$$(3.15) \qquad (s^B[f\bar{a}], \, t^B[f\bar{a}]) = (f(s^A[\bar{a}]), \, f(t^A[\bar{a}])) \in R^B.$$

Hence $B \vDash \phi[f\bar{a}]$ by (3.11) again. Essentially the same proof works for every atomic formula ϕ.

For the converse, again we take a typical example. Assume that (3.12) holds for all atomic ϕ and sequences $\bar{a}$. Suppose $(a_0, a_1) \in R^A$. Then writing $\bar{a}$ for (a_0, a_1), we have $A \vDash R(x_0, x_1)[\bar{a}]$. Then (3.12) implies $B \vDash R(x_0, x_1)[f\bar{a}]$, which by (3.11) implies that $(fa_0, fa_1) \in R^B$ as required. Thus f is a homomorphism.

(c) is proved like (b), but using (2.4) in place of (2.2).   □

A variant of Theorem 1.3.1(c) is often useful. By a **negated atomic formula** of L we mean a string $\neg\phi$ where ϕ is an atomic formula of L. We read the symbol $\neg$ as 'not' and we define

$$(3.16) \qquad \text{`} A \vDash \neg\phi[\bar{a}] \text{' holds} \qquad \text{iff} \qquad \text{`} A \vDash \phi[\bar{a}] \text{' doesn't hold.}$$

where A is any L-structure, ϕ is an atomic formula and $\bar{a}$ is a sequence from A. A **literal** is an atomic or negated atomic formula; it's a **closed literal** if it contains no variables.

Corollary 1.3.2. *Let A and B be L-structures and f a map from* dom(A) *to* dom(B). *Then f is an embedding if and only if, for every literal* $\phi(\bar{x})$ *of L and sequence* $\bar{a}$ *from A,*

(3.17) $A \vDash \phi[\bar{a}] \Rightarrow B \vDash \phi[f\bar{a}]$.

Proof. Immediate from (c) of the theorem and (3.16).   □

The term algebra

The Delphic oracle said 'Know thyself'. Terms can do this – they can describe themselves. Let L be any signature and X a set of variables. We define the **term algebra of L with basis** X to be the following L-structure A. The domain of A is the set of all terms of L whose variables are in X. We put

(3.18) $c^A = c$ for each constant c of L,

(3.19) $F^A(\bar{t}) = F(\bar{t})$ for each n-ary function symbol F of L and
                                      n-tuple $\bar{t}$ of elements of dom(A),

(3.20) R^A is empty   for each relation symbol R of L.

The term algebra of L with basis X is also known as the **absolutely free L-structure with basis** X.

Exercises for section 1.3

1. Let B be an L-structure and Y a set of elements of B. Show that $\langle Y \rangle_B$ consists of those elements of B which have the form $t^B[\bar{b}]$ for some term $t(\bar{x})$ of L and some tuple $\bar{b}$ of elements of Y. [Use the construction of $\langle Y \rangle_B$ in the proof of Theorem 1.2.3.]

2. (a) If $t(x, y, z)$ is $F(G(x, z), x)$, what are $t(z, y, x)$, $t(x, z, z)$, $t(F(x, x)$, $G(x, x), x)$, $t(t(a, b, c), b, c)$?
(b) Let A be the structure $\langle \mathbb{N}, +, \cdot, 0, 1 \rangle$ where $\mathbb{N}$ is the set of natural numbers 0, 1, ... Let $\phi(x, y, z, u)$ be the atomic formula $x + z = y \cdot u$ and let $t(x, y)$ be the term $y \cdot y$. Which of the following are true? $A \vDash \phi[0, 1, 2, 3]$; $A \vDash \phi[1, 5, t^A[4, 2], 1]$; $A \vDash \phi[9, 1, 16, 25]$; $A \vDash \phi[56, t^A[9, t^A[0, 3]], t^A[5, 7], 1]$. [Answers: $F(G(z, x), z)$, $F(G(x, z), x)$, $F(G(F(x, x), x), F(x, x))$, $F(G(F(G(a, c), a), c), F(G(a, c), a))$; no, yes, yes, no.]

3. Let A be an L-structure, $\bar{a} = (a_0, a_1, \ldots)$ a sequence of elements of A and $\bar{s} = (s_0, s_1, \ldots)$ a sequence of closed terms of L such that, for each i, $s_i^A = a_i$. Show (a) for each term $t(\bar{x})$ of L, $t(\bar{s})^A = t^A[\bar{a}]$, (b) for each atomic formula $\phi(\bar{x})$ of L, $A \vDash \phi(\bar{s}) \Leftrightarrow A \vDash \phi[\bar{a}]$.

4. Let A and B be L-structures, $\bar{a}$ a sequence of elements which generate A and f a map from $\mathrm{dom}\,(A)$ to $\mathrm{dom}\,(B)$. Show that f is a homomorphism if and only if for every atomic formula $\phi(\bar{x})$ of L, $A \vDash \phi[\bar{a}]$ implies $B \vDash \phi[f\bar{a}]$.

5. (Unique parsing lemma). Let L be a signature and t a term of L. Show that t can be constructed in only one way. [In other words, (a) if t is a constant then t is not also of the form $F(\bar{s})$, (b) if t is $F(s_0, \ldots, s_{m-1})$ and $G(r_0, \ldots, r_{n-1})$ then $F = G$, $m = n$ and for each $i < n$, $s_i = r_i$. For each occurrence σ of a symbol in t, define $^\#(\sigma)$ to be the number of occurrences of '(' to the left of σ, minus the number of occurrences of ')' to the left of σ; use $^\#(\sigma)$ to identify the commas connected with F in t.]

6. Let A be the term algebra of L with basis X. Show that for each term $t(\bar{x})$ of L and each tuple $\bar{s}$ of elements of $\mathrm{dom}\,(A)$, $t^A[\bar{s}] = t(\bar{s})$.

7. Let A be the term algebra of signature L with basis X, and B any L-structure. Show that for each map $f \colon X \to \mathrm{dom}\,(B)$ there is a unique homomorphism $g \colon A \to B$ which agrees with f on X. Show also that if $t(\bar{x})$ is any term in $\mathrm{dom}\,(A)$ then $t^B[f\bar{x}] = g(t)$.

1.4 Parameters and diagrams

The conventions for interpreting variables are one of the more irksome parts of model theory. We can avoid them, at a price. Instead of interpreting a variable as a name of the element b, we can add a new constant for b to the signature. The price we pay is that the language changes every time another element is named. When constants are added to a signature, the new constants and the elements they name are called **parameters**.

Suppose for example that A is an L-structure, $\bar{a}$ is a sequence of elements of A, and we want to name the elements in $\bar{a}$. Then we choose a sequence $\bar{c}$ of distinct new constant symbols, of the same length as $\bar{a}$, and we form the signature $L(\bar{c})$ by adding the constants $\bar{c}$ to L. In the notation of (2.5), $(A, \bar{a})$ is an $L(\bar{c})$-structure, and each element a_i is $c_i^{(A, \bar{a})}$.

Likewise if B is another L-structure and $\bar{b}$ a sequence of elements of B of the same length as $\bar{c}$, then there is an $L(\bar{c})$-structure $(B, \bar{b})$ in which these same constants c_i name the elements of $\bar{b}$. The next lemma is about this situation. It comes straight out of the definitions, and it is often used silently.

Lemma 1.4.1. *Let A, B be L-structures and suppose $(A, \bar{a})$, $(B, \bar{b})$ are $L(\bar{c})$-structures. Then a homomorphism $f \colon (A, \bar{a}) \to (B, \bar{b})$ is the same thing as a homomorphism $f \colon A \to B$ such that $f\bar{a} = \bar{b}$. Likewise an embedding $f \colon (A, \bar{a}) \to (B, \bar{b})$ is the same thing as an embedding $f \colon A \to B$ such that $f\bar{a} = \bar{b}$.* $\qquad\square$

In the situation above, if $t(\bar{x})$ is a term of L then $t^A[\bar{a}]$ and $t(\bar{c})^{(A,\bar{a})}$ are the same element; and if $\phi(\bar{x})$ is an atomic formula then $A \vDash \phi[\bar{a}] \Leftrightarrow (A, \bar{a}) \vDash \phi(\bar{c})$. These two notations – the square-bracket and the parameter notation – serve the same purpose, and it's a burden on one's patience to keep them separate. The following compromise works well in practice. We use the elements a_i as *constants naming themselves*. The expanded signature is then $L(\bar{a})$, and we write $t^A(\bar{a})$ and $A \vDash \phi(\bar{a})$ with an easy conscience. One should take care when $\bar{a}$ contains repetitions, or when two separate $L(\bar{c})$-structures are under discussion. To play safe I retain the square-bracket notation for the next few sections.

Let $\bar{a}$ be a sequence of elements of A. We say that $\bar{a}$ **generates** A, in symbols $A = \langle \bar{a} \rangle_A$, if A is generated by the set of all elements in $\bar{a}$. Suppose that A is an L-structure, $(A, \bar{a})$ is an $L(\bar{c})$-structure and $\bar{a}$ generates A. Then (cf. Exercise 1 below) every element of A is of the form $t^{(A,\bar{a})}$ for some closed term t of $L(\bar{c})$; so every element of A has a name in $L(\bar{c})$. The set of all closed literals of $L(\bar{c})$ which are true in $(A, \bar{a})$ is called the (Robinson) **diagram** of A, in symbols diag(A). The set of all atomic sentences of $L(\bar{c})$ which are true in $(A, \bar{a})$ is called the **positive diagram** of A, in symbols diag$^+(A)$.

Note that diag(A) and diag$^+(A)$ are not uniquely determined, because in general there are many ways of choosing $\bar{a}$ and $\bar{c}$ so that $\bar{a}$ generates A. But this never matters. There is always at least one possible choice of $\bar{a}$ and $\bar{c}$: simply list all the elements of A without repetition.

Diagrams and positive diagrams should be thought of as a generalisation of the multiplication tables of groups. If we know either diag(A) or diag$^+(A)$, we know A up to isomorphism. The name 'diagram' is due to Abraham Robinson, who was the first model theorist to use diagrams systematically. It's a slightly unfortunate name, because it invites confusion with diagrams in the sense of pictures with arrows, as in category theory. Actually the next result, which we shall use many times, is about diagrams in both senses.

Lemma 1.4.2 (*Diagram lemma*). *Let A and B be L-structures, $\bar{c}$ a sequence of constants, and $(A, \bar{a})$ and $(B, \bar{b})$ $L(\bar{c})$-structures. Then* (a) *and* (b) *are equivalent.*
(a) *For every atomic sentence ϕ of $L(\bar{c})$, if $(A, \bar{a}) \vDash \phi$ then $(B, \bar{b}) \vDash \phi$.*
(b) *There is a homomorphism $f: \langle \bar{a} \rangle_A \to B$ such that $f\bar{a} = \bar{b}$.*
The homomorphism f in (b) *is unique if it exists; it is an embedding if and only if*
(c) *for every atomic sentence ϕ of $L(\bar{c})$, $(A, \bar{a}) \vDash \phi \Leftrightarrow (B, \bar{b}) \vDash \phi$.*

Proof. Assume (a). Since the inclusion map embeds $\langle \bar{a} \rangle_A$ in A, Theorem 1.3.1(c) says that in (a) we can replace A by $\langle \bar{a} \rangle_A$. So without loss we can assume that $A = \langle \bar{a} \rangle_A$. By Lemma 1.4.1, to prove (b) it suffices to find a homomorphism $f: (A, \bar{a}) \to (B, \bar{b})$. We define f as follows. Since $\bar{a}$ generates A, each element of A is of the form $t^{(A, \bar{a})}$ for some closed term t of $L(\bar{c})$. Put

$$(4.1) \qquad\qquad f(t^{(A, \bar{a})}) = t^{(B, \bar{b})}.$$

The definition is sound, since $s^{(A, \bar{a})} = t^{(A, \bar{a})}$ implies $(A, \bar{a}) \vDash s = t$, so $(B, \bar{b}) \vDash s = t$ by (a) and hence $s^{(B, \bar{b})} = t^{(B, \bar{b})}$. Then f is a homomorphism by (a) and Theorem 1.3.1(b), which proves (b). Since any homomorphism f from $(A, \bar{a})$ to $(B, \bar{b})$ must satisfy (4.1), f is unique in (b). The converse (b) $\Rightarrow$ (a) follows at once from Theorem 1.3.1(b).

The argument for embeddings and (c) is similar, using Theorem 1.3.1(c). $\square$

Lemma 1.4.2 doesn't mention Robinson diagrams outright, so let me make the connection explicit. Suppose $\bar{a}$ generates A. Then the implication (a) $\Rightarrow$ (b) in the lemma says that A can be mapped homomorphically to a reduct of B whenever $B \vDash \mathrm{diag}^+(A)$. Similarly the last part of the lemma says that if $B \vDash \mathrm{diag}(A)$ then A can be embedded in a reduct of B.

Exercises for section 1.4

1. Let B be an L-structure. If $\bar{b}$ is a sequence of elements of B and $\bar{c}$ are parameters such that $(B, \bar{b})$ is an $L(\bar{c})$-structure, show that $\langle \bar{b} \rangle_B$ consists of those elements of B which have the form $t^{(B, \bar{b})}$ for some closed term t of $L(\bar{c})$.

2. Let A, B be L-structures and $\bar{a}$ a sequence of elements of A. Let g be a map from the elements of $\bar{a}$ to $\mathrm{dom}(B)$ such that for every atomic sentence ϕ of $L(\bar{a})$, $(A, \bar{a}) \vDash \phi$ implies $(B, g\bar{a}) \vDash \phi$. Show that g has a unique extension to a homomorphism $g': \langle \bar{a} \rangle_A \to B$. (In practice one often identifies g and g'.)

3. Let $(A, \bar{a})$ and $(B, \bar{b})$ be $L(\bar{c})$-structures which satisfy exactly the same atomic sentences of $L(\bar{c})$. Suppose also that the L-structures A, B are generated by $\bar{a}$, $\bar{b}$ respectively. Show that there is an isomorphism $f: A \to B$ such that $f\bar{a} = \bar{b}$.

1.5 Canonical models

In the previous section we saw how one can translate any structure into a set of atomic sentences. It turns out that there is a good route back home again: we can convert any set of atomic sentences into a structure.

Let L be a signature, A an L-structure and T the set of all atomic sentences of L which are true in A. Then T has the following two properties.

(5.1) For every closed term t of L, the atomic sentence $t = t$ is in T.

(5.2)   If $\phi(x)$ is an atomic formula of L and the equation $s = t$ is in
         T, then $\phi(s) \in T$ if and only if $\phi(t) \in T$.

Any set T of atomic sentences which satisfies (5.1) and (5.2) will be said to
be **=-closed** (in L).

Lemma 1.5.1. *Let T be an =-closed set of atomic sentences of L. Then there is
an L-structure A such that*
(a) *T is the set of all atomic sentences of L which are true in A,*
(b) *every element of A is of the form t^A for some closed term t of L.*

Proof. Let X be the set of all closed terms of L. We define a relation $\sim$ on
X by

(5.3) $s \sim t$ iff $s = t \in T$.

We claim that $\sim$ is an equivalence relation. (i) By (5.1), $\sim$ is reflexive. (ii)
Suppose $s \sim t$; then $s = t \in T$. But if $\phi(x)$ is the formula $x = s$, then $\phi(s)$ is
$s = s$ which is in T by (5.1); so by (5.2), T also contains $\phi(t)$, which is $t = s$.
Hence $t \sim s$. (iii) Suppose $s \sim t$ and $t \sim r$. Then let $\phi(x)$ be $s = x$. By
assumption both $\phi(t)$ and $t = r$ are in T, so by (5.2) T also contains $\phi(r)$,
which is $s = r$. Hence $s \sim r$. This proves the claim.

For each closed term t let $t^\sim$ be the equivalence class of t under $\sim$, and
let Y be the set of all equivalence classes $t^\sim$ with $t \in X$. We shall define an
L-structure A with $\mathrm{dom}\,(A) = Y$.

First, for each constant c of L we put $c^A = c^\sim$. Next, if $0 < n < \omega$ and F is
an n-ary function symbol of L, we define F^A by

(5.4) $F^A(s_0^\sim, \ldots, s_{n-1}^\sim) = (F(s_0, \ldots, s_{n-1}))^\sim.$

One must check that (5.4) is a sound definition. Suppose $s_i \sim t_i$ for each
$i < n$. Then by n applications of (5.2) to the sentence $F(s_0, \ldots, s_{n-1}) =
F(s_0, \ldots, s_{n-1})$, which is in T by (5.1), we find that the equation $F(s_0, \ldots,
s_{n-1}) = F(t_0, \ldots, t_{n-1})$ is in T. Hence $F(s_0, \ldots, s_{n-1})^\sim = F(t_0, \ldots, t_{n-1})^\sim.$
This shows that the definition (5.4) is sound.

Finally we define the relation R^A, where R is any n-ary relation symbol of
L, by

(5.5) $(s_0^\sim, \ldots, s_{n-1}^\sim) \in R^A$ iff $R(s_0, \ldots, s_{n-1}) \in T$.

(5.5) is justified in the same way as (5.4), and it completes the description of
the L-structure A.

Now it is easy to prove by induction on the complexity of t, using (3.6)
and (3.7) from section 1.3, that for every closed term t of L,

(5.6) $t^A = t^\sim.$

From this we infer that if s and t are any closed terms of L then

(5.7) $A \models s = t$ $\Leftrightarrow$ $s^A = t^A$ $\Leftrightarrow$ $s^\sim = t^\sim$ $\Leftrightarrow$ $s = t \in T$.

Together with a similar argument for atomic sentences of the form $R(t_0, . . ., t_{n-1})$, using (3.11) of section 1.3, this shows that T is the set of all atomic sentences of L which are true in A. Also by (5.6) every element of A is of the form t^A for some closed term t of L. $\qquad\square$

Now if T is any set of atomic sentences of L, there is a least set U of atomic sentences of L which contains T and is =-closed in L. We call U the =-**closure** of T in L. Any L-structure which is a model of U must also be a model of T since $T \subseteq U$.

Theorem 1.5.2. *For any signature L, if T is a set of atomic sentences of L then there is an L-structure A such that*
(a) $A \vDash T$,
(b) *every element of A is of the form t^A for some closed term t of L,*
(c) *if B is an L-structure and $B \vDash T$ then there is a unique homomorphism $f: A \to B$.*

Proof. Apply the lemma to the =-closure U of T to get the L-structure A. Then (a) and (b) are clear. Now $A = \langle \varnothing \rangle_A$ by (b). So (c) will follow from the diagram lemma (Lemma 1.4.2) if we can show that every atomic sentence true in A is true in all models B of T. By the choice of A, every atomic sentence true in A is in U. The set of all atomic sentences true in a model B of T is an =-closed set containing T, and so it must contain the =-closure of T, which is U. $\qquad\square$

By clause (c), the model A of T in Theorem 1.5.2 is unique up to isomorphism. (See Exercise 1.2.2.) We call it the **canonical model** of T. Note that it will be the empty L-structure if and only if L has no constant symbols.

Sometimes – for example in logic programming – one doesn't include equations as atomic formulas. In this situation the canonical model is much easier to construct, because there is no need to factor out an equivalence relation. Thus we get what has become known as the **Herbrand universe** of a set of atomic sentences.

Here is an example at the opposite extreme, where all the atomic sentences are equations.

Example 1: *Adding roots of polynomials to a field.* Let F be a field and $p(X)$ an irreducible polynomial over F in the indeterminate X. We can regard $F[X]$ as a structure in the signature of rings with constants added for X and all the elements of F. Let T be the set of all equations which are true in $F[X]$. (For example if a is $b \cdot (c + d)$ in $F[X]$, then T contains the equation '$a = b \cdot (c + d)$'. Also rings satisfy the law $1 \cdot x = x$, so T contains the equation $1 \cdot t = t$ for every closed term t.) Then T is a set of atomic sentences, and the equation '$p(X) = 0$' is another atomic sentence. Let C be the canonical model of the set $T \cup \{p(X) = 0\}$. Then C is a homomorphic image of $F[X]$, because it is a model of T and every element is named by a closed term. In particular C is a ring; moreover '$p(X) = 0$' holds in C, and so every element of the ideal I in $F[X]$ generated by $p(X)$ goes to 0 in C. Let θ be any root of p; then the field extension $F[\theta]$ is a model of $T \cup \{p(X) = 0\}$ too, with X read as a name of θ. So by Theorem 1.5.2(c), $F[\theta]$ is a homomorphic image of C. Since $F[\theta] = F[X]/I$, it follows that C is isomorphic to $F[\theta]$.

Exercises for section 1.5

1. Show that the property of being $=$-closed is not altered if we change 'if and only if' in (5.2) to '$\Rightarrow$'.

2. Let T be a set of closed literals of signature L. Show that (a) is equivalent to (b). (a) Some L-structure is a model of T. (b) If $\neg \phi$ is a negated atomic sentence in T, then ϕ is not in the $=$-closure of the set of atomic sentences in T.

3. Let T be a finite set of closed literals of signature L. Show that (a) of Exercise 2 is equivalent to (c) Some finite L-structure is a model of T. [Suppose $A \vDash T$. Let X be the set of all closed terms which occur as parts of sentences in T, including those inside other terms. Choose $a_0 \in \text{dom}(A)$. Define a structure A' with domain $\{t^A : t \in X\} \cup \{a_0\}$ so that $A' \vDash T$.]

Further reading

Readers who feel they need a fuller background in elementary logic will find there are several excellent texts available. To name two:

Cori, R. and Lascar, D., *Logique Mathématique, cours et exercises*. Paris: Masson 1994 (two volumes).
Ebbinghaus, H. D., Flum, J. & Thomas, W. *Mathematical logic*. New York: Springer-Verlag 1984.

Two interesting papers on the historical and philosophical background to model theory are:

Demopoulos, W. Frege, Hilbert, and the conceptual structure of model theory. *History and Philosophy of Logic*, **15** (1994), 211–225.
Hintikka, J. On the development of the model-theoretical tradition in logical theory. *Synthese*, **77** (1988), 1–16.

2

Classifying structures

I must get into this stone world now.
Ratchel, striae, relationships of tesserae,
  Innumerable shades of grey . . .
I try them with the old Norn words – hraun
Duss, rønis, queedaruns, kollyarum . . .
  Hugh MacDiarmid, On a raised beach.

Now that we have structures in front of us, the most pressing need is to start classifying them and their features. Classifying is a kind of defining. Most mathematical classification is by axioms or defining equations – in short, by formulas. This chapter could have been entitled 'The elementary theory of mathematical classification by formulas'.

Notice three ways in which mathematicians use formulas. First, a mathematician writes the equation '$y = 4x^2$'. By writing this equation one names a set of points in the plane, i.e. a set of ordered pairs of real numbers. As a model theorist would put it, the equation *defines a 2-ary relation on the reals*. We study this kind of definition in section 2.1.

Or second, a mathematician writes down the laws

(*)   For all x, y and z, $x \leqslant y$ and $y \leqslant z$ imply $x \leqslant z$;
        for all x and y, exactly one of $x \leqslant y$, $y \leqslant x$, $x = y$ holds.

By doing this one *names a class of relations*, namely those relations $\leqslant$ for which (*) is true. Section 2.2 lists some more examples of this kind of naming. They cover most branches of algebra.

Third, a mathematician defines a **homomorphism** from a group G to a group H to be a map from G to H such that $x = y \cdot z$ implies $f(x) = f(y) \cdot f(z)$. Here the equation $x = y \cdot z$ *defines a class of maps*. See section 2.5 for more examples.

Many of the founding fathers of model theory wanted to understand how language works in mathematics, and how it ought to. (One might name Frege, Padoa, Russell, Gödel and Tarski among others.) But a successful branch of mathematics needs more than just a desire to correct and catalogue things. It needs a programme of problems which are interesting and not quite too hard to solve. The first systematic programme of model theory was known as 'elimination of quantifiers'. Its aim was to find, in any concrete mathematical situation, the simplest set of formulas that would give all the classifications that one needs. Thoralf Skolem set this programme moving in 1919. One recent offshoot of the programme is the study of unification in computer science.

2.1 Definable subsets

We begin with a single structure A. What are the interesting sets of elements of A? More generally, what are the interesting relations on elements of A?

Example 1: *Algebraic curves.* Regard the field $\mathbb{R}$ of reals as a structure. An algebraic curve in the real plane is a set of ordered pairs of elements of $\mathbb{R}$ given by an equation $p(x, y) = 0$, where p is a polynomial with coefficients from $\mathbb{R}$. The parabola $y = x^2$ is perhaps the most quoted example; this equation can be written without naming any elements of $\mathbb{R}$ as parameters.

Example 2: *Recursive sets of natural numbers.* For this we use the structure $\mathbb{N} = (\omega, 0, 1, +, \cdot, <)$ of natural numbers. Any recursive subset X of ω can be defined, for example by an algorithm for computing whether any given number is in X. But unlike Example 1, the definition will usually be much too complicated to be written as an atomic formula. There is no need to use parameters in this case, since every element of $\mathbb{N}$ is named by a closed term of the signature of $\mathbb{N}$.

Example 3: *Connected components of graphs.* Let G be a graph as in Example 1 of section 1.1, and g an element of G. The **connected component** of g in G is the smallest set of Y of vertices of G such that (1) $g \in Y$ and (2) if $a \in Y$ and a is joined to b by an edge, then $b \in Y$. This description defines Y, using g as a parameter. But again there is generally no hope of expressing the definition as an atomic formula. Also it will probably be hopeless to try to define Y without mentioning any element as a parameter.

We shall describe some simple relations on a structure, and then we shall describe how to generate more complicated relations from these. Each definable relation will be defined by a formula, and the formula will show how the relation is built up from simpler relations.

This approach pays dividends in several ways. First, the formulas give a way of describing the relations. Second, we can prove theorems about all definable sets by induction on the complexity of their defining formulas. Third, we can prove theorems about those relations which are defined by formulas of particular kinds (as for example in section 2.4 below).

For our starting point, we take the relations expressed by atomic formulas.

Some notation will be helpful here. Given an L-structure A and an atomic formula $\phi(x_0, \ldots, x_{n-1})$ of L, we write $\phi(A^n)$ for the set of n-tuples $\{\bar{a}: A \vDash \phi(\bar{a})\}$. For example if R is a relation symbol of the signature L, then the relation R^A is of the form $\phi(A^n)$: take $\phi(x_0, \ldots, x_{n-1})$ to be the formula $R(x_0, \ldots, x_{n-1})$.

We allow parameters too. Let $\psi(x_0, \ldots, x_{n-1}, \bar{y})$ be an atomic formula of L and $\bar{b}$ a tuple from A. Then $\psi(A^n, \bar{b})$ means the set $\{\bar{a}: A \vDash \psi(\bar{a}, \bar{b})\}$. For example if A consists of the real numbers and $\psi(x, y)$ is the formula $x > y$, then $\psi(A, 0)$ is the set of all positive reals.

To build up more complicated relations, we introduce a formal language $L_{\infty\omega}$ based on the signature L, as follows. (Read on for a couple of pages to see what the subscripts $_\infty$ and $_\omega$ mean.)

Building a language

Let L be a signature. The language $L_{\infty\omega}$ will be infinitary, which means that some of its formulas will be infinitely long. It's a language in a formal or set-theoretic sense. The symbols of $L_{\infty\omega}$ are those of L together with some logical symbols, variables and punctuation signs. The logical symbols are the following:

(1.1) $=$ 'equals',

        $\neg$ 'not',

        $\bigwedge$ 'and',

        $\bigvee$ 'or',

        $\forall$ 'for all elements . . . .',

        $\exists$ 'there is an element . . . .'.

We define the **terms**, the **atomic formulas** and the **literals** of $L_{\infty\omega}$ to be the same as those of L. The class of **formulas** of $L_{\infty\omega}$ is defined to be the smallest class X such that

(1.2)   all atomic formulas of L are in X,

(1.3)   if ϕ is in X then the expression $\neg\phi$ is in X, and if $\Phi \subseteq X$ then the expressions $\bigwedge\Phi$ and $\bigvee\Phi$ are both in X,

(1.4)   if ϕ is in X and y is a variable then $\forall y\, \phi$ and $\exists y\, \phi$ are both in X.

The formulas which go into the making of a formula ϕ are called the **subformulas** of ϕ. The formula ϕ is counted as a subformula of itself; its **proper subformulas** are all its subformulas except itself.

The quantifiers $\forall y$ ('for all y') and $\exists y$ ('there is y') bind variables just as in elementary logic. Just as in elementary logic, we can distinguish between free and bound occurrences of variables. The **free variables** of a formula ϕ

are those which have free occurrences in ϕ. We sometimes introduce a formula ϕ as $\phi(\bar{x})$, for some sequence $\bar{x}$ of variables; this means that the variables in $\bar{x}$ are all distinct, and the free variables of ϕ all lie in $\bar{x}$. Then $\phi(\bar{s})$ means the formula that we get from ϕ by putting the terms s_i in place of the free occurrences of the corresponding variables x_i. This extends the notation that we introduced in section 1.3 for atomic formulas.

Likewise for any L-structure A and sequence $\bar{a}$ of elements of A, we extend the notation '$A \vDash \phi[\bar{a}]$' or '$A \vDash \phi(\bar{a})$' ('$\bar{a}$ satisfies ϕ in A') to all formulas $\phi(\bar{x})$ of $L_{\infty\omega}$ as follows, by induction on the construction of ϕ. These definitions are meant to fit the intuitive meanings of the symbols as given in (1.1) above.

(1.5)   If ϕ is atomic, then '$A \vDash \phi[\bar{a}]$' holds or fails just as in (3.10) and (3.11) of section 1.3.

(1.6) $A \vDash \neg\phi[\bar{a}]$ iff it is not true that $A \vDash \phi[\bar{a}]$.

(1.7) $A \vDash \bigwedge\Phi[\bar{a}]$ iff for every formula $\psi(\bar{x}) \in \Phi$, $A \vDash \psi[\bar{a}]$.

(1.8) $A \vDash \bigvee\Phi[\bar{a}]$ iff for at least one formula $\psi(\bar{x}) \in \Phi$, $A \vDash \psi[\bar{a}]$.

(1.9)   Suppose ϕ is $\forall y\, \psi$, where ψ is $\psi(y, \bar{x})$. Then $A \vDash \phi[\bar{a}]$ iff for all elements b of A, $A \vDash \psi[b, \bar{a}]$.

(1.10)   Suppose ϕ is $\exists y\, \psi$, where ψ is $\psi(y, \bar{x})$. Then $A \vDash \phi[\bar{a}]$ iff for at least one element b of A, $A \vDash \psi[b, \bar{a}]$.

If $\bar{x}$ is an n-tuple of variables, $\phi(\bar{x}, \bar{y})$ is a formula of $L_{\infty\omega}$ and $\bar{b}$ is a sequence of elements of A whose length matches that of $\bar{y}$, we write $\phi(A^n, \bar{b})$ for the set $\{\bar{a}: A \vDash \phi(\bar{a}, \bar{b})\}$. Then $\phi(A^n, \bar{b})$ is the **relation defined in** A by the formula $\phi(\bar{x}, \bar{b})$.

Example 3 *continued.* x_0 is in the same component as g if either x_0 is g, or x_0 is joined by an edge to g (in symbols $R(x_0, g)$), or there is x_1 such that $R(x_0, x_1)$ and $R(x_1, g)$, or there are x_1 and x_2 such that $R(x_0, x_1)$, $R(x_1, x_2)$ and $R(x_2, g)$, or $\dots$. In other words, the connected component of g is defined by the formula

(1.11) $$\bigvee(\{x_0 = g\} \cup \{\exists x_1 \dots \exists x_n \bigwedge (\{R(x_i, x_{i+1}):$$
$$i < n\} \cup \{R(x_n, g)\}): n < \omega\})$$

with parameter g. The formula (1.11) is not easy to read, but it is quite precise.

Levels of language

We can define the **complexity** of a formula ϕ, comp(ϕ), so that it is greater than the complexity of any proper subformula of ϕ. Using ordinals, one possible definition is

(1.12) $\text{comp}(\phi) = \sup\{\text{comp}(\psi) + 1 : \psi \text{ is a proper subformula of } \phi\}$.

But the exact notion of complexity never matters. What does matter is that we shall now be able to prove theorems about relations definable in $L_{\infty\omega}$, by using induction on the complexity of the formulas that define them.

In fact complexity is not always the most useful measure. The subscripts $_{\infty\omega}$ suggest other classifications. The second subscript, $_\omega$, means that we can put only finitely many quantifiers together in a row. By the same token, $L_{\infty 0}$ is the language consisting of those formulas of $L_{\infty\omega}$ in which no quantifiers occur; we call such formulas **quantifier-free**. Every atomic formula is quantifier-free. (Thus the curves in Example 1 above were defined by quantifier-free formulas.)

The first subscript $_\infty$ means that we can join together arbitrarily many formulas by $\bigwedge$ or $\bigvee$. The **first-order language** of L, in symbols $L_{\omega\omega}$, consists of those formulas in which $\bigwedge$ and $\bigvee$ are only used to join together finitely many formulas at a time, so that the whole formula is finite. More generally if κ is any regular cardinal (such as the first uncountable cardinal ω_1), then $L_{\kappa\omega}$ is the same as $L_{\infty\omega}$ except that $\bigwedge$ and $\bigvee$ are only used to join together fewer than κ formulas at a time. For example the formula (1.11) lies in $L_{\omega_1\omega}$, since $\bigvee$ is taken over a countable set and $\bigwedge$ is taken over finite sets.

Thus we can pick out many smaller languages inside $L_{\infty\omega}$, by choosing subclasses of the class of formulas of $L_{\infty\omega}$. In fact $L_{\infty\omega}$ itself is much too large for everyday use; most of this book will be concerned with the first-order level $L_{\omega\omega}$, making occasional forays into $L_{\omega_1\omega}$.

We shall use L as a symbol to stand for languages as well as signatures. Since a language determines its signature, there is no ambiguity if we talk about L-structures for a language L. Also if L is a first-order language, it is clear what is meant by $L_{\infty\omega}$, $L_{\kappa\omega}$ etc.; these are infinitary languages extending L. If a set X of parameters is added to L, forming a new language $L(X)$, we shall refer to the formulas of $L(X)$ as **formulas of L with parameters from X**.

Let L be a first-order language and A an L-structure. If $\phi(\bar{x})$ is a first-order formula, then a set or relation of the form $\phi(A^n)$ is said to be **first-order definable without parameters**, or more briefly $\varnothing$-**definable** (pronounced 'zero-definable'.) A set or relation of the form $\psi(A^n, \bar{b})$, where $\psi(\bar{x}, \bar{y})$ is a first-order formula and $\bar{b}$ is a tuple from some set X of elements of A, is said to be X-**definable** and **first-order definable with parameters**. (When people say simply 'first-order definable', you should check whether they allow parameters. Some do, some don't.)

The following abbreviations are standard:

(1.13) $\quad x \neq y$ $\qquad$ for $\neg x = y$,

$\qquad (\phi_1 \wedge \ldots \wedge \phi_n)$ $\qquad$ for $\bigwedge\{\phi_1, \ldots, \phi_n\}$ (finite conjunction),

$\qquad (\phi_1 \vee \ldots \vee \phi_n)$ $\qquad$ for $\bigvee\{\phi_1, \ldots, \phi_n\}$ (finite disjunction),

$$\bigwedge_{i \in I} \phi_i \qquad\qquad\qquad \text{for } \bigwedge \{\phi_i : i \in I\},$$

$$\bigvee_{i \in I} \phi_i \qquad\qquad\qquad \text{for } \bigvee \{\phi_i : i \in I\},$$

$(\phi \rightarrow \psi)$ for $(\neg \phi) \vee \psi$ ('if ϕ then ψ'),

$(\phi \leftrightarrow \psi)$ for $(\phi \rightarrow \psi) \wedge (\psi \rightarrow \phi)$ ('ϕ iff ψ'),

$\forall x_1 \ldots x_n$ or $\forall \bar{x}$ for $\forall x_1 \ldots \forall x_n$,

$\exists x_1 \ldots x_n$ or $\exists \bar{x}$ for $\exists x_1 \ldots \exists x_n$,

$\perp$ for $\bigvee \varnothing$ (empty disjunction, false everywhere).

Normal conventions for dropping brackets are in force: brackets around $(\phi \wedge \psi)$ or $(\phi \vee \psi)$ can be omitted when either $\rightarrow$ or $\leftrightarrow$ stands immediately outside these brackets. Thus $\phi \wedge \psi \rightarrow \chi$ always means $(\phi \wedge \psi) \rightarrow \chi$, not $\phi \wedge (\psi \rightarrow \chi)$.

With these conventions, formula (1.11) would normally be written

$$(1.14) \qquad x_0 = g \vee \bigvee_{n < \omega} \exists x_1 \ldots x_n \left(\left(\bigwedge_{i < n} R(x_i, x_{i+1}) \right) \wedge R(x_n, g) \right),$$

which is slightly easier to read.

A family of languages which differ from each other only in signature is called a **logic**. Thus **first-order logic** consists of the languages $L_{\omega\omega}$ as L ranges over all signatures.

Remarks on variables

Mostly we shall be interested in formulas $\phi(x_0, \ldots, x_{n-1})$ with just finitely many free variables. Every first-order formula has finitely many free variables, since it only has finite length.

We say that two formulas $\phi(x_0, \ldots, x_{n-1})$ and $\psi(y_0, \ldots, y_{n-1})$ are **equivalent** in the L-structure A if $\phi(A^n) = \psi(A^n)$, or equivalently, if $A \vDash \forall \bar{x}(\phi(\bar{x}) \leftrightarrow \psi(\bar{x}))$. Thus two formulas are equivalent in A if and only if they define the same relation in A. Likewise sets of formulas $\Phi(\bar{x})$ and $\Psi(\bar{y})$ are **equivalent in** A if $\bigwedge \Phi(A^n) = \bigwedge \Psi(A^n)$.

Of course these definitions depend on the listing of variables. For example if $\phi(x, y)$ and $\psi(y, x)$ are both $x < y$, then we shouldn't expect $\phi(x, y)$ to be equivalent to $\psi(y, x)$, since $\psi(A^2)$ will be $\phi(A^2)$ back to front.

The formula $\phi(x_0, \ldots, x_{n-1})$ and the formula $\phi(y_0, \ldots, y_{n-1})$ are equivalent in any structure. Likewise $\forall y \, R(x, y)$ is equivalent to $\forall z \, R(x, z)$. Thus we have too many formulas trying to do the same job. To cut down the clutter, we shall say that one formula is a **variant** of another formula if the two formulas differ only in the choice of variables, i.e. if each can be got from the other by a consistent replacement of variables.

Variance is an equivalence relation on the class of formulas. We shall always take the **cardinality** of a first-order language L, $|L|$, to be the number of equivalence classes of formulas of L under the relation of being variants. This agrees with the definition of $|L|$ for a signature L in section 1.2; see Exercise 7 below.

These dreary matters of syntax do have one important consequence. In a given structure A, two different first-order formulas can define the same relation; so the family of first-order definable relations on A may be much less rich than the language used to define them. How can we tell exactly what the different first-order definable relations on A are? There is no uniform machinery for answering this question; often it takes months of research and inspiration. I close this section with two very different kinds of example.

Few definable subsets: minimality

Lemma 2.1.1. *Let L be a signature, A an L-structure, X a set of elements of A and Y a relation on $\mathrm{dom}(A)$. Suppose Y is definable by some formula of signature L with parameters from X. Then for every automorphism f of A, if f fixes X pointwise (i.e. $f(a) = a$ for all a in X), then f fixes Y setwise (i.e. for every element a of A, $a \in Y \Leftrightarrow fa \in Y$).*

Proof. For formulas of $L_{\infty\omega}$ this can be proved by induction on complexity; see Exercise 8. But I urge the reader to see the lemma as a fundamental insight about mathematical structure, which must apply equally well to formulas in other logics besides $L_{\infty\omega}$. $\qquad\square$

For example, we have the following.

Theorem 2.1.2. *Let L be the empty signature and A an L-structure so that A is simply a set. Let X be any subset of A, and let Y be a subset of $\mathrm{dom}(A)$ which is definable in A by a formula of some logic of signature L, using parameters from X. Then Y is either a subset of X, or the complement in $\mathrm{dom}(A)$ of a subset of X.*

Proof. Immediate from the lemma. $\qquad\square$

Note that in this theorem, all finite subsets of X and their complements in A can be defined by first-order formulas with parameters in X. The set $\{a_0, \ldots, a_{n-1}\}$ is defined by the formula $x = a_0 \vee \ldots \vee x = a_{n-1}$ (which is $\bot$ if the set is empty); negate this formula to get a definition of the complement.

The situation in Theorem 2.1.2 is commoner than one might expect. We say that a structure A is **minimal** if A is infinite but the only subsets of $\mathrm{dom}(A)$ which are first-order definable with parameters are either finite or cofinite (i.e. complements of finite sets). More generally a set $X \subseteq \text{dom}(A)$ which is first-order definable with parameters is said to be **minimal** if X is infinite, and for every set Z which is first-order definable in A with parameters, either $X \cap Z$ or $X \backslash Z$ is finite. See section 9.2 below.

Many definable subsets: arithmetic

The definable subsets of the natural numbers have been very closely analysed, because of their importance for recursion theory.

We take $\mathbb{N}$ as in Example 2; let L be its signature. We write $(\forall x < y)\phi$, $(\exists x < y)\phi$ as shorthand for $\forall x(x < y \to \phi)$ and $\exists x(x < y \wedge \phi)$ respectively. The quantifiers $(\forall x < y)$ and $(\exists x < y)$ are said to be **bounded**. We define a hierarchy of first-order formulas of L, as follows.

(1.15)   A first-order formula of L is said to be a Π_0^0 formula, or equivalently a Σ_0^0 formula, if all quantifiers in it are bounded.

(1.16)   A formula is said to be a Π_{k+1}^0 formula if it is of form $\forall \bar{x} \, \psi$ for some Σ_k^0 formula ψ. (The tuple $\bar{x}$ may be empty.)

(1.17)   A formula is said to be a Σ_{k+1}^0 formula if it is of form $\exists \bar{x} \, \psi$ for some Π_k^0 formula ψ. (The tuple $\bar{x}$ may be empty.)

Thus for example a Σ_3^0 formula consists of three blocks of quantifiers, $\exists \bar{x} \forall \bar{y} \exists \bar{z}$, followed by a formula with only bounded quantifiers. Because the blocks are allowed to be empty, every Π_k^0 formula is also a Σ_{k+1}^0 formula and a Π_{k+1}^0 formula; the higher classes gather up the lower ones.

Let $\bar{x}$ be $(x_0, \ldots, x_{n-1})$. A set R of n-tuples of natural numbers is called a Π_k^0 relation (resp. a Σ_k^0 relation) if it is of the form $\phi(\mathbb{N}^n)$ for some Π_k^0 formula (resp. some Σ_k^0 formula) $\phi(\bar{x})$. We say that R is a Δ_k^0 relation if it is both a Π_k^0 relation and a Σ_k^0 relation. A relation is said to be **arithmetical** if it is Σ_k^0 for some k (so the arithmetical relations are exactly the first-order definable ones).

Intuitively the hierarchy measures how many times we have to run through the entire set of natural numbers if we want to check by (1.5)–(1.10) whether a particular tuple belongs to the relation R. An important theorem of Kleene says that the Δ_1^0 relations are exactly the recursive ones, and the Σ_1^0 relations are exactly the recursively enumerable ones. Another theorem of Kleene says that for each $k < \omega$ there is a relation R which is Σ_{k+1}^0 but neither Σ_k^0 nor Π_k^0. So the hierarchy keeps growing.

Exercises for section 2.1

1. By exhibiting suitable formulas, show that the set of even numbers is a Σ_0^0 set in $\mathbb{N}$. Show the same for the set of prime numbers.

2. Let A be the partial ordering (in a signature with $\leqslant$) whose elements are the positive integers, with $m \leqslant^A n$ iff m divides n. (a) Show that the set $\{1\}$ and the set of primes are both $\varnothing$-definable in A. (b) A number n is **square-free** if there is no prime p such that p^2 divides n. Show that the set of square-free numbers is $\varnothing$-definable in A.

3. Let A be the graph whose vertices are all the sets $\{m, n\}$ of exactly two natural numbers, with a joined to b iff $a \cap b \neq \varnothing \wedge a \neq b$. Show that A is not minimal, but it has infinitely many minimal subsets.

An L-structure is said to be **O-minimal** *(read 'Oh-minimal' – the O is for Ordering) if L contains a symbol $\leqslant$ which linearly orders $\mathrm{dom}\,(A)$ in such a way that every subset of $\mathrm{dom}\,(A)$ which is first-order definable with parameters is a union of finitely many intervals of the forms (a, b), $\{a\}$, $(-\infty, b)$, (a, ∞) where a, b are elements of A. A theory is* **O-minimal** *if all its models are.*

4. Let A be a linear ordering with the order-type of the rationals (see Example 2 of section 1.1 for the signature). Show that A is O-minimal. [Use Lemma 2.1.1; Example 3 in section 3.2 below may help.]

5. Let A be an infinite-dimensional vector space over a finite field. Show that A is minimal, and that the only $\varnothing$-definable sets in A are $\varnothing$, $\{0\}$ and $\mathrm{dom}\,(A)$. *In Exercise 2.7.9 we shall remove the condition that the field of scalars is finite.*

6 (Time-sharing with formulas). Let $\bar{x}$ be a k-tuple of variables and let $\phi_i(\bar{x}, \bar{y})$ $(i < n)$ be formulas of a first-order language L. Show that there is a formula $\psi(\bar{x}, \bar{y}, \bar{w})$ of L such that for every L-structure A with at least two elements, and every tuple $\bar{a}$ in A, the set of all relations of the form $\psi(A^k, \bar{a}, \bar{b})$ with $\bar{b}$ in A is exactly the set of all relations of the form $\phi_i(A^k, \bar{a})$ with $i < n$. [Let $\bar{w}$ be $(w_0, \ldots, w_n)$ and arrange that $\psi(A^k, \bar{a}, c_0, \ldots, c_n)$ is $\phi_i(A^k, \bar{a})$ when $c_n = c_i$ and $c_n \neq c_j$ $(i \neq j)$.]

7. Show that if L is a first-order language then $|L|$ is equal to $\omega +$ (the number of symbols in the signature of L).

8. Prove Lemma 2.1.1 for formulas of $L_{\infty\omega}$ with finitely many free variables, by induction on the complexity of the formulas.

9. Let L be the signature of abelian groups and p a prime. Let A be the direct sum of infinitely many copies of $\mathbb{Z}(p^2)$, the cyclic group of order p^2. Show (a) the subgroup of elements of order $\leqslant p$ is $\varnothing$-definable and minimal, (b) the set of elements of order p^2 is $\varnothing$-definable but not minimal.

10. Let G be a group, and let us call a subgroup H **definable** if H is first-order definable in G with parameters. Suppose that G satisfies the descending chain condition on definable groups of finite index in G. Show that there is a unique

smallest definable subgroup of finite index in G. Show that this subgroup is in fact $\emptyset$-definable, and deduce that it is a characteristic subgroup of G. *This subgroup is known as $G°$, by analogy with the connected component $G°$ in an algebraic group – which is in fact a special case.*

2.2 Definable classes of structures

Until the mid 1920s, formal languages were used for the most part either in a purely syntactic way, or for talking about definable sets and relations in a single structure. The chief exception was geometry, where Hilbert and others had used formal axioms to classify geometric structures. Today it's a commonplace that we can classify structures by asking what formal axioms are true in them – and so we can speak of *definable (or axiomatisable) classes of structures*. Model theory studies such classes.

We begin with some definitions. A **sentence** is a formula with no free variables. A **theory** is a set of sentences. (Strictly one should say 'class', since a theory in $L_{\infty\omega}$ could be a proper class. But normally theories are sets.)

If ϕ is a sentence of $L_{\infty\omega}$ and A is an L-structure, then clauses (1.5)–(1.10) in the previous section define a relation '$A \vDash \phi[]$', i.e. 'the empty sequence satisfies ϕ in A', taking $\bar{a}$ to be the empty sequence. We omit [] and write simply '$A \vDash \phi$'. We say that A is a **model** of ϕ, or that ϕ is **true in** A, when '$A \vDash \phi$' holds. Given a theory T in $L_{\infty\omega}$, we say that A is a **model of** T, in symbols $A \vDash T$, if A is a model of every sentence in T.

Let T be a theory in $L_{\infty\omega}$ and $\mathbf{K}$ a class of L-structures. We say that T **axiomatises $\mathbf{K}$**, or is a **set of axioms** for $\mathbf{K}$, if $\mathbf{K}$ is the class of all L-structures which are models of T. Obviously this determines $\mathbf{K}$ uniquely, and so we can write $\mathbf{K} = \text{Mod}(T)$ to mean that T axiomatises $\mathbf{K}$. Note that T is also a theory in $L^+_{\infty\omega}$ where L^+ is any signature containing L, and that $\text{Mod}(T)$ in L^+ is a different class from $\text{Mod}(T)$ in L. So the notion of 'model of T' depends on the signature. But we can leave it to the context to find a signature; if no signature is mentioned, choose the smallest L such that T is in $L_{\infty\omega}$.

Likewise if T is a theory, we say that a theory U **axiomatises** T (or is **equivalent to** T) if $\text{Mod}(U) = \text{Mod}(T)$. In particular if A is an L-structure and T is a first-order theory, we say that T **axiomatises** A if the first-order sentences true in A are exactly those which are true in every model of T. (The next section will examine this notion more closely.)

Let L be a language and $\mathbf{K}$ a class of L-structures. We define the L-**theory** of $\mathbf{K}$, $\text{Th}_L(\mathbf{K})$, to be the set (or class) of all sentences ϕ of L such that $A \vDash \phi$ for every structure A in $\mathbf{K}$. We omit the subscript $_L$ when L is first-order: the **theory** of $\mathbf{K}$, $\text{Th}(\mathbf{K})$, is the set of all *first-order* sentences which are true in every structure in $\mathbf{K}$.

We say that **K** is *L*-**definable** if **K** is the class of all models of some sentence in *L*. We say that **K** is *L*-**axiomatisable**, or **generalised *L*-definable**, if **K** is the class of models of some theory in *L*. For example **K** is **first-order definable** if **K** is the class of models of some first-order sentence, or equivalently, of some finite set of first-order sentences. Note that **K** is generalised first-order definable if and only if **K** is the class of all *L*-structures which are models of Th (**K**).

First-order definable and first-order axiomatisable classes are also known as EC and EC$_\Delta$ classes respectively. The E is for Elementary and the delta for Intersection (cf. Exercise 2 below – the German for intersection is Durchschnitt).

When one writes theories, there is no harm in using standard mathematical abbreviations, so long as they can be seen as abbreviations of genuine terms or formulas. For example we write

(2.1) $x + y + z$ for $(x + y) + z$,

  $x - y$ for $x + (-y)$,

  n for $1 + \ldots + 1$ (n times),   where n is a positive integer,

  nx for $\begin{cases} x + \ldots + x \ (n \text{ times}), & \text{where } n \text{ is a positive integer,} \\ 0 & \text{where } n \text{ is } 0, \\ -(-n)x & \text{where } n \text{ is a negative integer,} \end{cases}$

  xy for $x \cdot y$,

  x^n for $x \ldots x$ (n times), where n is a positive integer,

  $x \leqslant y$ for $x < y \vee x = y$,

  $x \geqslant y$ for $y \leqslant x$.

The following notation is useful too; it allows us to say 'There are exactly n elements x such that ...', for any $n < \omega$. Let $\phi(x, \bar{z})$ be a formula. Then we define $\exists_{\geqslant n} x \, \phi$ ('At least n elements x satisfy ϕ') as follows, by induction on n.

(2.2) $\exists_{\geqslant 0} x \, \phi$ is $\forall x \, x = x$.

  $\exists_{\geqslant 1} x \, \phi$ is $\exists x \, \phi$.

  $\exists_{\geqslant n+1} x \, \phi$ is $\exists x (\phi(x, \bar{z}) \wedge \exists_{\geqslant n} y (\phi(y, \bar{z}) \wedge y \neq x))$ (for $n \geqslant 1$).

Then we put $\exists_{\leqslant n} x \, \phi$ for $\neg \exists_{\geqslant n+1} x \, \phi$, and finally $\exists_{=n} x \, \phi$ is $\exists_{\geqslant n} x \, \phi \wedge \exists_{\leqslant n} x \, \phi$. So for example the first-order sentence $\exists_{=n} x \, x = x$ expresses that there are exactly n elements.

Axioms for particular structures

Even in geometry, axioms were first used for describing a particular structure, not for defining a class of structures. When a theory T is written down in order to describe a particular structure A, we say that A is the **intended model** of T. It often happens – as it did in geometry – that people decide to take an interest in the unintended models too.

Example 1: *The term algebra.* Let L be an algebraic signature, X a set of variables and A the term algebra of L with basis X (see section 1.3). Then we can describe A by the set of all sentences of the following forms.

(2.3) $c \neq d$ where c, d are distinct constants.

(2.4) $\forall \bar{x}\, F(\bar{x}) \neq c$ where F is a function symbol and c a constant.

(2.5) $\forall \bar{x}\bar{y}\, F(\bar{x}) \neq G(\bar{y})$ where F, G are distinct function symbols.

(2.6) $\forall x_0 \ldots x_{n-1}y_0 \ldots y_{n-1}\, (F(x_0, \ldots, x_{n-1})$

$$= F(y_0, \ldots, y_{n-1}) \to \bigwedge_{i<n} x_i = y_i).$$

(2.7) $\forall x_0 \ldots x_{n-1}\, t(x_0, \ldots, x_{n-1}) \neq x_i$ where $i < n$ and t is any term containing x_i but distinct from x_i.

(2.8)   [Use this axiom only when L is finite.] Write $\mathrm{Var}(x)$ for the formula $\bigwedge\{x \neq c : c$ a constant of $L\} \wedge \bigwedge\{\forall \bar{y}\, x \neq F(\bar{y}): F$ a function symbol of $L\}$. Then if X has finite cardinality n, we add the axiom $\exists_{=n}x\, \mathrm{Var}(x)$. If X is infinite, we add the infinitely many axioms $\exists_{\geqslant n}x\, \mathrm{Var}(x)$ $(n < \omega)$.

These axioms are all first-order. Each of them says something which is obviously true of A.

One can show that these axioms (2.3)–(2.8) axiomatise A. (This is by no means obvious. See section 2.7 below). Unfortunately they don't suffice to pin down the structure A itself, even up to isomorphism. For example let L consist of one 1-ary function symbol F and one constant c, and let X be empty. Then (2.3)–(2.8) reduce to the following.

(2.9) $\forall x\, F(x) \neq c$. $\forall xy\, (F(x) = F(y) \to x = y)$.
         $\forall x\, F(F(F(\ldots (F(x) \ldots) \neq x$ (for any positive number of Fs).
         $\forall x(x = c \vee \exists y\, x = F(y))$.

We can get a model B of (2.9) by taking the intended term model A and adding all the integers as new elements, putting $F^B(n) = n + 1$ for each integer n:

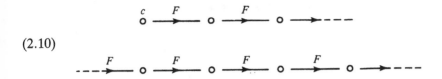

(2.10)

This model B is clearly not isomorphic to A. One can think of new elements of B as terms that can be analysed into smaller terms infinitely often.

Example 2: *The first-order Peano axioms.* This is another example of a first-order theory with an intended model. Gödel studied a close variant of it. For him it was a set of sentences which are true of the natural number structure $\mathbb{N}$ (cf. Example 2 in section 2.1); his Incompleteness Theorem says that the theory fails to axiomatise $\mathbb{N}$.

(2.11) $\forall x\; x + 1 \neq 0.$

(2.12) $\forall xy(x + 1 = y + 1 \rightarrow x = y).$

(2.13) $\forall \bar{z}(\phi(0, \bar{z}) \wedge \forall x(\phi(x, \bar{z}) \rightarrow \phi(x + 1, \bar{z})) \rightarrow \forall x\; \phi(x, \bar{z}))$

$\qquad\qquad\qquad\qquad\qquad$ for each first-order formula $\phi(x, \bar{z})$.

(2.14) $\forall x\; x + 0 = x;\; \forall xy\; x + (y + 1) = (x + y) + 1.$

(2.15) $\forall x\; x \cdot 0 = 0;\; \forall xy\; x \cdot (y + 1) = x \cdot y + x.$

(2.16) $\forall x \neg(x < 0);\; \forall xy(x < (y + 1) \leftrightarrow x < y \vee x = y).$

Clause (2.13) is an example of an **axiom schema**, i.e. a set of axioms consisting of all the sentences of a certain pattern. This schema expresses that if X is a set which is first-order definable with parameters, and (1) $0 \in X$ and (2) if $n \in X$ then $n + 1 \in X$, then every number is in X. This is the **first-order induction schema**. The axioms (2.14)–(2.16) are the **recursive definitions** of $+$, $\cdot$ and $<$; given the meanings of 0 and 1 and the function $x \mapsto x + 1$, there is only one way of defining $+$, $\cdot$ and $<$ in $\mathbb{N}$ so as to make these axioms true.

The axioms (2.11)–(2.16) are known as **first-order Peano arithmetic**, or P for short. It's natural to ask whether P has any models besides the intended one. In Chapter 6 we shall see that the compactness theorem (Theorem 5.1.1) gives the answer Yes at once. Models of P which are not isomorphic to the intended one are known as **nonstandard models**. Like randomly discovered poisonous plants, they turn out to have important and wholly benign applications that nobody dreamed of beforehand. Unfortunately there is no space in this book to discuss 'nonstandard methods'; but see the references at the end of this chapter.

A list of axiomatisable classes

There follows a list of some classes which are definable or axiomatisable. The list is for reference, not for light reading.

In most cases the sentences given in the list are just the standard definition of the class, thrown into formal symbols. We shall refer to these sentences as the **theory of** the class: thus the axioms (2.21) form the **theory of left** R-**modules**. Besides providing examples, the list shows what signatures are commonly used for various classes. For example rings normally have signature $+$, $\cdot$, $-$ (1-ary), 0 and 1. (A ring has 1 unless otherwise stated.)

(2.17)   **Groups (multiplicative)**:
$\forall xyz\ (xy)z = x(yz)$, $\forall x\ x \cdot 1 = x$, $\forall x\ x \cdot x^{-1} = 1$.

(2.18)   **Groups of exponent** n (n **a fixed positive integer**):
(2.17) together with $\forall x\ x^n = 1$.

(2.19)   **Abelian groups (additive)**:
$\forall xyz\ (x + y) + z = x + (y + z)$, $\forall x\ x + 0 = x$, $\forall x\ x - x = 0$,
$\forall xy\ x + y = y + x$.

(2.20)   **Torsion-free abelian groups**:
(2.19) together with $\forall x\ (nx = 0 \to x = 0)$ for each positive integer n.

(2.21)   **Left** R-**modules where** R **is a ring**:
As in vector spaces (Example 4 of section 1.1) the module elements are the elements of the structures. Each ring element r is used as a 1-ary function symbol, so that $r(x)$ represents rx. The axioms are (2.19) together with

$\forall xy\ r(x + y) = r(x) + r(y)$	for all $r \in R$,
$\forall x\ (r + s)(x) = r(x) + s(x)$	for all $r,s \in R$,
$\forall x\ (rs)(x) = r(s(x))$	for all $r,s \in R$,
$\forall x\ 1(x) = x$.	

(2.22)   **Rings**:
(2.19) together with
$\forall xyz\ (xy)z = x(yz)$,
$\forall x\ x1 = x$,
$\forall x\ 1x = x$,
$\forall xyz\ x(y + z) = xy + xz$,
$\forall xyz\ (x + y)z = xz + yz$.

(2.23)   **Von Neumann regular rings**:
(2.22) together with $\forall x \exists y\ xyx = x$.

(2.24)   **Fields**:
(2.22) together with $\forall xy\ xy = yx$, $0 \neq 1$,
$\forall x(x \neq 0 \to \exists y\ xy = 1)$.

(2.25) **Fields of characteristic p (p prime):**
(2.24) together with $p = 0$.

(2.26) **Algebraically closed fields:**
(2.24) together with
$$\forall x_1 \ldots x_n \exists y \; y^n + x_1 y^{n-1} + \ldots + x_{n-1} y + x_n = 0,$$
for each positive integer n.

(2.27) **Real-closed fields:**
(2.24) together with
$$\forall x_1 \ldots x_n \; x_1^2 + \ldots + x_n^2 \neq -1 \quad \text{(for each positive integer } n),$$
$$\forall x \exists y \; (x = y^2 \vee -x = y^2),$$
$$\forall x_1 \ldots x_n \exists y \; y^n + x_1 y^{n-1} + \ldots + x_{n-1} y + x_n = 0$$
$$\text{(for all odd } n).$$

(2.28) **Lattices:**
$$\forall x \; x \wedge x = x, \; \forall xy \; x \wedge y = y \wedge x,$$
$$\forall xy \; (x \wedge y) \vee y = y, \; \forall xyz \; (x \wedge y) \wedge z = x \wedge (y \wedge z),$$
$$\forall x \; x \vee x = x, \; \forall xy \; x \vee y = y \vee x,$$
$$\forall xy \; (x \vee y) \wedge y = y, \; \forall xyz \; (x \vee y) \vee z = x \vee (y \vee z).$$

(In lattices we write $x \leqslant y$ as an abbreviation of $x \wedge y = x$. Note that in sentences about lattices, the symbols $\wedge$ and $\vee$ have two meanings: the lattice meaning and the logical meaning. Brackets can help to keep them distinct.)

(2.29) **Boolean algebras:**
(2.28) together with
$$\forall xyz \; x \wedge (y \vee z) = (x \wedge y) \vee (x \wedge z),$$
$$\forall xyz \; x \vee (y \wedge z) = (x \vee y) \wedge (x \vee z),$$
$$\forall x \; x \vee x^* = 1, \; \forall x \; x \wedge x^* = 0, \; 0 \neq 1.$$

(2.30) **Atomless boolean algebras:**
(2.29) together with $\forall x \exists y (x \neq 0 \rightarrow 0 < y \wedge y < x)$.
($y < x$ is shorthand for $y \leqslant x \wedge y \neq x$.)

(2.31) **Linear orderings:**
$$\forall x \; x \not< x, \; \forall xy (x = y \vee x < y \vee y < x),$$
$$\forall xyz (x < y \wedge y < z \rightarrow x < z).$$

(2.32) **Dense linear orderings without endpoints:**
(2.31) together with $\forall xy (x < y \rightarrow \exists z (x < z \wedge z < y))$,
$$\forall x \exists z \; z < x, \; \forall x \exists z \; x < z.$$

All the classes above are generalised first-order definable. Here is a class with an infinitary definition:

(2.33)   **Locally finite groups**:
         (2.17) together with
         $\forall x_1 \ldots x_n$

$$\bigvee_{m<\omega} (\exists y_1 \ldots y_m \bigwedge_{t(\bar{x}) \text{ a term}} (t(\bar{x}) = y_1 \vee \ldots \vee t(\bar{x}) = y_m)).$$

Exercises for section 2.2

1. Let L be a first-order language and T a theory in L. Show: (a) if T and U are theories in L then $T \subseteq U$ implies $\mathrm{Mod}\,(U) \subseteq \mathrm{Mod}\,(T)$, (b) if $\mathbf{J}$ and $\mathbf{K}$ are classes of L-structures then $\mathbf{J} \subseteq \mathbf{K}$ implies $\mathrm{Th}\,(\mathbf{K}) \subseteq \mathrm{Th}\,(\mathbf{J})$, (c) $T \subseteq \mathrm{Th}\,(\mathrm{Mod}\,(T))$ and $\mathbf{K} \subseteq \mathrm{Mod}\,(\mathrm{Th}\,(\mathbf{K}))$, (d) $\mathrm{Th}\,(\mathrm{Mod}\,(T)) = T$ if and only if T is of the form $\mathrm{Th}\,(\mathbf{K})$, and likewise $\mathrm{Mod}\,(\mathrm{Th}\,(\mathbf{K})) = \mathbf{K}$ if and only if $\mathbf{K}$ is of the form $\mathrm{Mod}\,(T)$.

2. Let L be a first-order language and for each $i \in I$ let $\mathbf{K}_i$ be a class of L-structures. Show that $\mathrm{Th}\,(\bigcup_{i \in I} \mathbf{K}_i) = \bigcap_{i \in I} \mathrm{Th}\,(\mathbf{K}_i)$.

3. Let L be a first-order language and for each $i \in I$ let T_i be a theory in L. (a) Show that $\mathrm{Mod}\,(\bigcup_{i \in I} T_i) = \bigcap_{i \in I} \mathrm{Mod}\,(T_i)$. In particular if T is any theory in L, $\mathrm{Mod}\,(T) = \bigcap_{\phi \in T} \mathrm{Mod}\,(\phi)$. (b) Show that the statement $\mathrm{Mod}\,(\bigcap_{i \in I} T_i) = \bigcup_{i \in I} \mathrm{Mod}\,(T_i)$ holds when I is finite and each T_i is of the form $\mathrm{Th}\,(\mathbf{K}_i)$ for some class $\mathbf{K}_i$. It can fail if *either of these conditions is dropped*.

4. Let L be any signature containing a 1-ary relation symbol P and a k-ary relation symbol R. (a) Write down a sentence of $L_{\omega_1 \omega}$ expressing that at most finitely many elements x have the property $P(x)$. (b) When $n < \omega$, write down a sentence of $L_{\omega\omega}$ expressing that at least n k-tuples $\bar{x}$ of elements have the property $R(\bar{x})$. *One abbreviates this sentence to* $\exists_{\geqslant n} \bar{x}\, R(\bar{x})$.

5. Let L be a first-order language and A a finite L-structure. Show that every model of $\mathrm{Th}\,(A)$ is isomorphic to A. [Caution: L may have infinitely many symbols.]

6. For each of the following classes, show that it can be defined by a single first-order sentence. (a) Nilpotent groups of class k ($k \geqslant 1$). (b) Commutative rings with identity. (c) Integral domains. (d) Commutative local rings (i.e. commutative rings with a single maximal ideal). (e) Ordered fields. (f) Distributive lattices.

7. For each of the following classes, show that it can be defined by a set of first-order sentences. (a) Divisible abelian groups. (b) Fields of characteristic 0. (c) Formally real fields. (d) Separably closed fields.

8. Show that the class of simple groups is definable by a sentence of $L_{\omega_1 \omega}$.

9. Let L be a signature with a symbol $<$, and T the theory in L which expresses that $<$ is a linear ordering. (a) Define, by induction on the ordinal α, a formula $\theta_\alpha(x)$ of $L_{\infty\omega}$ which expresses (in any model of T) 'The order-type of the set of predecessors of x is α'. [The idea of (2.2) may help.] (b) Write down a set of axioms in $L_{\infty\omega}$ for the class of orderings of order-type α. Check that if α is infinite and of cardinality κ, your axioms can be written as a single sentence of $L_{\kappa^+\omega}$.

2.3 Some notions from logic

The work of the two previous sections allows us to define several important notions from logic. Any general text of logic (see the references at the end of Chapter 1) will give background information about these notions. We begin with some definitions that will be in force throughout the book, and we finish the section with an important lemma about constructing models. The definitions are worth reading but they are not worth memorising – remember that this book has an index.

Truth and consequences

Let L be a signature, T a theory in $L_{\infty\omega}$ and ϕ a sentence of $L_{\infty\omega}$. We say that ϕ is a **consequence** of T, or that T **entails** ϕ, in symbols $T \vdash \phi$, if every model of T is a model of ϕ. (In particular if T has no models then T entails ϕ.)

Warning: we don't require that if $T \vdash \phi$ then there is a proof of ϕ from T. In any case, with infinitary languages it's not always clear what would constitute a proof. Some writers use '$T \vdash \phi$' to mean that ϕ is deducible from T in some particular formal proof calculus, and they write '$T \vDash \phi$' for our notion of entailment (a notation which clashes with our '$A \vDash \phi$'). For first-order logic the two kinds of entailment coincide by the completeness theorem for the proof calculus in question.

We say that ϕ is **valid**, or is a **logical theorem**, in symbols $\vdash \phi$, if ϕ is true in every L-structure. We say that ϕ is **consistent** if ϕ is true in some L-structure. Likewise we say that a theory T is **consistent** if it has a model.

We say that two theories S and T in $L_{\infty\omega}$ are **equivalent** if they have the same models, i.e. if $\text{Mod}(S) = \text{Mod}(T)$. We also have a relativised notion of equivalence: when T is a theory in $L_{\infty\omega}$ and $\phi(\bar{x})$, $\psi(\bar{x})$ are formulas of $L_{\infty\omega}$, we say that ϕ is **equivalent to** ψ **modulo** T if for every model A of T and every sequence $\bar{a}$ from A, $A \vDash \phi(\bar{a}) \Leftrightarrow A \vDash \psi(\bar{a})$. Thus $\phi(\bar{x})$ is equivalent to $\psi(\bar{x})$ modulo T if and only if $T \vdash \forall \bar{x}(\phi \leftrightarrow \psi)$. (This sentence is not in $L_{\infty\omega}$ if ϕ and ψ have infinitely many free variables, but the sense is clear.) There is a metatheorem saying that if ϕ is equivalent to ψ modulo T, and χ' comes from

χ by putting ψ in place of ϕ somewhere inside χ, then χ' is equivalent to χ modulo T. Results like this are proved for first-order logic in elementary texts, and the proofs for other languages are no different. We can generalise relative equivalence and talk of two sets of formulas $\Phi(\bar{x})$ and $\Psi(\bar{x})$ being **equivalent modulo** T, meaning that $\bigwedge \Phi$ is equivalent to $\bigwedge \Psi$ modulo T.

A special case is where T is empty: $\phi(\bar{x})$ and $\psi(\bar{x})$ are said to be **logically equivalent** if they are equivalent modulo the empty theory. In the terminology of section 2.1, this is the same as saying that they are equivalent in every L-structure. The reader will know some examples: $\neg \forall x\, \phi$ is logically equivalent to $\exists x \neg \phi$, and $\exists x \bigvee_{i \in I} \psi_i$ is logically equivalent to $\bigvee_{i \in I} \exists x\, \psi_i$.

As another example, a formula ϕ is said to be a **boolean combination** of formulas in a set Φ if ϕ is in the smallest set X such that (1) $\Phi \cup \{\perp\} \subseteq X$ and (2) X is closed under $\wedge$, $\vee$ and $\neg$. We say that ϕ is in **disjunctive normal form over** Φ if ϕ is a finite disjunction of finite conjunctions of formulas in Y, where Y is Φ together with the negations of all formulas in Φ. By convention the empty disjunction $\perp$ and the empty conjunction $\neg \perp$ count as formulas in disjunctive normal form. *Every boolean combination* $\phi(\bar{x})$ *of formulas in a set* Φ *is logically equivalent to a formula* $\psi(\bar{x})$ *in disjunctive normal form over* Φ. (The same is true if we replace $\wedge$ and $\vee$ by $\bigwedge$ and $\bigvee$ respectively, dropping the word 'finite'; in this case we speak of **infinite boolean combinations** and **infinitary disjunctive normal form**.)

A formula is **prenex** if it consists of a string of quantifiers (possibly empty) followed by a quantifier-free formula. *Every first-order formula is logically equivalent to a prenex first-order formula.* (The result fails if one allows signatures to contain 0-ary relation symbols; see Exercise 7 below. This is one of the two embarrassing consequences of the fact that we allow structures to have empty domains. The other embarrassment is that $\vdash \exists x\, x = x$ holds if and only if the signature contains a constant. In practice these points never matter.)

Lemma 2.3.1. *Let T be a theory in a first-order language L, and Φ a set of formulas of L. Suppose*
(a) *every atomic formula of L is in Φ,*
(b) *Φ is closed under boolean combinations, and*
(c) *for every formula $\psi(\bar{x}, y)$ in Φ, $\exists y\, \psi$ is equivalent modulo T to a formula $\phi(\bar{x})$ in Φ.*
Then every formula $\chi(\bar{x})$ of L is equivalent modulo T to a formula $\phi(\bar{x})$ in Φ. (If (c) is weakened by requiring that $\bar{x}$ is non-empty, then the same conclusion holds provided $\bar{x}$ in $\chi(\bar{x})$ is also non-empty.)

Proof. By induction on the complexity of χ, using the fact that $\forall y\, \phi$ is equivalent to $\neg \exists x \neg \phi$. $\qquad \square$

An n-**type** of a theory T is a set $\Phi(\bar{x})$ of formulas, with $\bar{x} = (x_0, \ldots, x_{n-1})$, such that for some model A of T and some n-tuple $\bar{a}$ of elements of A, $A \vDash \phi(\bar{a})$ for all ϕ in Φ. We say then that A **realises** the n-type Φ, and that $\bar{a}$ **realises** Φ in A. We say that A **omits** Φ if no tuple in A realises Φ. A set Φ is a **type** if it is an n-type for some $n < \omega$. (These notions are central to model theory. See section 5.2 below for discussion and examples.)

Usually we work in a language L which is smaller than $L_{\infty\omega}$, for example a first-order language. Then all formulas in a type will automatically be assumed to come from L.

Let L be a language and A, B two L-structures. We say that A is L-**equivalent** to B, in symbols $A \equiv_L B$, if for every sentence ϕ of L, $A \vDash \phi \Leftrightarrow B \vDash \phi$. This means that A and B are indistinguishable by means of L. Two structures A and B are said to be **elementarily equivalent**, $A \equiv B$, if they are first-order equivalent. We write $\equiv_{\infty\omega}$, $\equiv_{\kappa\omega}$ for equivalence in $L_{\infty\omega}$, $L_{\kappa\omega}$ respectively.

If L is a language and A is an L-structure, the L-**theory** of A, $\mathrm{Th}_L(A)$, is the class of all sentences of L which are true in A. Thus $A \equiv_L B$ if and only if $\mathrm{Th}_L(A) = \mathrm{Th}_L(B)$. The **complete theory** of A, $\mathrm{Th}(A)$ without a language L specified, always means the complete first-order theory of A.

There is another usage of the word 'complete'. Let L be a first-order language and T a theory in L. We say that T is **complete** if T has models and any two of its models are elementarily equivalent. This is equivalent to saying that for every sentence ϕ of L, exactly one of ϕ and $\neg\phi$ is a consequence of T. Of course if A is any L-structure then $\mathrm{Th}(A)$ is complete in this sense; the compactness theorem implies that any complete theory in L is equivalent to a theory of the form $\mathrm{Th}(A)$ for some L-structure A.

We say that a theory T is **categorical** if T is consistent and all models of T are isomorphic. It will appear in section 5.1 that the only categorical first-order theories are the complete theories of finite structures, so the notion is not too useful. Instead we make the following definitions. Let λ be a cardinal. We say that a class **K** of L-structures is λ-**categorical** if there is, up to isomorphism, exactly one structure in **K** which has cardinality λ. Likewise a theory T is λ-**categorical** if the class of all its models is λ-categorical.

In a loose but very convenient turn of phrase, we say that a single structure A is λ-**categorical** if $\mathrm{Th}(A)$ is λ-categorical. This is a fairly recent change in terminology, and it reflects a shift of interest from theories to individual structures.

In section 1.4 we remarked that if $(A, \bar{a})$ is an $L(\bar{c})$-structure with A an L-structure, then for every atomic formula $\phi(\bar{x})$ of L, $A \vDash \phi[\bar{a}]$ if and only if $(A, \bar{a}) \vDash \phi(\bar{c})$. This remains true for all formulas $\phi(\bar{x})$ of $L_{\infty\omega}$, and it justifies us in using the compromise notation $A \vDash \phi(\bar{a})$ to represent either. (Thus $\bar{a}$ are either elements of A satisfying $\phi(\bar{x})$, or added constants naming themselves in the true sentence $\phi(\bar{a})$.)

Lemma 2.3.2 (Lemma on constants). *Let L be a signature, T a theory in $L_{\infty\omega}$ and $\phi(\bar{x})$ a formula in $L_{\infty\omega}$. Let $\bar{c}$ be a sequence of distinct constants which are not in L. Then $T \vdash \phi(\bar{c})$ if and only if $T \vdash \forall \bar{x}\, \phi$.*

Proof. Exercise.   □

Lastly, suppose the signature L has just finitely many symbols. Then we can identify each of these symbols with a natural number, and each term and first-order formula of L with a natural number. If this is done in a reasonable way, then syntactic operations such as forming the conjunction of two formulas, or substituting a term for a free variable in some formula, become recursive functions. In this situation we say we have a **recursive language**. Then it makes sense to talk of **recursive sets of terms** and **recursive sets of formulas**; these notions don't depend on the choice of coding. A theory T in L is said to be **decidable** if the set of its consequences is recursive. With a little more care, these definitions also make sense when the signature of L is an infinite recursive set.

Hintikka sets

Each structure A has a first-order theory $\mathrm{Th}(A)$. Does each first-order theory have a model? Clearly not. In fact a theorem of Church implies that there is no algorithm to determine whether or not a given first-order sentence has a model.

Nevertheless we found in section 1.5 that a set of atomic sentences always has a model. That fact can be generalised, as we shall see now. The results below are important for constructing models; we shall use them in Chapters 5 and 6 below.

Consider an L-structure A which is generated by its constant elements. Let T be the class of all sentences of $L_{\infty\omega}$ which are true in A. Then T has the following properties.

(3.1)   For every atomic sentence ϕ of L, if $\phi \in T$ then $\neg\phi \notin T$.

(3.2)   For every closed term t of L, the sentence $t = t$ is in T.

(3.3)   If $\phi(x)$ is an atomic formula of L, s and t are closed terms of L and $s = t \in T$, then $\phi(s) \in T$ if and only if $\phi(t) \in T$.

(3.4)   If $\neg\neg\phi \in T$ then $\phi \in T$.

(3.5)   If $\bigwedge\Phi \in T$ then $\Phi \subseteq T$; if $\neg\bigwedge\Phi \in T$ then there is $\psi \in \Phi$ such that $\neg\psi \in T$.

(3.6)   If $\bigvee\Phi \in T$ then there is $\psi \in \Phi$ such that $\psi \in T$. (In particular $\perp \notin T$.) If $\neg\bigvee\Phi \in T$ then $\neg\psi \in T$ for all $\psi \in \Phi$.

(3.7)   Let ϕ be $\phi(x)$. If $\forall x\, \phi \in T$ then $\phi(t) \in T$ for every closed term t of L; if $\neg\forall x\, \phi \in T$ then $\neg\phi(t) \in T$ for some closed term t of L.

(3.8)   Let ϕ be $\phi(x)$. If $\exists x\, \phi \in T$ then $\phi(t) \in T$ for some closed term t of L; if $\neg\exists x\, \phi \in T$ then for every closed term t of L, $\neg\phi(t) \in T$.

A theory T with the properties (3.1)–(3.8) is called a **Hintikka set** for L.

Theorem 2.3.3. *Let L be a signature and T a Hintikka set for L. Then T has a model in which every element is of the form t^A for some closed term t of L. In fact the canonical model of the set of atomic sentences in T is a model of T.*

Proof. Write U for the set of atomic sentences in T, and let A be the canonical model of U. We assert that for every sentence ϕ of $L_{\infty\omega}$,

(3.9)   if $\phi \in T$ then $A \vDash \phi$, and if $\neg\phi \in T$ then $A \vDash \neg\phi$.

(3.9) is proved as follows, by induction on the construction of ϕ, using the definition of $\vDash$ in clauses (1.5)–(1.10) of section 2.1.

By (3.2) and (3.3), U is $=$-closed in L (see section 1.5). Hence if ϕ is atomic, (3.9) is immediate by (3.1) and the definition of A.

If ϕ is of the form $\neg\psi$ for some sentence ψ, then by induction hypothesis (3.9) holds for ψ. This immediately gives the first half of (3.9) for ϕ. For the second half, suppose $\neg\phi \in T$; then $\psi \in T$ by (3.4) and hence $A \vDash \psi$ by (3.9) for ψ. But then $A \vDash \neg\phi$.

Suppose next that ϕ is $\forall x\, \psi$. If $\phi \in T$ then by (3.7), $\psi(t) \in T$ for every closed term t of L, so $A \vDash \psi(t)$ by induction hypothesis. Since every element of the canonical model is named by a closed term, this implies $A \vDash \forall x\, \psi$. If $\neg\phi \in T$ then by (3.7) again, $\neg\psi(t) \in T$ for some closed term t, and hence $A \vDash \neg\psi(t)$. Therefore $A \vDash \neg\forall x\, \psi$.

It should now be clear how the remaining cases go. From (3.9) it follows that A is a model of T. $\square$

Theorem 2.3.3 reduces the problem of finding a model to the problem of finding a particular kind of theory. The next result shows where we might look for theories of the right kind.

Theorem 2.3.4. *Let L be a first-order language. Let T be a theory in L such that*
(a) *every finite subset of T has a model,*
(b) *for every sentence ϕ of L, either ϕ or $\neg\phi$ is in T,*
(c) *for every sentence $\exists x\, \psi(x)$ in T there is a closed term t of L such that $\psi(t)$ is in T.*
Then T is a Hintikka set for L.

Proof. First we claim that

(3.10)   if U is a finite subset of T and ϕ is a sentence of L such that
$U \vdash \phi$, then $\phi \in T$.

For let U and ϕ be a counterexample. Then $\phi \notin T$, so by (b), $\neg \phi \in T$. It
follows by (a) that there is a model of $U \cup \{\neg \phi\}$, contradicting the
assumption that $U \vdash \phi$. This proves the claim, using just (a) and (b).

Now from (a) we deduce (3.1), and from the claim (3.10) we deduce (3.2),
(3.3), (3.4), the first halves of (3.5) and (3.7) and the second halves of (3.6)
and (3.8).

Suppose Φ is a finite set $\{\psi_0, \ldots, \psi_{n-1}\}$, and $\bigvee \Phi \in T$ but $\psi_i \notin T$ for all
$i < n$. Then by (b), $\neg \psi_i \in T$ for all $i < n$, and so by (a) the set
$\{\bigvee \Phi, \neg \psi_0, \ldots, \neg \psi_{n-1}\}$ has a model; which is absurd. So (3.6) holds.
Analogous arguments prove the second half of (3.5).

Finally (c) implies the first half of (3.8), and by (3.10) this implies the
second half of (3.7).   □

Exercises for section 2.3

1. Show that a theory T in a first-order language L is closed under taking
consequences if and only if $T = \mathrm{Th}\,(\mathrm{Mod}\,(T))$.

2. Let T be the theory of vector spaces over a field K. (Take T to be (2.21) from the
previous section, with $R = K$.) Show that T is λ-categorical whenever λ is an infinite
cardinal $> |K|$.

3. Let $\mathbb{N}$ be the natural number structure $(\omega, 0, 1, +, \cdot, <)$; let its signature be L.
Write a sentence of $L_{\omega_1 \omega}$ whose models are precisely the structures isomorphic to $\mathbb{N}$.
(*So there are categorical sentences of $L_{\omega_1 \omega}$ with infinite models.*)

4. Prove the lemma on constants (Lemma 2.3.2).

5. Show that if L is a first-order language with finitely many relation, function and
constant symbols, then there is an algorithm to determine, for any finite set T of
quantifier-free sentences of L, whether or not T has a model. [Use Exercises 1.5.2
and 1.5.3.]

6. For each $n < \omega$ let L_n be a signature and Φ_n a Hintikka set for L_n. Suppose that
for all $m < n < \omega$, $L_m \subseteq L_n$ and $\Phi_m \subseteq \Phi_n$. Show that $\bigcup_{n < \omega} \Phi_n$ is a Hintikka set for
the signature $\bigcup_{n < \omega} L_n$.

7. Let L be a first-order language. (a) Show that if there is an empty L-structure A
and ϕ is a prenex sentence which is true in A, then ϕ begins with a universal
quantifier. (b) Show (without assuming that every structure is non-empty) that every

formula $\phi(\bar{x})$ of L is logically equivalent to a prenex formula $\psi(\bar{x})$ of L. [You only need a new argument when $\bar{x}$ is empty.] (c) Sometimes it is convenient to allow L to contain 0-ary relation symbols (i.e. sentence letters) p; we interpret them so that for each L-structure A, p^A is either truth or falsehood, and in the definition of $\vDash$ we put $A \vDash p \Leftrightarrow p^A = $ truth. Show that in such a language L there can be a sentence which is not logically equivalent to a prenex sentence.

8. Let L be a first-order language. An L-structure A is said to be **locally finite** if every finitely generated substructure of A is finite. (a) Show that there is a set Ω of quantifier-free types (i.e. types consisting of quantifier-free formulas) such that for every L-structure A, A is locally finite if and only if A omits every type in Ω. (b) Show that if L has finite signature, then we can choose the set Ω in (a) to consist of a single type.

2.4 Maps and the formulas they preserve

Let $f: A \to B$ be a homomorphism of L-structures and $\phi(\bar{x})$ a formula of $L_{\infty\omega}$. We say that f **preserves** ϕ if for every sequence $\bar{a}$ of elements of A,

(4.1) $A \vDash \phi(\bar{a}) \Rightarrow B \vDash \phi(f\bar{a})$.

In this terminology, Theorem 1.3.1 and its corollary said that homomorphisms preserve atomic formulas, and that a homomorphism is an embedding if and only if it preserves literals. (Set theorists speak of a formula ϕ being **absolute** under f if (4.1) holds with $\Rightarrow$ replaced by $\Leftrightarrow$; thus atomic formulas are absolute under embeddings.)

The notion of preservation can be used two ways round. In this section we classify formulas in terms of the maps which preserve them. The next section will classify maps in terms of the formulas which they preserve.

Classifying formulas by maps

Our main results will say that certain types of map preserve all formulas with certain syntactic features. Later (see sections 5.4 and 8.3) we shall be able to show that in a broad sense these results are best possible for first-order formulas.

A formula ϕ is said to be an $\forall_1$ formula (pronounced 'A1 formula'), or **universal**, if it is built up from quantifier-free formulas by means of $\bigwedge$, $\bigvee$ and universal quantification (at most). It is said to be an $\exists_1$ formula (pronounced 'E1 formula'), or **existential**, if it is built up from quantifier-free formulas by means of $\bigwedge$, $\bigvee$ and existential quantification (at most).

This definition is the bottom end of a hierarchy. We shall barely need the higher reaches of the hierarchy, but here it is for the record.

(4.2)   Formulas are said to be $\forall_0$, and $\exists_0$, if they are quantifier-free.

(4.3)   A formula is an $\forall_{n+1}$ formula if it is in the smallest class of
         formulas which contains the $\exists_n$ formulas and is closed under
         $\bigwedge, \bigvee$ and adding universal quantifiers at the front.

(4.4)   A formula is an $\exists_{n+1}$ formula if it is in the smallest class of
         formulas which contains the $\forall_n$ formulas and is closed under
         $\bigwedge, \bigvee$ and adding existential quantifiers at the front.

$\forall_2$ formulas are sometimes known as $\forall\exists$ formulas. Note that just like the
arithmetical hierarchy in section 2.1, the classes of formulas increase as we go
up: every quantifier-free formula is $\forall_1$ and $\exists_1$, and all $\forall_1$ or $\exists_1$ formulas are
$\forall_2$. (Some writers use this classification only for prenex formulas.)

   If a formula is formed from other formulas by means of just $\wedge$ and $\vee$, we
say it is a **positive boolean combination** of these other formulas. If just $\bigwedge$ and
$\bigvee$ are used, we talk of a **positive infinite boolean combination**. Note that the
class of $\forall_n$ formulas and the class of $\exists_n$ formulas of $L_{\infty\omega}$, for any $n < \omega$, are
both closed under positive infinite boolean combinations.

Theorem 2.4.1. *Let $\phi(\bar{x})$ be an $\exists_1$ formula of signature L and $f: A \to B$ an
embedding of L-structures. Then f preserves ϕ.*

Proof. We first show that if $\phi(\bar{x})$ is a quantifier-free formula of L and $\bar{a}$ is a
sequence of elements of A, then

(4.5) $A \vDash \phi(\bar{a}) \Leftrightarrow B \vDash \phi(f\bar{a})$.

This is proved by induction on the complexity of ϕ. If ϕ is atomic, we have it
by Theorem 1.3.1(c). If ϕ is $\neg\psi$, $\bigwedge\Phi$ or $\bigvee\Phi$, then the result follows by
induction hypothesis and (1.6)–(1.8) of section 2.1.

   We prove the theorem by showing that for every $\exists_1$ formula $\phi(\bar{x})$ and
every sequence $\bar{a}$ of elements of A,

(4.6) $A \vDash \phi(\bar{a}) \Rightarrow B \vDash \phi(f\bar{a})$.

For quantifier-free ϕ this follows from (4.5), and $\bigwedge$ and $\bigvee$ raise no new
questions. There remains the case where $\phi(\bar{x})$ is $\exists y\, \psi(y, \bar{x})$; cf. (1.10) of
section 2.1. If $A \vDash \phi(\bar{a})$ then for some element c of A, $A \vDash \psi(c, \bar{a})$. So by
induction hypothesis, $B \vDash \psi(fc, f\bar{a})$ and hence $B \vDash \phi(f\bar{a})$ as required. $\square$

   We say that a formula $\phi(\bar{x})$ is **preserved in substructures** if whenever A
and B are L-structures, A is a substructure of B and $\bar{a}$ is a sequence of
elements of A such that $B \vDash \phi(\bar{a})$, then $A \vDash \phi(\bar{a})$ too. We say that a theory T
is an $\forall_1$ theory if all the sentences in T are $\forall_1$ formulas.

Corollary 2.4.2. (a) $\forall_1$ *formulas are preserved in substructures.*
   (b) *If T is an $\forall_1$ theory then the class of models of T is closed under taking
substructures.*

Proof. (a) Every $\forall_1$ formula is logically equivalent to the negation of an $\exists_1$ formula. (b) follows at once.   □

Part (b) of the corollary can be checked against the theories listed in section 2.2. It depends on the choice of language, of course. In the signature with just the symbol $\cdot$, a substructure of a group need not be a group; so there is no $\forall_1$ axiomatisation of the class of groups in this signature, and one has to make do with $\forall_2$ axioms instead. (Cf. Exercise 4 for similar examples.)

A formula of $L_{\infty\omega}$ is said to be **positive** if $\neg$ never occurs in it (and so by implication $\rightarrow$ and $\leftrightarrow$ never occur in it either – but it can contain $\bot$). We call a formula $\exists_1^+$ or **positive existential** if it is both positive and existential. The proof of the next theorem uses no new ideas.

Theorem 2.4.3. *Let $\phi(\bar{x})$ be a formula of signature L and $f: A \rightarrow B$ a homomorphism of L-structures.*
(a) *If ϕ is an $\exists_1^+$ formula then f preserves ϕ.*
(b) *If ϕ is positive and f is surjective, then f preserves ϕ.*
(c) *If f is an isomorphism then f preserves ϕ.*   □

There are innumerable other similar results for other types of homomorphism. See for example Exercises 5, 6, 9, 10.

Chains

Let L be a signature and $(A_i: i < \gamma)$ a sequence of L-structures. We call $(A_i: i < \gamma)$ a **chain** if for all $i < j < \gamma$, $A_i \subseteq A_j$. If $(A_i: i < \gamma)$ is a chain, then we can define another L-structure B as follows. The domain of B is $\bigcup_{i<\gamma} \mathrm{dom}\,(A_i)$. For each constant c, c^{A_i} is independent of the choice of i, so we can put $c^B = c^{A_i}$ for any $i < \gamma$. Likewise if F is an n-ary function symbol of L and $\bar{a}$ is an n-tuple of elements of B, then $\bar{a}$ is in $\mathrm{dom}\,(A_i)$ for some $i < \gamma$, and without ambiguity we can define $F^B(\bar{a})$ to be $F^{A_i}(\bar{a})$. Finally if R is an n-ary relation symbol of L, we put $\bar{a} \in R^B$ if $\bar{a} \in R^{A_i}$ for some (or all) A_i containing $\bar{a}$. By construction, $A_i \subseteq B$ for every $i < \gamma$. We call B the **union** of the chain $(A_i: i < \gamma)$, in symbols $B = \bigcup_{i<\gamma} A_i$.

We say that a formula $\phi(\bar{x})$ of L is **preserved in unions of chains** if whenever $(A_i: i < \gamma)$ is a chain of L-structures, $\bar{a}$ is a sequence of elements of A_0 and $A_i \vDash \phi(\bar{a})$ for all $i < \gamma$, then $\bigcup_{i<\gamma} A_i \vDash \phi(\bar{a})$.

Theorem 2.4.4. *Let $\psi(\bar{y}, \bar{x})$ be an $\exists_1$ formula of signature L with $\bar{y}$ finite. Then $\forall \bar{y}\, \psi$ is preserved in unions of chains of L-structures.*

Proof. Let $(A_i : i < \gamma)$ be a chain of L-structures and $\bar{a}$ a sequence of elements of A_0 such that $A_i \vDash \forall \bar{y} \, \psi(\bar{y}, \bar{a})$ for all $i < \gamma$. Put $B = \bigcup_{i < \gamma} A_i$. To show that $B \vDash \forall \bar{y} \, \psi(\bar{y}, \bar{a})$, let $\bar{b}$ be any tuple of elements of B. Since $\bar{b}$ is finite, there is some $i < \gamma$ such that $\bar{b}$ lies in A_i. By assumption, $A_i \vDash \psi(\bar{b}, \bar{a})$. Since $A_i \subseteq B$, it follows from Theorem 2.4.1 that $B \vDash \psi(\bar{b}, \bar{a})$. $\qquad\square$

Any $\forall_2$ first-order formula can be brought to the form $\forall \bar{y} \, \psi$ with ψ existential. So Theorem 2.4.4 says in particular that all $\forall_2$ first-order formulas are preserved in unions of chains. For example the axioms (2.32) in section 2.2 (for dense linear orderings without endpoints) are $\forall_2$ first-order, and it follows at once that the union of a chain of dense linear orderings without endpoints is a dense linear ordering without endpoints.

Exercises for section 2.4

1. Let L be a first-order language. (a) Suppose ϕ is an $\forall_1$ sentence of L and A is an L-structure. Show that $A \vDash \phi$ if and only if $B \vDash \phi$ for every finitely generated substructure B of A. (b) Show that if A and B are L-structures, and every finitely generated substructure of A is embeddable in B, then every $\forall_1$ sentence of L which is true in B is true in A too.

2. Suppose the first-order language L has just finitely many relation symbols and constants, and no function symbols. Show that if A and B are L-structures such that every $\forall_1$ sentence of L which is true in B is true in A too, then every finitely generated substructure of A is embeddable in B.

3. Let L be a first-order language and T a theory in L, such that every $\forall_1$ formula $\phi(\bar{x})$ of L is equivalent modulo T to an $\exists_1$ formula $\psi(\bar{x})$. Show that every formula $\phi(\bar{x})$ of L is equivalent modulo T to an $\exists_1$ formula $\psi(\bar{x})$. [Put ϕ into prenex form and peel away the quantifier blocks, starting at the inside.]

4. In section 2.2 above there are axiomatisations of several important classes of structure. Show that, using the signatures given in section 2.2, it is not possible to write down sets of axioms of the following forms: (a) a set of $\exists_1$ axioms for the class of groups; (b) a set of $\forall_1$ axioms for the class of atomless boolean algebras; (c) a single $\exists_2$ first-order axiom for the class of dense linear orderings without endpoints.

*Let L be a signature containing the 2-ary relation symbol $<$. If A and B are L-structures, we say that B is an **end-extension** of A if $A \subseteq B$ and whenever a is an element of A and $B \vDash b < a$ then b is an element of A too. We say that an embedding $f : A \to B$ of L-structures is an **end-embedding** if B is an end-extension of the image of f. We define $(\forall x < y)$ and $(\exists x < y)$ as in section 2.1: $(\forall x < y)\phi$ is $\forall x(x < y \to \phi)$ and $(\exists x < y)\phi$ is $\exists x(x < y \land \phi)$, and the quantifiers $(\forall x < y)$ and $(\exists x < y)$ are said to be **bounded**.*

5. Let L be as above. A Π_0^0 formula is one in which all quantifiers are bounded. A Σ_1^0 formula is a formula in the smallest class of formulas which contains the Π_0^0 formulas and is closed under $\bigwedge$, $\bigvee$ and existential quantification. Show that end-embeddings preserve Σ_1^0 formulas.

6. Let L be a signature containing a 1-ary symbol P. By a P-**embedding** we mean an embedding $e: A \to B$, where A and B are L-structures, such that e maps P^A onto P^B. Let Φ be the smallest class of formulas of $L_{\infty\omega}$ such that (i) every quantifier-free formula is in Φ, (ii) Φ is closed under $\bigwedge$ and $\bigvee$, (iii) if ϕ is in Φ and x is a variable then $\exists x\, \phi$ and $\forall x(Px \to \phi)$ are in Φ. Show that every P-embedding preserves all the formulas in Φ.

Let L be a signature. By a **descending chain of L-structures** *we mean a sequence $(A_i: i < \gamma)$ of L-structures such that $A_j \subseteq A_i$ whenever $i < j < \gamma$.*

7. Show that if $(A_i: i < \gamma)$ is a descending chain of L-structures, then there is a unique L-structure B which is a substructure of each A_j and has domain $\bigcap_{i<\gamma} \mathrm{dom}\,(A_i)$. (We call this structure the **intersection** of the chain, in symbols $\bigcap_{i<\gamma} A_i$.)

We say that a formula ϕ of $L_{\infty\omega}$ is **preserved in intersections of descending chains** *if for every descending chain $(A_i: i < \gamma)$ of L-structures and every tuple $\bar{a}$ of elements of $\bigcap_{i<\gamma} A_i$, if $A_j \vDash \phi(\bar{a})$ for every $j < \gamma$ then $\bigcap_{i<\gamma} A_i \vDash \phi(\bar{a})$.*

8. (a) Show that, if ϕ is a formula of $L_{\infty\omega}$ of the form $\forall \bar{x} \exists_{=n} y\, \psi(\bar{x}, y, \bar{z})$, where ψ is quantifier-free, then ϕ is preserved in intersections of descending chains of L-structures. (b) Write a set of first-order axioms of this form for the class of real-closed fields. (c) Can axioms of this form be found for the class of dense linear orderings without endpoints?

Let $\mathbf{K}$ be the class of all L-structures that carry a partial ordering (named by the relation symbol $<$) which has no maximal element. When A and B are structures in $\mathbf{K}$, we say that A is a **cofinal substructure** *of B (and that B is a* **cofinal extension** *of A) if $A \subseteq B$ and for every element b of B there is an element a in A such that $B \vDash b < a$. A formula $\phi(\bar{x})$ of L is* **preserved in cofinal substructures** *if whenever A is a cofinal substructure of B and $\bar{a}$ is a tuple in A such that $B \vDash \phi(\bar{a})$, then $A \vDash \phi(\bar{a})$.*

9. Let L be a signature and Φ the smallest class of formulas of $L_{\infty\omega}$ such that (1) all literals of L are in Φ, (2) Φ is closed under $\bigwedge$ and $\bigvee$, and (3) if $\phi(x, \bar{y})$ is any formula in Φ, then so are the formulas $\forall x\, \phi$ and $\exists z \forall x(z < x \to \phi)$. Show that every formula in Φ is preserved in cofinal substructures.

We say that a relation symbol R is **positive in** *the formula ϕ of $L_{\infty\omega}$ if ϕ is in the smallest class X of formulas such that (1) every literal of L which doesn't contain R is in X, (2) every atomic formula of L is in X, and (3) X is closed under $\bigwedge$, $\bigvee$ and quantification. (For example R is positive in $\forall x(Qx \to Rx)$, but not in $\forall x(Rx \to Qx)$.)*

10. Let L be a signature and L^+ the signature got by adding to L a new n-ary relation symbol P. Let $\bar{x}$ be an n-tuple of variables and $\phi(\bar{x})$ a formula of L^+ in which

P is positive. Let A be an L-structure; suppose X and Y are n-ary relations on dom(A) with $X \subseteq Y$. It is clear that the identity map on A forms an embedding $e: (A, X) \to (A, Y)$ of L^+-structures. (a) Show that e preserves ϕ. (b) For any n-ary relation X on dom(A) we define $\pi(X)$ to be the relation $\{\bar{a}: (A, X) \vDash \phi(\bar{a})\}$. Show that if $X \subseteq Y$ then $\pi(X) \subseteq \pi(Y)$.

2.5 Classifying maps by formulas

Let L be a signature, $f: A \to B$ a homomorphism of L-structures and Φ a class of formulas of $L_{\infty\omega}$. We call f a **Φ-map** if f preserves all the formulas in Φ.

By far the most important example is when Φ is the class of all first-order formulas. A homomorphism which preserves all first-order formulas must be an embedding (by Corollary 1.3.2); we call such a map an **elementary embedding**.

We say that B is an **elementary extension** of A, or that A is an **elementary substructure** of B, in symbols $A \preccurlyeq B$, if $A \subseteq B$ and the inclusion map is an elementary embedding. (This map is then described as an **elementary inclusion**.) We write $A \prec B$ when A is a proper elementary substructure of B.

Note that $A \preccurlyeq B$ implies $A \equiv B$. But there are examples to show that $A \subseteq B$ and $A \equiv B$ together don't imply $A \preccurlyeq B$; see Exercise 2.

Theorem 2.5.1 (*Tarski–Vaught criterion for elementary substructures*). *Let L be a first-order language and let A, B be L-structures with $A \subseteq B$. Then the following are equivalent.*

(a) *A is an elementary substructure of B.*

(b) *For every formula $\psi(\bar{x}, y)$ of L and all tuples $\bar{a}$ from A, if $B \vDash \exists y\, \psi(\bar{a}, y)$ then $B \vDash \psi(\bar{a}, d)$ for some element d of A.*

Proof. Let $f: A \to B$ be the inclusion map. (a) $\Rightarrow$ (b): If $B \vDash \exists y\, \psi(\bar{a}, y)$ then $A \vDash \exists y\, \psi(\bar{a}, y)$ since f is elementary; hence there is d in A such that $A \vDash \psi(\bar{a}, d)$, and we reach $B \vDash \psi(\bar{a}, d)$ by applying f again. (b) $\Rightarrow$ (a): take the proof of Theorem 2.4.1 in the previous section; our condition (b) is exactly what is needed to make that proof show that f is elementary. $\square$

Theorem 2.5.1 by itself isn't very useful for detecting elementary substructures in nature (though see Exercises 3, 4 below). Its main use is for constructing elementary substructures, as in Exercise 5.

If Φ is a set of formulas, we say that a chain $(A_i: i < \gamma)$ of L-structures is a **Φ-chain** when each inclusion map $A_i \subseteq A_j$ is a Φ-map. In particular an **elementary chain** is a chain in which the inclusions are elementary.

Theorem 2.5.2 (Tarski–Vaught theorem on unions of elementary chains). *Let* $(A_i: i < \gamma)$ *be an elementary chain of L-structures. Then* $\bigcup_{i<\gamma} A_i$ *is an elementary extension of each A_j $(j < \gamma)$.*

Proof. Put $A = \bigcup_{i<\gamma} A_i$. Let $\phi(\bar{x})$ be a first-order formula of signature L. We show by induction on the complexity of ϕ that for every $j < \gamma$ and every tuple $\bar{a}$ of elements of A_j,

(5.1) $\qquad\qquad\qquad A_j \vDash \phi(\bar{a}) \Leftrightarrow A \vDash \phi(\bar{a}).$

When ϕ is atomic, we have (5.1) by Theorem 1.3.1(c). The cases $\neg \psi, (\psi \wedge \chi)$ and $(\psi \vee \chi)$ are straightforward. Suppose then that ϕ is $\exists y\, \psi(\bar{x}, y)$. If $A \vDash \phi(\bar{a})$ then there is some b in A such that $A \vDash \psi(\bar{a}, b)$. Choose $k < \gamma$ so that b is in $\mathrm{dom}(A_k)$ and $k \geqslant j$. Then $A_k \vDash \psi(\bar{a}, b)$ by induction hypothesis, hence $A_k \vDash \phi(\bar{a})$. Then $A_j \vDash \phi(\bar{a})$ as required, since the chain is elementary. This proves right to left in (5.1); the other direction is easier. The argument for $\forall y\, \psi(\bar{x}, y)$ is similar. $\qquad\square$

Lemma 2.5.3 (Elementary diagram lemma). *Suppose L is a first-order language, A and B are L-structures, $\bar{c}$ is a tuple of distinct constants not in L, $(A, \bar{a})$ and $(B, \bar{b})$ are $L(\bar{c})$-structures, and $\bar{a}$ generates A. Then the following are equivalent.*
(a) *For every formula $\phi(\bar{x})$ of L, if $(A, \bar{a}) \vDash \phi(\bar{c})$ then $(B, \bar{b}) \vDash \phi(\bar{c})$.*
(b) *There is an elementary embedding $f: A \to B$ such that $f\bar{a} = \bar{b}$.*

Proof. (b) clearly implies (a). For the converse, define f as in the proof of Lemma 1.4.2. If $\bar{a}'$ is any tuple of elements of A and $\phi(\bar{z})$ is any formula of L, then by choosing a suitable sequence $\bar{x}$ of variables, we can write $\phi(\bar{z})$ as $\psi(\bar{x})$ so that $\phi(\bar{a}')$ is the same formula as $\psi(\bar{a})$. Then $A \vDash \phi(\bar{a}')$ implies $A \vDash \psi(\bar{a})$, which by (a) implies $B \vDash \psi(f\bar{a})$ and hence $B \vDash \phi(f\bar{a}')$. So f is an elementary embedding. $\qquad\square$

We define the **elementary diagram** of an L-structure A, in symbols $\mathrm{eldiag}(A)$, to be $\mathrm{Th}(A, \bar{a})$ where $\bar{a}$ is any sequence which generates A. By (a) $\Rightarrow$ (b) in the lemma we have the following fact, which will be used constantly for constructing elementary extensions: *if D is a model of the elementary diagram of the L-structure A, then there is an elementary embedding of A into the reduct $D|L$.*

The notion of a Φ-map has other applications.

Example 1: *Pure extensions.* This example is familiar to abelian group theorists and people who work with modules. Let A and B be left R-modules, and

A a submodule of *B*. We say that *A* is **pure** in *B*, or that *B* is a **pure extension** of *A*, if the following holds:

(5.2)   for every finite set *E* of equations with parameters in *A*, if *E* has a solution in *B* then *E* already has a solution in *A*.

Now the statement that a certain finite set of equations with parameters $\bar{a}$ has a solution can be written $\exists \bar{x}(\psi_1(\bar{x}, \bar{a}) \wedge \ldots \wedge \psi_k(\bar{x}, \bar{a}))$ with $\psi_1, \ldots, \psi_k$ atomic; a first-order formula of this form is said to be **positive primitive**, or **p.p.** for short. So we can define a **pure** embedding to be one which preserves the negations of all positive primitive formulas.

Exercises for section 2.5

Our first exercise is a slight refinement of Theorem 2.5.1.

1. Let *L* be a first-order language and *B* an *L*-structure. Suppose *X* is a set of elements of *B* such that, for every formula $\psi(\bar{x}, y)$ of *L* and all tuples $\bar{a}$ of elements of *X*, if $B \vDash \exists y\, \psi(\bar{a}, y)$ then $B \vDash \psi(\bar{a}, d)$ for some element *d* in *X*. Show that *X* is the domain of an elementary substructure of *B*.

2. Give an example of a structure *A* with a substructure *B* such that $A \cong B$ but *B* is not an elementary substructure of *A*. [Take *A* to be $(\omega, <)$.]

3. Let *B* be an *L*-structure and *A* a substructure with the following property: if $\bar{a}$ is any tuple of elements of *A* and *b* is an element of *B*, then there is an automorphism *f* of *B* such that $f\bar{a} = \bar{a}$ and $fb \in \mathrm{dom}(A)$. Show that if $\phi(\bar{x})$ is any formula of $L_{\infty\omega}$ and $\bar{a}$ a tuple in *A*, then $A \vDash \phi(\bar{a}) \Leftrightarrow B \vDash \phi(\bar{a})$. (In particular $A \preccurlyeq B$.)

4. Suppose *B* is a vector space and *A* is a subspace of infinite dimension. Show that $A \preccurlyeq B$.

The next exercise will be refined in section 3.1 below.

5. Let *L* be a countable first-order language and *B* an *L*-structure of infinite cardinality μ. Show that for every infinite cardinal $\lambda < \mu$, *B* has an elementary substructure of cardinality λ. [Choose a set *X* of λ elements of *B*, and close off *X* so that Exercise 1 applies.]

6. Let $(A_i : i < \gamma)$ be a chain of structures such that for all $i < j < \gamma$, A_i is a pure substructure of A_j. Show that each structure $A_j \, (j < \gamma)$ is a pure substructure of the union $\bigcup_{i<\gamma} A_i$.

The next two exercises are not particularly hard, but they use things from later sections.

7. Show that if *n* is a positive integer, then there are a first-order language *L* and *L*-structures *A* and *B* such that $A \subseteq B$ and for every *n*-tuple $\bar{a}$ in *A* and every formula $\phi(x_0, \ldots, x_{n-1})$ of *L*, $A \vDash \phi(\bar{a}) \Leftrightarrow B \vDash \phi(\bar{a})$, but *B* is not an elementary extension of *A*. [Use Fraïssé limits (Chapter 6) to construct a countable structure C^+

in which a $(2n + 2)$-ary relation symbol E defines a random equivalence relation on the sets of $n + 1$ elements, and a $(2n + 2)$-ary relation symbol R randomly arranges these equivalence classes in a linear ordering whose first element is defined by the $(n + 1)$-ary relation symbol P. Form C by dropping the symbol P from C^+. Find an embedding $e \colon C \to C$ whose image misses the first element of the ordering.]

8. Suppose R is a ring and A, B are left R-modules with $A \subseteq B$, and, for every element a and A and every p.p. formula $\phi(x)$ without parameters, $A \vDash \phi(a) \Leftrightarrow B \vDash \phi(a)$. Show that A is pure in B. [Show by induction on n that if $\phi(x_0, \ldots, x_n)$ is a pure formula and $\bar{a}$ an $(n + 1)$-tuple in A such that $B \vDash \phi(\bar{a})$ then $A \vDash \phi(\bar{a})$. If $B \vDash \phi(\bar{a})$ then $B \vDash \exists x_n \, \phi(\bar{a} | n, x_n)$, so by induction hypothesis there is c in A such that $A \vDash \phi(\bar{a} | n, c)$, so by subtraction (justify this!) $B \vDash \phi(0, \ldots, 0, a_n - c)$; hence $A \vDash \phi(0, \ldots, 0, a_n - c)$, and addition yields $A \vDash \phi(\bar{a})$.]

9. Let R be a ring and L the language of left R-modules. Let M be a left R-module and $\phi(x_0, \ldots, x_{n-1})$ a p.p. formula of L. (a) Show that $\phi(M^n)$ is a subgroup of M^n regarded as an abelian group. *Groups of this form are known as the* **p.p.-definable subgroups** *of M^n. See Exercise 2.7.12 for their importance.* (b) Show that if $\phi(\bar{x}, \bar{y})$ is a p.p. formula of L and $\bar{b}$ a tuple from M, then $\phi(M^n, \bar{b})$ is either empty or a coset of the p.p-definable subgroup $\phi(M^n, 0, \ldots, 0)$. Show that both possibilities can occur.

2.6 Translations

In model theory as elsewhere, there can be several ways of saying the same thing. Say it in the wrong way and you may not get the results you intended.

This section is an introduction to some kinds of paraphrase that model theorists find useful. *These paraphrases never alter the class of definable relations on a structure – they only affect formulas which can be used to define them.*

1: Unnested formulas

Let L be a signature. By an **unnested atomic formula** of signature L we mean an atomic formula of one of the following forms:

(6.1) $x = y$;

(6.2) $c = y$ for some constant c of L;

(6.3) $F(\bar{x}) = y$ for some function symbol F of L;

(6.4) $R\bar{x}$ for some relation symbol R of L.

We call a formula **unnested** if all of its atomic subformulas are unnested.

Unnested formulas are handy when we want to make definitions or proofs by induction on the complexity of formulas. The atomic case becomes particularly simple: we never need to consider any terms except variables, constants and terms $F(\bar{x})$ where F is a function symbol. There will be examples in sections 3.3 (back-and-forth games) and 4.3 (interpretations).

Theorem 2.6.1 *Let L be a signature. Then every atomic formula $\phi(\bar{x})$ of L is logically equivalent to unnested first-order formulas $\phi^\forall(\bar{x})$ and $\phi^\exists(\bar{x})$ of signature L such that $\phi^\forall$ is an $\forall_1$ formula and $\phi^\exists$ is an $\exists_1$ formula.*

Proof by example. The formula $F(G(x), z) = c$ is logically equivalent to

(6.5) $\forall uw (G(x) = u \wedge F(u, z) = w \to c = w)$

and to

(6.6) $\exists uw (G(x) = u \wedge F(u, z) = w \wedge c = w)$.   □

Corollary 2.6.2. *Let L be a first-order language. Then every formula $\phi(\bar{x})$ of L is logically equivalent to an unnested formula $\psi(\bar{x})$ of L. More generally every formula of $L_{\infty\omega}$ is logically equivalent to an unnested formula of $L_{\infty\omega}$.*

Proof. Use the theorem to replace all atomic subformulas by unnested first-order formulas.   □

If ϕ in the corollary is an $\exists_1$ formula, then by choosing wisely between $\theta^\forall$ and $\theta^\exists$ for each atomic subformula θ of ϕ, we can arrange that ψ in the corollary is an $\exists_1$ formula too. In fact we can always choose ψ to lie in the same place in the $\forall_n$, $\exists_n$ hierarchy (see (4.2)–(4.4) in section 2.4 above) as ϕ, unless ϕ is quantifier-free.

2: Definitional expansions and extensions

Let L and L^+ be signatures with $L \subseteq L^+$, and let R be a relation symbol of L^+. Then an **explicit definition of R in terms of L** is a sentence of the form

(6.7) $\forall \bar{x} \, (R\bar{x} \leftrightarrow \phi(\bar{x}))$

where ϕ is a formula of L. Likewise if c is a constant and F is a function symbol of L^+, **explicit definitions of c, F in terms of L** are sentences of the form

(6.8) $\forall y \, (c = y \leftrightarrow \phi(y))$,

         $\forall \bar{x} y \, (F(\bar{x}) = y \leftrightarrow \psi(\bar{x}, y))$

where ϕ, ψ are formulas of L. Note that the sentences in (6.8) have consequences in the language L. They imply respectively

(6.9) $\exists_{=1} y \, \phi(y)$,

         $\forall \bar{x} \exists_{=1} y \, \psi(\bar{x}, y)$.

We call the sentences (6.9) the **admissibility conditions** of the two explicit definitions in (6.8).

Explicit definitions have two main properties, as follows.

Theorem 2.6.3 (*Uniqueness of definitional expansions*). *Let L and L^+ be signatures with $L \subseteq L^+$. Let A and B be L^+-structures, R a relation symbol of L^+ and θ an explicit definition of R in terms of L. If A and B are both models of θ, and $A|L = B|L$, then $R^A = R^B$. Similarly for constants and function symbols.*

Proof. Immediate.   □

Theorem 2.6.4 (*Existence of definitional expansions*). *Let L and L^+ be signatures with $L \subseteq L^+$. Suppose that for each symbol S of $L^+ \backslash L$, θ_S is an explicit definition of S in terms of L; let U be the set of these definitions.*

(a) *If C is any L-structure which satisfies the admissibility conditions (if any) of the definitions θ_S, then we can expand C to form an L^+-structure C^+ which is a model of U.*

(b) *Every formula $\chi(\bar{x})$ of signature L^+ is equivalent modulo U to a formula $\chi^*(\bar{x})$ of signature L.*

(c) *If χ and all the sentences θ_S are first-order, then so is χ^*.*

Proof. The definitions tell us exactly how to interpret the symbols S in C^+, so we have (a). For (b), use Theorem 2.6.1 to replace every atomic formula in χ by an unnested formula, and observe that the explicit definitions translate each unnested atomic formula directly into a formula of signature L. Then (c) is clear too.   □

A structure C^+ as in Theorem 2.6.4(a) is called a **definitional expansion** of C. If L and L^+ are signatures with $L \subseteq L^+$, and T is a theory of signature L, then a **definitional extension** of T to L^+ is a theory equivalent to $T \cup \{\theta_S : S \text{ a symbol in } L^+ \backslash L\}$ where for each symbol S in $L^+ \backslash L$,

(6.10) θ_S is an explicit definition of S in terms of L, and

(6.11)   if S is a constant or function symbol and χ is the admissibility condition for θ_S then $T \vdash \chi$.

Theorems 2.6.3 and 2.6.4 tell us that if T^+ is a definitional extension of T to L^+, then every model C of T has a unique expansion C^+ which is a model of T^+, and C^+ is a definitional expansion of C.

Let T^+ be a theory in the language L^+, and L a language $\subseteq L^+$. We say that a symbol S of L^+ is **explicitly definable in T^+ in terms of L** if T^+ entails some explicit definition of S in terms of L. So, up to equivalence of theories,

T^+ is a definitional extension of a theory T in L if and only if (1) T and T^+ have the same consequences in L and (2) every symbol of L^+ is explicitly definable in T^+ in terms of L.

Definitional extensions are useful for replacing complicated formulas by simple ones. For example in set theory they allow us to write $\phi(x \cup y)$ instead of the less readable formula $\exists z(\phi(z) \wedge \forall t(t \in z \leftrightarrow t \in x \vee t \in y))$.

Warning – particularly for software developers using first-order logic. It's important that the symbols being defined in (6.7) and (6.8) don't occur in the formulas ϕ, ψ. If the symbols were allowed to occur in ϕ or ψ, Theorems 2.6.3 and 2.6.4 would both fail.

There is a deathtrap here. One sometimes meets things called 'definitions', which look like explicit definitions except that the symbol being defined occurs on both sides of the formula. These may be implicit definitions, which are harmless – at least in first-order logic; see Beth's theorem, Theorem 5.5.4 below. On the other hand if they have purple fins underneath and pink spots on top, they are almost certainly *recursive definitions*. The recursive definitions of plus and times in arithmetic, (2.14)–(2.15) in section 2.2, are a typical example; one can rewrite them to look dangerously like explicit definitions. *In general, recursive definitions define symbols on a particular structure, not on all models of a theory. There is no guarantee that they can be translated into explicit definitions in a first-order theory.* The notion of a recursive definition lies outside model theory and I shall say no more about it here.

Before we leave definitional extensions, here is an extended example which will be useful in the next section.

Suppose L_1 and L_2 are signatures; to simplify the next definition let us assume they are disjoint. Let T_1 and T_2 be first-order theories of signature L_1, L_2 respectively. Then we say that T_1 and T_2 are **definitionally equivalent** if there is a first-order theory T in signature $L_1 \cup L_2$ which is a definitional extension both of T_1 and of T_2.

When theories T_1 and T_2 are definitionally equivalent as above, we can turn a model A_1 of T_1 into a model A_2 of T_2 by first expanding A_1 to a model of T and then restricting to the language L_2; we can get back to A_1 from A_2 by doing the same in the opposite direction. In this situation we say that the structures A_1 and A_2 are **definitionally equivalent**.

Example 1: *Term algebras in another language.* In Example 1 of section 2.2 we wrote down some sentences (2.3)–(2.7) which are true in every term algebra of a fixed algebraic signature L. Call these sentences T_1 and let L_1

be their first-order language. Let L_2 be the first-order language whose signature consists of the following symbols:

(6.12) 1-ary relation symbols Is_c (for each constant c of L_1) and Is_F (for each function symbol F of L_1);

(6.13) a 1-ary function F_i for each function symbol F of L_1 and each $i < \text{arity}(F)$.

We claim that T_1 is definitionally equivalent to the following theory T_2 in L_2.

(6.14) $\exists_{=1} y\, \text{Is}_c(y)$ for each constant symbol c of L.

(6.15) $\forall x_0 \ldots x_{n-1} \exists_{=1} y\, (\text{Is}_F(y) \wedge \bigwedge_{i<n} F_i(y) = x_i)$

                                                    for each function symbol F of L.

(6.16) $\forall x\, \neg(\text{Is}_c(x) \wedge \text{Is}_d(x))$ where c, d are distinct constant or function symbols.

(6.17) $\forall x(\neg \text{Is}_F(x) \rightarrow F_i(x) = x)$ for each function symbol F_i.

(6.18) $\forall x(t(F_i(x)) = x \rightarrow \neg \text{Is}_F(x))$ for each function symbol F_i and term $t(y)$ of L_2.

To show this, we must write down explicit definitions U_1 of the symbols of L_2 in terms of L_1, and explicit definitions U_2 of the symbols of L_1 in terms of L_2, in such a way that T_i implies the admissibility conditions for U_i ($i = 1, 2$), and $T_1 \cup U_1$ is equivalent to $T_2 \cup U_2$. Here are the definitions of L_2 in terms of L_1, where c is any constant of L_1, F is any function symbol of L_1 with arity n and $i < n$.

(6.19) $\forall y\, (\text{Is}_c(y) \leftrightarrow y = c)$.

$\forall y\, (\text{Is}_F(y) \leftrightarrow \exists \bar{x}\, F\bar{x} = y)$.

$\forall xy\, (F_i x = y \leftrightarrow (\exists y_0 \ldots y_{i-1} y_{i+1} \cdots y_{n-1}$

$\qquad F(y_0, \ldots, y_{i-1}, y, y_{i+1}, \ldots, y_{n-1}) = x)$

$\qquad \vee (x = y \wedge \neg \exists \bar{y}\, F\bar{y} = x))$.

And here are the explicit definitions of L_1 in terms of L_2, where c is any constant of L_1 and F any function symbol of L_1:

(6.20) $\forall y(y = c \leftrightarrow \text{Is}_c(y))$;

$\forall x_0 \ldots x_{n-1} y$

$$(F(x_0, \ldots, x_{n-1}) = y \leftrightarrow (\text{Is}_F(y) \wedge \bigwedge_{i<n} F_i(y) = x_i)).$$

These two theories T_1 and T_2 give opposite ways of looking at the term algebra: T_1 generates the terms from their components, while T_2 recovers the components from the terms. One curious feature is that T_2 uses only 1-ary function and relation symbols, while there is no bound on the arities of the symbols in T_1. The theory T_2 will be useful for analysing T_1 in the next section.

3: Atomisation

Here we have a theory T in a language L, and a set Φ of formulas of L which are not sentences. The object is to extend T to a theory T^+ in a larger language L^+ in such a way that every formula in Φ is equivalent modulo T^+ to an atomic formula. In fact the set of new sentences $T^+\backslash T$ will turn out to depend only on L and not on T.

This device is sometimes called Morleyisation. But it has been well known since 1920 when Skolem introduced it, and it has nothing particularly to do with Morley. So I thought it best to use a more descriptive name. 'Skolemisation' already means something else; see section 3.1 below.

Theorem 2.6.5 (Atomisation theorem). *Let L be a first-order language. Then there are a first-order language $L^\Theta \supseteq L$ and a theory Θ in L^Θ such that*
(a) *every L-structure A can be expanded in just one way to an L^Θ-structure A^Θ which is a model of Θ,*
(b) *every formula $\phi(\bar{x})$ of L^Θ is equivalent modulo Θ to a formula $\psi(\bar{x})$ of L, and also (when $\bar{x}$ is not empty) to an atomic formula $\chi(\bar{x})$ of L^Θ,*
(c) *every homomorphism between non-empty models of Θ is an elementary embedding,*
(d) *$|L^\Theta| = |L|$.*

Proof. For each formula $\phi(x_0, \ldots, x_{n-1})$ of L with $n > 0$, introduce a new n-ary relation symbol R_ϕ. Take L^Θ to be the first-order language got from L by adding all the symbols R_ϕ, and take Θ to be the set of all sentences of the form

$$(6.21) \qquad \forall \bar{x}(R_\phi \bar{x} \leftrightarrow \phi(\bar{x})).$$

Then (d) is immediate. The theory Θ is a definitional extension of the empty theory in L, so we have (a) and the first part of (b). The second part of (b) then follows by (6.21).

Now by (b), every formula of L which is not a sentence is equivalent modulo Θ to an atomic formula. If ϕ is a sentence of L then $\phi \wedge x = x$ is equivalent modulo Θ to an atomic formula $\chi(x)$; any homomorphism between non-empty models of Θ which preserves χ must also preserve ϕ. So (c) follows by Theorem 1.3.1(b).   □

One can apply the same technique to a particular set Φ of formulas of L, if one wants to study homomorphisms which preserve the formulas in Φ. If A and B are models of Θ, then every embedding (in fact every homomorphism) from A to B must preserve the formulas in Φ, by Theorem 1.3.1(b).

Theorem 2.6.6. *Let Θ be the theory constructed in the proof of Theorem 2.6.5. Then for every theory T in L^{Θ}, $T \cup \Theta$ is equivalent to an $\forall_2$ theory.*

Proof. By (b) of the theorem, every formula of L^{Θ} with at least one free variable is equivalent modulo Θ to an atomic formula of L^{Θ}. So $T \cup \Theta$ is equivalent to a theory $T' \cup \Theta$ where every sentence of T' is $\forall_1$ at worst. It suffices now to show that Θ itself is equivalent to an $\forall_2$ theory.

Let Θ' be the set of all sentences of the following forms:

(6.22) $\forall \bar{x}(\phi(\bar{x}) \leftrightarrow R_\phi(\bar{x}))$ where ϕ is an atomic formula of L;

(6.23) $\forall \bar{x}(R_\phi(\bar{x}) \wedge R_\psi(\bar{x}) \leftrightarrow R_{\phi \wedge \psi}(\bar{x}))$;

and likewise with $\vee$ for $\wedge$;

(6.24) $\forall \bar{x}(\neg R_\phi(\bar{x}) \leftrightarrow R_{\neg \phi}(\bar{x}))$;

(6.25) $\forall \bar{x}(\forall y\, R_{\phi(\bar{x},y)}(\bar{x}, y) \leftrightarrow R_{\forall y\, \phi(\bar{x},y)}(\bar{x}))$;

and likewise with $\exists$ for $\forall$.

After a slight rearrangement of the sentences (6.25), Θ' is an $\forall_2$ theory. We show that Θ is equivalent to Θ'. Clearly Θ implies all the sentences in Θ'. Conversely assume that Θ' holds. Then (6.21) follows by induction on the complexity of ϕ. $\square$

A first-order theory is said to be **model-complete** if every embedding between its models is elementary. Atomisation shows that we can turn any first-order theory into a model-complete theory in a harmless way. But the real interest of the notion of model-completeness is that a number of theories in algebra have this property without any prior tinkering. We shall think about this in section 7.3.

Exercises for section 2.6

We can eliminate function symbols in favour of relation symbols:

1. Let L be a signature. Form a signature L^r from L as follows: for each positive n and each n-ary function symbol F of L, introduce an $(n + 1)$-ary relation symbol R_F. If A is an L-structure, let A^r be the L^r-structure got from A by interpreting each R_F as the relation $\{(\bar{a}, b): A \vDash (F\bar{a} = b)\}$ (the **graph** of the function F^A). (a) Define a translation $\phi \mapsto \phi^r$ from formulas of L to formulas of L^r, which is independent of A. Formulate and prove a theorem about this translation and the structures A, A^r.

(b) Extend (a) so as to translate every formula of L into a formula which contains no function symbols and no constants.

*A formula ϕ is said to be **negation normal** if in ϕ the symbol $\neg$ never occurs except immediately in front of an atomic formula. (Recall that $\psi \to \chi$ is an abbreviation for $\neg \psi \vee \chi$.)*
2. Show that if L is a first-order language, then every formula $\phi(\bar{x})$ of L is logically equivalent to a negation normal formula $\phi^*(\bar{x})$ of L. (Your proof should adapt at once to show the same for $L_{\infty\omega}$ in place of L.) Show that if ϕ was unnested then ϕ^* can be chosen unnested.

3. Let L be a first-order language, R a relation symbol of L and ϕ a formula of L. Show that the following are equivalent. (a) R is positive in some formula of L which is logically equivalent to ϕ. (b) ϕ is logically equivalent to a formula of L in negation normal form in which R never has $\neg$ immediately before it.

Here is a more perverse rewriting, which depends on the properties of a particular theory.
4. Let T be the theory of linear orderings. For each positive integer n, write a first-order sentence which expresses (modulo T) 'There are at least n elements', and which uses only two variables, x and y.

5. Let T_0, T_1 and T_2 be first-order theories. Show that if T_2 is a definitional extension of T_1 and T_1 is a definitional extension of T_0, then T_2 is a definitional extension of T_0.

*The method of the following exercise is known as **Padoa's method**. It is not limited to first-order languages. Compare it with Lemma 2.1.1 above. See also the discussion after Theorem 5.5.4.*
6. Let L and L^+ be signatures with $L \subseteq L^+$; let T be a theory of signature L^+ and S a symbol of signature L^+. Suppose that there are two models A, B of T such that $A|L = B|L$ but $S^A \neq S^B$. Deduce that S is not explicitly definable in T in terms of L (and hence T is not a definitional extension of any theory of signature L).

7. Let L^+ be the first-order language of arithmetic with symbols $0, 1, +, \cdot$, and let T be the complete theory of the natural numbers in this language. Let L be the language L^+ with the symbol $+$ removed. Show that $+$ is not explicitly definable in T in terms of L.

*Let L and L^+ be first-order languages with $L \subseteq L^+$, and let T, T^+ be theories in L, L^+ respectively. We say that T^+ is a **conservative extension** of T if for every sentence ϕ of L, $T \vdash \phi \Leftrightarrow T^+ \vdash \phi$.*
8. (a) Show that if $T \subseteq T^+$ and every L-structure which is a model of T can be expanded to form a model of T^+, then T^+ is a conservative extension of T. In particular every definitional extension is conservative. (b) Prove that the converse of (a) fails. [Let T say that $<$ is a linear ordering with first element 0, every element has

an immediate successor and every element except 0 has an immediate predecessor. Let T^+ be Peano arithmetic. Show that T is complete in its language, so that T^+ is a conservative extension of T. Show that every countable model of T^+ has order-type either ω or $\omega + (\omega^* + \omega) \cdot \eta$ where ω^* is the reverse of ω and η is the order-type of the rationals; so not every countable model of T expands to a model of T^+.]

Even the definition of addition (a recursive definition, not an explicit one) can lead to new first-order consequences.

9. Let L be the language with constant symbol 0 and 1-ary function symbol S; let L^+ be L with a 2-ary function symbol $+$ added. Let T^+ be the theory $\forall x\, x + 0 = x$, $\forall xy\, x + Sy = S(x + y)$. Show that T^+ is not a conservative extension of the empty theory in L.

10. Show that the theory of boolean algebras is definitionally equivalent to the theory of commutative rings with $\forall x\, x^2 = x$.

2.7 Quantifier elimination

The first systematic programme for model theory appeared in the decade after the first world war. This programme is known as **elimination of quantifiers**. Let me summarise it.

Take a first-order language L and a class **K** of L-structures. The class **K** might be, for example, the class of all dense linear orderings, or it might be the singleton $\{\mathbb{R}\}$ where $\mathbb{R}$ is the field of real numbers. We say that a set Φ of formulas of L is an **elimination set** for **K** if

(7.1)   for every formula $\phi(\bar{x})$ of L there is a formula $\phi^*(\bar{x})$ which is
         a boolean combination of formulas in Φ, and ϕ is equivalent
         to ϕ^* in every structure in **K**.

The programme can be stated briefly: given **K**, *find an elimination set for* **K**. There are analogous programmes for other languages, but the first-order case is the most interesting.

Of course there always is at least one elimination set Φ for any class **K** of L-structures: take Φ to be the set of all formulas of L. But with care and attention we can often find a much more revealing elimination set than this.

For example, here are two results which we owe to Tarski's Warsaw seminar in the late 1920s. (A linear ordering is **dense** if for all elements $x < y$ there is z such that $x < z < y$; cf. (2.31) and (2.32) in section 2.2.)

Theorem 2.7.1. *Let L be the first-order language whose signature consists of the 2-ary relation symbol $<$, and let **K** be the class of all dense linear orderings. Let Φ consist of formulas of L which express each of the following.*

(7.2)   *There is a first element.*
          There is a last element.
          x is the first element.
          x is the last element.
          $x < y$.

Then Φ is an elimination set for **K**.

Proof. This will be Exercise 1 below.   □

Theorem 2.7.2. *Let L be the first-order language of rings, whose symbols are $+, -, \cdot, 0, 1$. Let* **K** *be the class of real-closed fields. Let Φ consist of the formulas*

(7.3) $$\exists y \, y^2 = t(x)$$

where t ranges over all terms of L not containing the variable y. Then Φ is an elimination set for **K**. *(Note that (7.3) expresses $t(x) \geqslant 0$.)*

Proof. We shall see an algebraic proof of this in Theorem 7.4.4 below.   □

The name 'quantifier elimination' refers either to the process of reducing a formula to a boolean combination of formulas in Φ, or to the process of discovering the appropriate set Φ in the first place. One should distinguish the method of quantifier elimination from the *property of quantifier elimination*, which is a property that some theories have. A theory T **has quantifier elimination** if the set of quantifier-free formulas forms an elimination set for the class of all models of T. (Cf. section 7.4 below, and note that some of the formulas in (7.2) and (7.3) are not quantifier-free.)

The point of quantifier elimination

Suppose we have an elimination set Φ for the class **K**. What does it tell us?

(a) *Classification up to elementary equivalence.* Suppose A and B are structures in the class **K**, and A is not elementarily equivalent to B. Then there is some boolean combination of sentences in Φ which is true in A but false in B. It follows at once that some sentence ϕ in Φ is true in one of A and B but false in the other. The conclusion is that we can classify the structures in **K**, up to elementary equivalence, by looking to see which *sentences* in Φ are true in these structures.

If the sentences in Φ each express some 'algebraic' property of structures (a vague notion this, but sharp enough to be helpful), then we have reduced elementary equivalence in **K** to a purely 'algebraic' notion. For example Tarski showed that two algebraically closed fields are elementarily equivalent if and only if they have the same characteristic.

(b) *Completeness proofs*. As a special case of (a), suppose that **K** is the class Mod (T) of all models of some first-order theory T. Suppose that all the sentences in Φ are either deducible from T or inconsistent with T. Then it follows that all models of T are elementarily equivalent, and so T is a complete theory.

Theorem 2.7.2 is a case in point. The sentences in Φ can all be written as $s = t$ or $s \leqslant t$ where s, t are closed terms of L. But every real-closed field has characteristic 0. In fields of characteristic 0, each closed term t has an integer value independent of the choice of the field, and so we can prove or refute the sentences $s = t, s \leqslant t$ from the axioms for real-closed fields. Thus Theorem 2.7.2 shows that the theory of real-closed fields is complete.

(c) *Decidability proofs*. This is a special case of (b) in turn. Suppose L is a recursive language (see section 2.3 above). The theory T in L is decidable if and only if there is an algorithm to determine whether any given sentence of L is a consequence of T. The **decision problem** for a theory T in L is the problem of finding such an algorithm (or showing that there isn't one).

Now suppose that **K** is Mod (T) and the map $\phi \mapsto \phi^*$ in (7.1) is recursive. Suppose also that we have an algorithm which tells us, for any sentence ψ in the elimination set Φ, either that ψ is provable from T or that it is refutable from T. Then by putting everything together, we derive an algorithm for determining which sentences are consequences of T; this is a positive solution of the decision problem for T. Again real-closed fields are a case in point.

(d) *Description of definable relations*. Suppose Φ is an elimination set for **K** and A is a structure in **K**. Let D be the set of all relations on A which have the form $\psi(A^n)$ for some formula $\psi(x_0, \ldots, x_{n-1})$ in Φ. Then the $\varnothing$-definable relations on A are precisely the boolean combinations of relations in D.

(e) *Description of elementary embeddings*. If Φ is an elimination set for **K**, then the elementary maps between structures in **K** are precisely those homomorphisms which preserve ψ and $\neg\psi$ for every formula ψ in Φ. For example, by Theorem 2.7.1, every embedding between dense linear orderings with no endpoints is elementary.

Points (a), (d) and (e) were vital for the future of model theory. What they said was that in certain important classes of structure, the natural model-theoretic classifications could be paraphrased into straightforward algebraic notions. This allowed logicians and algebraists to talk to each other and merge their methods.

The main snag of quantifier elimination is that the method proceeds entirely on the level of deducibility from a set of axioms. This makes it heavily syntactic, and it can prevent us using good algebraic information about the class **K**. In particular the method doesn't allow us to exploit anything we know about maps between structures in **K**. For example, to prove the following result of Tarski by the method of quantifier elimination we would need to undertake some rather heavy study of equations; but the more structural argument of Example 2 in section 7.4 below makes it almost a triviality.

Theorem 2.7.3. *The theory of algebraically closed fields has quantifier elimination.*   □

For this reason the example done in detail below is not one of the well-known algebraic results from Tarski's school. Most of those can be handled by smoother methods today. Instead I choose an example where the structures are syntactic objects themselves, so that the method meshes well with the problem.

But first a brief word on strategy. We have a first-order language L and a class **K** of L-structures. We also have a theory T which is a candidate for an axiomatisation of **K**, and a set of formulas Φ which is a candidate for an elimination set. If **K** is defined as Mod(T), then of course T does axiomatise **K**. But if **K** was given and T is a guess at an axiomisation, we may find during the course of the quantifier elimination that we have to adjust T.

The following straightforward lemma very much eases the burden of showing that Φ is an elimination set. We write Φ^- for the set $\{\neg \phi : \phi \in \Phi\}$.

Lemma 2.7.4. *Suppose that*

(7.4)   *every atomic formula of L is in Φ, and*

(7.5)   *for every formula $\theta(\bar{x})$ of L which is of form $\exists y \bigwedge_{i<n} \psi_i(\bar{x}, y)$*

   with each ψ_i in $\Phi \cup \Phi^-$, there is a formula $\theta^*(\bar{x})$ of L which
   (i) *is a boolean combination of formulas in Φ, and*
   (ii) *is equivalent to θ in every structure in **K**.*

*Then Φ is an elimination set for **K**.*

Proof. See Lemma 2.3.1.   □

So to find an elimination set, we must discover a way of getting rid of the quantifier $\exists y$ in (7.5). Hence the name 'quantifier elimination'. As Lemma 2.7.4 suggests, we start with an arbitrary finite subset $\Theta(y, \bar{x})$ of $\Phi \cup \Phi^-$, and we aim to find a boolean combination $\psi(\bar{x})$ of formulas in Φ so that $\exists y \bigwedge \Theta$ is equivalent to ψ modulo T. Typically the move from Θ to ψ takes several steps, depending on what kinds of formula appear in Θ. If we run into a dead end, we can add sentences to T and formulas to Φ until the process moves again.

Example: term algebras

We consider the class **K** of term algebras of an algebraic signature L_1: see Example 1 in section 2.2 and Example 1 of section 2.6. The theory T_1, which consists of the sentences (2.3)–(2.7) of section 2.2, is true in every algebra in **K**. Our elimination will be easier to carry out if we switch to the language L_2 and the theory T_2 of section 2.6. Since T_2 is definitionally equivalent to T_1, everything can be translated back into the language of T_1 if needed.

If L_1 (and hence L_2) has finite signature, we can write for each positive integer k a sentence α_k of L_2 which says 'There are at least k elements satisfying all $\neg \mathrm{Is}_c$ and $\neg \mathrm{Is}_F$'. Let β be the sentence $\exists x\, x = x$.

Theorem 2.7.5. *Let* **K** *be the class of term algebras in the signature* L_2 *described above. Let* Φ *be the set of atomic formulas of* L_2, *together with the sentences* α_k *if* L_1 *has finite signature, and the sentence* β *if* L_1 *has no constants. Then* Φ *is an elimination set for the class of all models of* T_2, *and hence for* **K**.

Proof. Our task is the following. We have a finite set $\Theta_0(\bar{x}, y)$ which consists of literals of L_2 (and possibly some sentences β, α_k or their negations), and we must eliminate the quantifier $\exists y$ from the formula $\exists y \bigwedge \Theta_0$. We can assume without loss that

(7.6)   No sentence α_k or β or its negation is in Θ_0.

Reason: The variable y is not free in α_k, and so $\exists y(\alpha_k \wedge \psi)$ is logically equivalent to $\alpha_k \wedge \exists y\, \psi$. Likewise with β.

Also we can assume without loss the following:

(7.7)   there is no formula ψ such that both ψ and $\neg \psi$ are in Θ_0; moreover $y \neq y$ is not in Θ_0.

Reason: Otherwise $\exists y \bigwedge \Theta_0$ reduces at once to $\bot$.

We can replace Θ_0 by a set Θ_1 which satisfies (7.6), (7.7) and

(7.8)   if t is a term of L_2 in which y doesn't occur, then the formulas $y = t$ and $t = y$ are not in Θ_1.

Reason: $\exists y(y = t \wedge \psi(y, \bar{x}))$ is logically equivalent to $\exists y(y = t \wedge \psi(t, \bar{x}))$, hence to $\psi(t, \bar{x}) \wedge \exists y\, y = t$, or equivalently $\psi(t, \bar{x})$. $\exists y\, y = t$ is equivalent to $t = t$.

We can replace Θ_1 by one or more sets Θ_2 which satisfy (7.6)–(7.8) and

(7.9)   the variable y never occurs inside another term.

Reason: Suppose a term $s(F_i(y))$ appears somewhere in Θ_1. Now $\exists y\, \psi$ is equivalent to $\exists y(\mathrm{Is}_F(y) \wedge \psi) \vee \exists y(\neg \mathrm{Is}_F(y) \wedge \psi)$, so we can suppose that exactly one of $\mathrm{Is}_F(y)$ and $\neg \mathrm{Is}_F(y)$ appears in Θ_1. If $\neg \mathrm{Is}_F(y)$ appears, we can replace $F_i(y)$ by y according to axiom (6.17). If $\mathrm{Is}_F(y)$ appears and F is

n-ary, we make the following changes. First if G is any function symbol of L_1 distinct from F, we replace any expression $G_j(y)$ in Θ_1 by y (again quoting (6.17)). We introduce n new variables $y_0, \ldots, y_{n-1}$, we replace each $F_j(y)$ by y_j and we add the formulas $F_j(y) = y_j$. Then $\exists y \bigwedge \Theta_1$ becomes equivalent to an expression $\exists y_0 \ldots y_{n-1} \exists y (\mathrm{Is}_F(y) \wedge \bigwedge_{j<n} F_j(y) = y_j \wedge \bigwedge \Theta_2)$ where Θ_2 satisfies (7.9). By (6.15) this reduces to $\exists y_0 \ldots y_{n-1} \bigwedge \Theta_2$. Here Θ_2 has more variables y_j to dispose of, but these all occur in shorter terms than those involving y in Θ_1. Hence we can deal with the variables $y_{n-1}, \ldots, y_0$ in turn, using an induction on the lengths of terms.

At this point Θ_2 consists of formulas of the forms $y \neq t$ or $t \neq y$ (where y doesn't occur in t), $y = y$, $\mathrm{Is}_c(y)$, $\neg\mathrm{Is}_c(y)$, $\mathrm{Is}_F(y)$ or $\neg\mathrm{Is}_F(y)$; as in (7.6), we can eliminate any literals in which y doesn't appear. We can replace Θ_2 by a set Θ_3 satisfying (7.6)–(7.9) and

(7.10)   there is no constant c such that $\mathrm{Is}_c(y)$ is in Θ_3, and no function symbol F such that $\mathrm{Is}_F(y)$ is in Θ_3.

Reason: $\exists y (\mathrm{Is}_c(y) \wedge y \neq t \wedge \neg\mathrm{Is}_F(y))$, say, is equivalent to $\neg\mathrm{Is}_c(t)$ by (6.14) and (6.16); $\exists y \, \mathrm{Is}_c(y)$ is equivalent to $\neg \bot$ by (6.14). Function symbols need a more complicated argument. Let F be an n-ary function symbol.

We claim T_2 implies that for any k elements ($k > 0$) there is an element distinct from all of them which satisfies $\mathrm{Is}_F(y)$. By (6.15), T_2 implies that if $\mathrm{Is}_F(x_0)$ then there is a unique x_1 such that $\mathrm{Is}_F(x_1)$ and $F_i(x_1) = x_0$ for all $i < n$; by (6.18), $x_0 \neq x_1$. Likewise by (6.15) there is x_2 such that $\mathrm{Is}_F(x_2)$ and $F_i(x_2) = x_1$ for all $i < n$, and then by (6.18) again, $x_2 \neq x_1$ and $x_2 \neq x_0$. Etc. etc.; this proves the claim. With (6.16), the claim allows the reduction to (7.10), unless the problem is to eliminate the quantifier from $\exists y \, \mathrm{Is}_F(y)$. When L_1 has at least one constant c, the formula $\exists y \, \mathrm{Is}_F(y)$ reduces to $\neg \bot$ by (6.15); but in general it is equivalent to β.

We are almost home. When L_1 has infinite signature, $\exists y \bigwedge \Theta_3$ reduces to β by (6.16). There remains only the case where L_1 has finitely many symbols. As at the start of the reason for (7.9), we can suppose that for each symbol S of L_1, Θ_3 contains either $\mathrm{Is}_S(y)$ or $\neg\mathrm{Is}_S(y)$, and we have seen how to deal with the first of these formulas. So assume henceforth that Θ_3 contains $\neg\mathrm{Is}_S(y)$ for each symbol S of L_1. By the same reasoning we can suppose that for each term t appearing in Θ_3 (even inside other terms), one of the formulas $\mathrm{Is}_S(t)$ and $\neg\mathrm{Is}_S(t)$ is in Θ_3. Also we can suppose that for each pair of terms s, t appearing in Θ_3, either $s = t$ or $s \neq t$ also appears in Θ_3. Now $\bigwedge \Theta_3$ asserts (among other things) that there are at least k distinct items, including y, which satisfy $\neg\mathrm{Is}_S(x)$ for all S. Such an element y can be found if and only if α_k holds; so $\exists y \bigwedge \Theta_3$ reduces to a conjunction of α_k and the formulas in Θ_3 which don't mention y.   □

Notice that if we had overlooked the formulas α_k or β, or one of the axioms that should have been in T_2, then this procedure would have shown up our mistake and suggested how to correct it.

Corollary 2.7.6. *Let* **K** *be (as above) the class of all term algebras of L_1, regarded as L_2-structures. If L_1 is infinite, or has at least one constant symbol, or no symbols at all, then* Th$(\mathbf{K})$ *is equivalent to T_2. If L_1 is finite and has function symbols but no constant symbols then* Th$(\mathbf{K})$ *is equivalent to $T_2 \cup \{\beta \to \alpha_1\}$.*

Proof. Certainly every sentence of T_2 is in Th$(\mathbf{K})$. In the other direction, the harder case is where L_1 is finite. Let ϕ be any sentence in Th$(\mathbf{K})$. By the theorem, ϕ is equivalent modulo T_2 to a boolean combination of sentences β, α_k. (The signature L_2 has no closed terms.) By logic, each α_{k+1} entails α_k and α_1 entails β. If L_1 has at least one constant symbol then β is provable from T_2 but there are no other implications between β and the α_k; so in this case ϕ must be provable from T_2.

If L_1 has function symbols but no constant symbol then the term algebra is empty unless α_1 holds; so $\beta \to \alpha_1$ lies in Th$(\mathbf{K})$. It is not a consequence of T_2, since we can construct a model of T_2 in which α_1 fails, by taking an infinitely decomposable 'term' as in (2.10) of section 2.2. Examples show that no other implications hold between β and the α_k. I leave to the reader the case where L_1 is empty. $\qquad\square$

Corollary 2.7.7. *The theory of term algebras of a given finite algebraic signature, in either the language L_1 or the language L_2 above, is decidable.*

Proof. Any sentence of L_1 can be translated effectively into a sentence ϕ of L_2 by the explicit definitions (6.20) of section 2.6. Then we can compute a sentence ϕ^* of L_2 which is equivalent to ϕ modulo T_2 and is a boolean combination of sentences in Φ, where Φ is as in Theorem 2.7.5. The argument for the previous corollary shows that we can effectively check whether ϕ^* is a consequence of T_2 or $T_2 \cup \{\beta \to \alpha_1\}$ as appropriate. $\qquad\square$

Exercises for section 2.7

For these exercises, be advised that the method of quantifier elimination is not intrinsically hard, but it does use up hours and paper.

1. Prove Theorem 2.7.1.

2. Let the signature L consist of finitely many 1-ary relation symbols $R_0, \ldots, R_{n-1}$. For each map $s: n \to 2$ let $\phi^s(x)$ be the conjunction $R_0^{s(0)}(x) \wedge \ldots \wedge R_{n-1}^{s(n-1)}(x)$, where R_i^j is R_i if $j = 1$ and $\neg R_i$ if $j = 0$. If **K** is the class of all L-structures, show that an elimination set for **K** is given by the formulas $\phi^s(x)$ and the sentences $\exists_{=k} x \, \phi^s(x)$ where $s: n \to 2$ and $k < \omega$.

3. Let L be the first-order language of boolean algebras (see (2.29) in section 2.2). Let Ω be a set, and let A be the power-set algebra of Ω, regarded as an L-structure with $\wedge$ for $\cap$, $\vee$ for $\cup$, * for complement in Ω, 0 for $\varnothing$ and 1 for Ω. Let **K** be $\{A\}$. For each positive integer k, write $\alpha_k(y)$ for the formula 'y has at least k elements'. (This can be written as 'There are at least k atoms $\leqslant y$', where an **atom** of a boolean algebra is an element $b > 0$ such that there is no element c with $b > c > 0$.) Let Φ be the set of all atomic formulas of L and all formulas of the form $\alpha_k(t)$ where t is a term of L. Show that Φ is an elimination set for **K**.

4. A boolean algebra B is said to be **atomic** if the supremum of the set of atoms in B is the top element of B. Let T be the theory of the class of atomic boolean algebras. Show (a) a boolean algebra B is atomic if and only if for every element $b > 0$ there is an atom $a \leqslant b$, (b) if B is an atomic boolean algebra then every formula $\phi(\bar{x})$ of the first-order language of boolean algebras is equivalent in B to a boolean combination of formulas which say exactly how many atoms are below elements $y_0 \wedge \ldots \wedge y_{n-1}$, where each y_i is either x_i or x_i^* (complement), (c) T is the theory of the class of finite boolean algebras, (d) T is decidable.

5. Let **K** be the class of boolean algebras B such that the set of atoms of B has a supremum in B. (a) Show that **K** is a first-order axiomatisable class. (b) Use the method of quantifier elimination to show that up to elementary equivalence there are exactly ω boolean algebras in **K**; describe these algebras.

6. Let L be the signature consisting of one 2-ary relation symbol $<$. Let **K** be the class of L-structures $(X, <')$ where X is a non-empty set and $<'$ is a linear ordering of X in which every element has an immediate predecessor and an immediate successor. (a) Use the method of quantifier elimination to show that any two structures in **K** are elementarily equivalent. (b) Give necessary and sufficient conditions for a map from A to B to be an elementary embedding, where A and B are in **K**. (c) Show that for every infinite cardinal λ there are 2^λ non-isomorphic structures in **K** with cardinality λ. [Construct models $\Sigma_{i<\lambda}((\omega^* + \omega) \cdot \rho_i)$, where each ρ_i is either ω or $(\omega^* + \omega)$; show that each ρ_i is recoverable from the model.]

We shall handle the next result differently in section 3.3.
7. Let the signature L consist of constants 0, 1 and a 2-ary function symbol $+$. Let **K** consist of one structure, namely the natural numbers $\mathbb{N}$ regarded as an L-structure in the obvious way. Using the method of elimination of quantifiers, find a set of axioms for $\mathrm{Th}(\mathbb{N})$ and show that $\mathrm{Th}(\mathbb{N})$ is a decidable theory. [One elimination set consists of equations and formulas which express '$t(\bar{x})$ is divisible by n' where n is a positive integer.]

8. Let the signature L consist of one 1-ary function symbol F and one constant symbol 0. Let $\mathbf{K}$ be the class of L-structures which obey the second-order induction axiom, namely, for every set X of elements, $((0 \in X \wedge \forall y(y \in X \to F(y) \in X)) \to \exists y(y \in X))$. Use the method of quantifier elimination to find (a) a set of axioms for the first-order theory $\mathrm{Th}(\mathbf{K})$ of $\mathbf{K}$, and (b) a classification of the models of $\mathbf{K}$, up to elementary equivalence.

9. Let K be a field; let $\mathbf{J}$ be the class of (left) vector spaces over K, in the language of left K-modules (i.e. with symbols $+, -, 0$ and for each scalar r a 1-ary function symbol $r(x)$ to represent multiplication of a vector by r; see Example 4 in section 1.1 above). (a) Show that the set of linear equations $r_0(x_0) + \ldots + r_{n-1}(x_{n-1}) = 0$, with $n < \omega$ and $r_0, \ldots, r_{n-1}$ scalars, together with the set of sentences $\exists_{=k} x\, x = x$ $(k < \omega)$, is an elimination set for $\mathbf{J}$. (b) Deduce that every infinite vector space in $\mathbf{J}$ is a minimal structure (in the sense of section 2.1).

An abelian group is **divisible** *if for every non-zero element b and every positive integer n there is an element c such that $nc = b$. It is* **ordered** *if it carries a linear ordering relation $<$ such that $a < b$ implies $a + c < b + c$ for all a, b, c.*
10. Use the method of quantifier elimination to axiomatise the class of non-trivial divisible ordered abelian groups. Show (a) all groups in this class are elementarily equivalent, (b) if A and B are non-trivial divisible ordered abelian groups and A is a subgroup of B then A is an elementary substructure of B.

11. Let T_1 be the theory of term algebras of a given algebraic signature L. (a) Show that if A is any finitely generated model of T_1, then A is isomorphic to a term algebra of L. (b) Show that if B is an L-structure then B is a model of T_1 if and only if every finitely generated substructure of B is isomorphic to a term algebra of L. *(Hence T_1 is known as the* **theory of locally free L-structures**.*)*

12. Let R be a ring and T the theory of left R-modules as in (2..21); let L be the language of T and M a left R-module. We use the notation of section 2.5 and Exercise 2.5.9. The exercise sketches Monk's proof of the **Baur–Monk theorem**, that the p.p. formulas together with the invariant sentences (defined below) form an elimination set for T. (a) Show that for all p.p. formulas $\phi(x)$ and $\psi(x)$ of L and every positive integer n there is a sentence $Inv(\phi, \psi, n)$ which expresses in M: the group $\phi(M)/(\phi(M) \cap \psi(M))$ has cardinality $\leq n$. *These sentences are called the* **invariant sentences**. (b) Suppose $\phi(x_0, \ldots, x_n)$ is a conjunction of p.p. formulas of L and negations of such formulas, and $\bar{d}$ is an n-tuple in M. Show that the claim $M \vDash \neg \exists x_n \phi(\bar{d})$ can be paraphrased as $(*)$: $G \subseteq \bigcup_{i<k}(H_i + \bar{b}_i)$, where $G, H_0, \ldots,$ H_{k-1} are certain p.p. subgroups of M^n, $\bar{b}_0, \ldots, \bar{b}_{n-1}$ are certain tuples from M, each H_i is a subgroup of G and k depends only on ϕ. (c) Using Corollary 5.6.4, show that we can put a finite bound (depending only on ϕ) on the indexes of the subgroups H_i in G. (d) We put $H = \bigcap_{i<k} H_i$, and for every union X of cosets of H we write $N(X)$ for the number of cosets of H in X. (By (c) this number is finite.) Show that $(*)$ can be paraphrased as $(**)$: $N(G) \leq \sum_{1 \leq j \leq k}(-1)^{j-1}\{\sum_{J \subseteq k, |J|=j} N(\bigcap_{i \in J}(H_i + \bar{b}_i))\}$. (e) Express $(**)$ by a boolean combination $\chi(\bar{d})$ of invariant sentences and

formulas $\phi(\bar{d})$ where $\phi(\bar{x})$ are p.p. formulas of L. Check that in any model M of T, $\chi(\bar{x})$ depends only on ϕ and not on M or $\bar{d}$.

Further reading

The original papers of Tarski, sometimes written with his students, are outstanding for their clarity. Two examples relevant to this chapter are:

> Tarski, A. & Vaught, R. L. Arithmetical extensions of relational systems. *Compositio Mathematica*, **13** (1957), 81–102.
>
> Tarski, A. *A decision method for elementary algebra and geometry*. Berkeley: University of California Press, 1951.

O-minimal theories (Exercise 2.1.4) have a central place in recent work on the model theory of fields. For a survey (at a more advanced level than this book), see

> van den Dries, L. O-minimal structures. In *Logic: from foundations to applications*, ed. Hodges, W. *et al*. pp. 143–85. Oxford: Oxford University Press, 1996.

Nonstandard models (see section 2.2) are the basis of *nonstandard analysis*; this is a way of doing analysis, quoting theorems from model theory to justify the use of infinitesimals. Keisler's book below is an undergraduate text using nonstandard analysis, while Cutland's collection surveys research.

> Keisler, H. J. *Foundations of infinitesimal analysis*. Boston: Prindle, Weber & Schmidt, 1976.
>
> Cutland, N. J. *Nonstandard analysis and its applications*. Cambridge: Cambridge University Press, 1988.

3

Structures that look alike

M Martin: J'ai une petite fille, ma petite fille, elle habite avec moi, chère Madame. Elle a deux ans, elle est blonde, elle a un oeil blanc et un oeil rouge, elle est très jolie, elle s'appelle Alice, chère Madame.

Mme Martin: Quelle bizarre coïncidence! moi aussi j'ai une petite fille, elle a deux ans, un oeil blanc et un oeil rouge, elle est très jolie et s'appelle Alice, cher Monsieur!

M Martin, *même voix traînante, monotone:* Comme c'est curieux et quelle coïncidence!

Eugène Ionesco, La cantatrice chauve, © *Editions GALLIMARD 1954.*

If we consider a first-order language L, the number of isomorphism types of L-structures is vastly greater than the number of theories in L (counted up to equivalence of theories). So there must be some huge family of non-isomorphic L-structures which it is impossible to tell apart by sentences of L.

In this chapter we prove a variety of theorems which have the general form: if some sentence is true here, then it must be true there too.

3.1 Theorems of Skolem

In earlier days there were people who disliked uncountable structures and wanted to show that they are unnecessary for mathematics. Thoralf Skolem was one such. He proved that for every infinite structure B of countable signature there is a countable substructure of B which is elementarily equivalent to B. He inferred from this that there are countable models of Zermelo–Fraenkel set theory, and hence countable models of the sentence 'There are uncountably many reals'. He hoped that this paradoxical result would scare people away from set-theoretical foundations. If anything it had the opposite effect.

The quickest way to prove Skolem's result is as follows. Let B be any infinite structure with countable signature. By Theorem 1.2.3 we can build a chain $(A_n: n < \omega)$ of countable substructures of B, in such a way that

(1.1)   for each first-order formula $\phi(y, \bar{x})$, each $n < \omega$ and each tuple $\bar{a}$ of elements of A_n such that $B \vDash \exists y \, \phi(y, \bar{a})$, if there

is b in B such that $B \vDash \phi(b, \bar{a})$ then there is such an element
b in A_{n+1}.

Put $A = \bigcup_{n<\omega} A_n$. Clearly A is countable. Also A is an elementary substructure of B by the Tarski–Vaught criterion, Theorem 2.5.1, and so $A \equiv B$. (We took a similar route in Exercise 2.5.5 above.)

Skolem proceeded differently. He added functions to B in such a way that every substructure of B which is closed under these functions is automatically an elementary substructure, and then he invoked Theorem 1.2.3. The added functions are called *Skolem functions*.

Theories with Skolem functions

Suppose T is a theory in a first-order language L. Then a **skolemisation** of T is a theory $T^+ \supseteq T$ in a first-order language $L^+ \supseteq L$, such that

(1.2)   every L-structure which is a model of T can be expanded to a model of T^+, and

(1.3)   for every formula $\phi(\bar{x}, y)$ of L^+ with $\bar{x}$ non-empty, there is a term t of L^+ such that T^+ entails the sentence $\forall \bar{x}(\exists y\, \phi(\bar{x}, y) \to \phi(\bar{x}, t(\bar{x})))$.

The terms t of (1.3) (and the functions which they define in models of T^+) are called **Skolem functions** for T^+.

We say that T **has Skolem functions** (or that T is a **Skolem theory**) if T is a skolemisation of itself; in other words, if (1.3) holds with $L = L^+$ and $T = T^+$. Note that if T^+ is a skolemisation of T, then T^+ has Skolem functions. Note also that these notions depend on the language: if $L \subseteq L'$ and T is a Skolem theory in L, T will generally not be a Skolem theory in L'. On the other hand if T has Skolem functions and T' is a theory with $T \subseteq T'$, both in the first-order language L, then it's immediate that T' has Skolem functions too.

Suppose T is a theory which has Skolem functions, in a first-order language. Let A be an L-structure and X a set of elements of A. The **Skolem hull** of X is defined to be $\langle X \rangle_A$, the substructure of A generated by X.

Theorem 3.1.1. *Suppose T is a theory in a first-order language L, and T has Skolem functions.*

(a) *Modulo T, each formula $\phi(\bar{x})$ of L (with $\bar{x}$ not empty) is equivalent to a quantifier-free formula $\phi^*(\bar{x})$ of L.*

(b) *If A is an L-structure and a model of T, and X is a set of elements of A such that the Skolem hull $\langle X \rangle_A$ is non-empty, then $\langle X \rangle_A$ is an elementary substructure of A.*

Proof. In (1.3), the formula $\phi(\bar{x}, t(\bar{x}))$ logically implies $\exists y \, \phi(\bar{x}, y)$, so we could have written $\leftrightarrow$ in place of $\rightarrow$. Hence (a) follows at once from Lemma 2.3.1, taking Φ to be the set of quantifier-free formulas.

To prove (b), put $B = \langle X \rangle_A$. Let $\bar{b}$ be a tuple of elements of B and $\phi(\bar{x}, y)$ a formula of L such that $A \vDash \exists y \, \phi(\bar{b}, y)$. Then by (1.3) there is a term t such that $A \vDash \phi(\bar{b}, t(\bar{b}))$. But the element $t^A(\bar{b})$ is in B since B is closed under the functions of L. By the Tarski–Vaught criterion (Theorem 2.5.1) it follows that B is an elementary substructure of A. $\square$

Adding Skolem functions

Sadly, in a state of nature there are very few Skolem theories. They have to be constructed by artifice.

Theorem 3.1.2 *(Skolemisation theorem). Let L be a first-order language. Then there are a first-order language $L^\Sigma \supseteq L$ and a set Σ of sentences of L^Σ such that*
(a) *every L-structure A can be expanded to a model A^Σ of Σ,*
(b) *Σ is a Skolem theory in L^Σ,*
(c) *$|L^\Sigma| = |L|$.*

Proof. For each formula $\chi(\bar{x}, y)$ of L (where $\bar{x}$ is not empty), introduce a new function symbol $F_{\chi, \bar{x}}$ of the same arity as $\bar{x}$. The language L' will consist of L with these new function symbols added. The set $\Sigma(L)$ will consist of all the sentences

(1.4) $$\forall \bar{x} \, (\exists y \chi(\bar{x}, y) \rightarrow \phi(\bar{x}, F_{\chi, \bar{x}}(\bar{x}))).$$

We claim

(1.5) every L-structure A can be expanded to a model of $\Sigma(L)$.

If A is empty it is already a model of $\Sigma(L)$. If it is not empty, we expand it to an L'-structure A' as follows. Let $\chi(\bar{x}, y)$ be any formula of L with $\bar{x}$ non-empty, and let $\bar{a}$ be a tuple of elements of A. If there is an element b such that $A \vDash \chi(\bar{a}, b)$, choose one such element b and put $F_{\chi, \bar{x}}^{A'}(\bar{a}) = b$. (Here we generally need the axiom of choice.) If there is no such element, let $F_{\chi, \bar{x}}^{A'}(\bar{a})$ be, say, the first element in $\bar{a}$. Then certainly A' is a model of all the sentences in $\Sigma(L)$. Thus (1.5) is proved.

The theory Σ is built by iterating the construction of $\Sigma(L)$ ω times. We define a chain of languages $(L_n : n < \omega)$ and a chain of theories $(\Sigma_n : n < \omega)$ by induction on n. We put $L_0 = L$ and take Σ_0 to be the empty theory. Then we define L_{n+1} to be $(L_n)'$ as above, and we define $\Sigma_{n+1} = \Sigma_n \cup \Sigma(L_n)$. Finally we define $L^\Sigma = \bigcup_{n<\omega} L_n$ and $\Sigma = \bigcup_{n<\omega} \Sigma_n$. Now (a) is true by making repeated expansions as in (1.5). For (b), every formula χ of L^Σ lies in some Σ_n, so the required sentence (1.4) is in Σ_{n+1}. Finally (c) is clear. $\square$

The main application of the skolemisation theorem is to give us elementary substructures, as follows.

Corollary 3.1.3. *Let T be a theory in a first-order language L. Then T has a skolemisation T^+ in a first-order language L^+ with $|L^+| = |L|$.*

Proof. Put $T^+ = T \cup \Sigma$. In particular Σ is a skolemisation of the empty theory in L.   □

Corollary 3.1.4 *(Downward Löwenheim–Skolem theorem). Let L be a first-order language, A an L-structure, X a set of elements of A, and λ a cardinal such that $|L| + |X| \leq \lambda \leq |A|$. Then A has an elementary substructure B of cardinality λ with $X \subseteq \mathrm{dom}\,(B)$.*

Proof. Expand A to a model A^Σ of Σ in L^Σ. Let Y be a set of λ elements of A, with $X \subseteq Y$. Let B' be the Skolem hull $\langle Y \rangle_A$, and let B be the reduct $B'|L$. By Theorem 1.2.3, $|B| \leq |Y| + |L^\Sigma| = \lambda + |L| = \lambda = |Y| \leq |B|$. Since Σ is a Skolem theory, $B' \preccurlyeq A^\Sigma$ by Theorem 3.1.1(b). Hence $B \preccurlyeq A$.   □

Example 1: *Simple subgroups of simple groups.* Let G be an infinite simple group. We show that for every infinite cardinal $\lambda \leq |G|$, G has a subgroup of cardinality λ which is simple. The language of groups is countable, so that by the downward Löwenheim–Skolem theorem, G has an elementary substructure H of cardinality λ. Clearly H is a subgroup of G. To show that H is simple it suffices to prove that if a, b are two elements of H and $b \neq 1$, then a is in the normal subgroup of H generated by b. Since G is simple, this is certainly true with G in place of H. Suppose for example that

$$G \models \exists y \, \exists z (a = y^{-1}by \cdot z^{-1}b^{-1}z).$$

Since $H \preccurlyeq G$, the same sentence is true in H. Hence there are c, d in H such that $a = c^{-1}bc \cdot d^{-1}b^{-1}d$ as required.

Exercises for section 3.1

1. Let L be a first-order language and L' a language which comes from L by adding constants. Show that if T is a Skolem theory in L, then T is a Skolem theory in L' too (and hence so is any theory $T' \supseteq T$ in L').

2. Use the downward Löwenheim–Skolem theorem and the result of Example 3 in the next section, to show that if A and B are dense linear orderings without endpoints, then $A \equiv B$.

3. Show that, if T is a first-order theory which has Skolem functions, then T is model-complete. Give an example of a first-order theory which is model-complete but doesn't have Skolem functions.

4. Let L be a first-order language with at least one constant. Show that if T is a Skolem theory in L, then T has elimination of quantifiers.

The next exercise shows that not all the sentences (1.4) are needed for Theorem 3.1.2.
5. Suppose T is a theory in a first-order language L, and, for every quantifier-free formula $\phi(\bar{x}, y)$ of L with $\bar{x}$ non-empty, there is a term t of L such that T entails the sentence $\forall \bar{x}(\exists y\, \phi(\bar{x}, y) \rightarrow \phi(\bar{x}, t(\bar{x})))$. (a) Show that T has Skolem functions. [Use Lemma 2.3.1 to show that modulo T, every formula with at least one free variable is equivalent to a quantifier-free formula.] (b) Show that for any theory T' in L with $T \subseteq T'$, T' is equivalent to an $\forall_1$ theory.

We say that a structure A has **Skolem functions** *if* $\mathrm{Th}(A)$ *has Skolem functions.*
6. Let L be a first-order language and A an L-structure which has Skolem functions. Suppose X is a set of elements which generate A, and $<$ is a linear ordering of X (not necessarily expressible in L). Show that every element of A has the form $t^A(\bar{c})$ for some term $t(\bar{x})$ of L and some tuple $\bar{c}$ from X which is strictly increasing in the sense of $<$.

7. Let **K** be the class of boolean algebras which are isomorphic to power set algebras of sets. Show that **K** is not first-order axiomatisable. [Use Corollary 3.1.4.]

8. Let L be a finite relational signature and A an infinite L-structure. Suppose there is a simple group which acts transitively on A. *(This means that the automorphism group of A contains a subgroup G which is simple, such that if a, b are any two elements of A then some automorphism in G takes a to b.)* Show that A has a countable elementary substructure on which some simple group acts transitively.

3.2 Back-and-forth equivalence

Compare the two relations $\cong$ (isomorphism) and $\equiv$ (elementary equivalence) between structures. In one sense isomorphism is a more intrinsic property of structures, because it's defined directly in terms of structural properties, whereas $\equiv$ involves a language. But in another sense elementary equivalence is more intrinsic, because the existence of an isomorphism can depend on some subtle questions about the surrounding universe of sets.

We can sharpen this second point with the help of some set theory. If M is a transitive model of set theory containing vector spaces A and B of dimensions ω and ω_1 over the same countable field, then A and B are not isomorphic in M, but they are isomorphic in an extension of M got by 'collapsing the cardinal ω_1 down to ω'. By contrast the question whether two structures A and B are elementarily equivalent depends only on A and B, and not on the sets around them.

In the early 1950s, Roland Fraïssé discovered a family of equivalence relations which hover somewhere between $\cong$ and $\equiv$. His equivalence relations are purely structural – there are no languages involved. On the other hand they are independent of the surrounding universe of sets. The trick is to look at isomorphisms, but only between a finite number of elements at a time. In the next section we shall find that Fraïssé's equivalence relations do often give us a way of proving that two structures are elementarily equivalent. Sometimes they give us proofs of isomorphism too, as we shall see in a few pages.

Back-and-forth games

Let L be a signature and let A and B be L-structures. We imagine two people, called $\forall$ and $\exists$ (male and female respectively, say $\forall$belard and $\exists$loise), who are comparing these structures. To add a note of conflict we imagine that $\forall$ wants to prove that A is different from B, while $\exists$ tries to show that A is the same as B. So their conversation has the form of a game. Player $\forall$ wins if he manages to find a difference between A and B before the game finishes; otherwise player $\exists$ wins.

The game is played as follows. An ordinal γ is given, which is the length of the game. Usually but not always, γ is ω or a finite number. The game is played in γ steps. At the ith step of a play, player $\forall$ takes one of the structures A, B and chooses an element of this structure; then player $\exists$ chooses an element of the other structure. So between them they choose an element a_i of A and an element b_i of B. Apart from the fact that player $\exists$ must choose from the other structure from player $\forall$ at each step, both players have complete freedom to choose as they please; in particular either player can choose an element which was chosen at an earlier step. Player $\exists$ is allowed to know which element player $\forall$ has chosen, and more generally each player is allowed to see and remember all previous moves in the play. (As the game theorists would put it, this is a **game of perfect information**.) At the end of the play, sequences $\bar{a} = (a_i : i < \gamma)$ and $\bar{b} = (b_i : i < \gamma)$ have been chosen. The pair $(\bar{a}, \bar{b})$ is known as the **play**.

We count the play $(\bar{a}, \bar{b})$ as a **win for player** $\exists$, and we say that player $\exists$ **wins the play**, if there is an isomorphism $f: \langle \bar{a} \rangle_A \to \langle \bar{b} \rangle_B$ such that $f\bar{a} = \bar{b}$. A play which is not a win for player $\exists$ counts as a **win for player** $\forall$.

Example 1: *Rationals versus integers*. Suppose $\gamma \geq 2$. Let A be the additive group $\mathbb{Q}$ of rational numbers and let B be the additive group $\mathbb{Z}$ of integers. Then player $\forall$ can win by playing as follows. He chooses a_0 to be any non-zero element of $\mathbb{Q}$. Then player $\exists$ must choose b_0 to be a non-zero integer; otherwise she loses the game at once. Now there is some integer n which doesn't divide b_0 in $\mathbb{Z}$. Let player $\forall$ choose a_1 in $\mathbb{Q}$ so that $na_1 = a_0$. There is no way that player $\exists$ can choose an element b_1 of $\mathbb{Z}$ so that $nb_1 = b_0$. It follows that, if $\gamma \geqslant 2$, player $\forall$ can always arrange to win the game on $\mathbb{Q}$ and $\mathbb{Z}$.

Let us write $A \equiv_0 B$ to mean that for every atomic sentence ϕ of L, $A \vDash \phi \Leftrightarrow B \vDash \phi$. (Clearly it makes no difference if we replace 'atomic' by 'quantifier-free'.) Then

(2.1)   player $\exists$ wins the play $(\bar{a}, \bar{b})$ if and only if $(A, \bar{a}) \equiv_0 (B, \bar{b})$.

This is equivalent to our definition of a win for $\exists$, by Theorem 1.3.1(c).

The game we have just described is called the **Ehrenfeucht–Fraïssé game of length γ on A and B**, in symbols $\mathrm{EF}_\gamma(A, B)$.

The more A is like B, the better chance player $\exists$ has of winning these games. In fact if player $\exists$ knows an isomorphism $i: A \to B$ then she can be sure of winning every time. All she has to do is to follow the rule

(2.2)   choose $i(a)$ whenever player $\forall$ has just chosen an element a
         of A, and $i^{-1}(b)$ whenever player $\forall$ has just chosen b from
         B.

We can express this point more precisely by using a notion from game theory, viz. the notion of a **winning strategy**.

A **strategy** for a player in a game is a set of rules which tell the player exactly how to move, depending on what has happened earlier in the play. We say that the player **uses** the strategy σ in a play if each of his or her moves in the play obeys the rules of σ. We say that the strategy σ is a **winning strategy** if the player wins every play in which he or she uses σ. For example the rule in (2.2) is a winning strategy for player $\exists$.

We write $A \sim_\gamma B$ to mean that player $\exists$ has a winning strategy in the game $\mathrm{EF}_\gamma(A, B)$. Thus for example $\mathbb{Q} \nsim_2 \mathbb{Z}$ by Example 1 above.

Before we forget empty structures altogether, we should stipulate that for any positive ordinal γ, if at least one of A, B is empty, then $A \sim_\gamma B$ if and only if both are empty.

Lemma 3.2.1. *Let L be a signature and let A, B be L-structures.*

(a) *If $A \cong B$ then $A \sim_\gamma B$ for all ordinals γ.*

(b) *If $\beta < \gamma$ and $A \sim_\gamma B$ then $A \sim_\beta B$.*

(c) *If $A \sim_\gamma B$ and $B \sim_\gamma C$ then $A \sim_\gamma C$; in fact $\sim_\gamma$ is an equivalence relation on the class of L-structures.*

Proof. We have already proved (a) in the discussion above.

I leave (b) as an exercise and move on to (c). It's clear from the definition that $\sim_\gamma$ is reflexive and symmetric on the class of L-structures. (True, the definition of the game $\mathrm{EF}_\gamma(A, B)$ was phrased as if A and B must be different structures. But the reader can handle this.) We prove transitivity. Suppose $A \sim_\gamma B$ and $B \sim_\gamma C$, so that player $\exists$ has winning strategies σ and τ for $\mathrm{EF}_\gamma(A, B)$ and $\mathrm{EF}_\gamma(B, C)$ respectively. Let the two players sit down to a match of $\mathrm{EF}_\gamma(A, C)$. We have to find a winning strategy for player $\exists$.

Here we use a trick which is common in game theory. We make one of the players play a private game on the side, at the same time as the main game. In fact player $\exists$ will play two private games, one of $\mathrm{EF}_\gamma(A, B)$ and one of $\mathrm{EF}_\gamma(B, C)$. In the public game of $\mathrm{EF}_\gamma(A, C)$ she will proceed as follows. Every time player $\forall$ chooses an element a_i of A, she first imagines to herself that player $\forall$ has made this move in $\mathrm{EF}_\gamma(A, B)$; then σ tells her to pick an element b_i of B. Next she imagines that b_i was a choice of player $\forall$ in the private game of $\mathrm{EF}_\gamma(B, C)$, and she uses her strategy τ to choose a corresponding c_i in C. This element c_i will be her answer to a_i in the public game. If player $\forall$ chose an element c_i of C, then she would respond in the same way but moving in the other direction, from C through B to A.

At the end of the contest, the players have constructed sequences $\bar{a}$ from A, $\bar{b}$ from B and $\bar{c}$ from C. The play of the public game $\mathrm{EF}_\gamma(A, C)$ is $(\bar{a}, \bar{c})$. Now, in the private game $\mathrm{EF}_\gamma(A, B)$, player $\exists$ used her winning strategy σ, and so the play $(\bar{a}, \bar{b})$ is a win for $\exists$. Similarly $(\bar{b}, \bar{c})$ is a win for $\exists$ in $\mathrm{EF}_\gamma(B, C)$. So

(2.3) $$(A, \bar{a}) \equiv_0 (B, \bar{b}) \equiv_0 (C, \bar{c})$$

and hence $(A, \bar{a}) \equiv_0 (C, \bar{c})$. This shows that $(\bar{a}, \bar{c})$ is a win for $\exists$ in $\mathrm{EF}_\gamma(A, C)$, and hence the strategy we described for her is winning. Thus $A \sim_\gamma B$.   □

Back-and-forth systems

Two L-structures A and B are said to be **back-and-forth equivalent** if $A \sim_\omega B$, i.e. if player $\exists$ has a winning strategy for the game $\mathrm{EF}_\omega(A, B)$.

There is a useful criterion for two structures to be back-and-forth equivalent. A **back-and-forth system** from A to B is a set I of pairs $(\bar{a}, \bar{b})$ of tuples, with $\bar{a}$ from A and $\bar{b}$ from B, such that

(2.4)   if $(\bar{a}, \bar{b})$ is in I then $\bar{a}$ and $\bar{b}$ have the same length and $(A, \bar{a})$ $\equiv_0 (B, \bar{b})$,

(2.5) I is not empty,

(2.6)   for every pair $(\bar{a}, \bar{b})$ in I and every element c of A there is an element d of B such that the pair $(\bar{a}c, \bar{b}d)$ is in I, and

(2.7)   for every pair $(\bar{a}, \bar{b})$ in I and every element d of B there is an element c of A such that the pair $(\bar{a}c, \bar{b}d)$ is in I.

Note that by (2.4) and Theorem 1.3.1(c), if $(\bar{a}, \bar{b})$ is in I then there is an isomorphism $f: \langle \bar{a} \rangle_A \to \langle \bar{b} \rangle_B$ such that $f\bar{a} = \bar{b}$; f is unique since $\bar{a}$ generates $\langle \bar{a} \rangle_A$. We write I^* for the set of all such functions f corresponding to pairs of tuples in I. The conditions (2.4)–(2.7) imply some similar conditions on the set $J = I^*$:

(2.4′)   each $f \in J$ is an isomorphism from a finitely generated substructure of A to a finitely generated substructure of B,

(2.5′) J is not empty,

(2.6′)   for every $f \in J$ and c in A there is $g \supseteq f$ such that $g \in J$ and $c \in \text{dom } g$,

(2.7′)   for every $f \in J$ and d in B there is $g \supseteq f$ such that $g \in J$ and $d \in \text{im } g$.

And conversely, if J is any set obeying the conditions (2.4′)–(2.7′), then there is a back-and-forth system I such that $J = I^*$. Namely, take I to be the set of all pairs of tuples $(\bar{a}, \bar{b})$ such that $\bar{a}$ is from A, $\bar{b}$ is from B and J contains a map $f: \langle \bar{a} \rangle_A \to \langle \bar{b} \rangle_B$ such that $f\bar{a} = \bar{b}$.

Some writers refer to a set J satisfying (2.4′)–(2.7′) as a 'back-and-forth system from A to B'. The clash between their terminology and ours is quite harmless; the two notions are near enough the same.

Lemma 3.2.2. *Let L be a signature and let A, B be L-structures. Then A and B are back-and-forth equivalent if and only if there is a back-and-forth system from A to B.*

Proof. Suppose first that A is back-and-forth equivalent to B, so that player $\exists$ has a winning strategy σ for the game $\text{EF}_\omega(A, B)$. Then we define I to consist of the pairs of tuples which are of the form $(\bar{c}|n, \bar{d}|n)$ for some $n < \omega$ and some play $(\bar{c}, \bar{d})$ in which player $\exists$ uses σ.

The set I is a back-and-forth system from A to B. First, putting $n = 0$ in the definition of I, we see that I contains the pair of 0-tuples $(\langle \rangle, \langle \rangle)$. This establishes (2.5). Next, (2.6) and (2.7) express that σ tells player $\exists$ what to do at each step of the game. And finally (2.4) holds because the strategy σ is winning.

In the other direction, suppose that there exists a back-and-forth system I from A to B. Define the set I^* of maps as above, and choose an arbitrary well-ordering of I^*. Consider the following strategy σ for player $\exists$ in the game $\text{EF}_\omega(A, B)$:

(2.8)   at each step, if the play so far is $(\bar{a}, \bar{b})$ and player $\forall$ has just chosen an element c from A, find the first map f in I^* such that $\bar{a}$ and c are in the domain of f and $f\bar{a} = \bar{b}$, and then choose d to be fc; likewise in the other direction if player $\forall$ has just chosen an element d from B. (If there is no such map f, choose some arbitrarily assigned element of the appropriate structure.)

By (2.5')–(2.7'), if player $\exists$ follows this strategy then there always will be a map f in I^* as required. Suppose the resulting play is $(\bar{a}, \bar{b})$. Then by (2.4') and Theorem 1.3.1(c) we have $(A, \bar{a}) \equiv_0 (B, \bar{b})$, and so player $\exists$ wins. □

Example 2: *Algebraically closed fields.* Let A and B be algebraically closed fields of the same characteristic and infinite transcendence degree. We shall show that A is back-and-forth equivalent to B. Let J be the set of all isomorphisms $e: A' \to B'$ where A', B' are finitely generated subfields of A, B respectively. (A **finitely generated subfield** of A is the smallest subfield of A containing some given finite set of elements of A. It need not be finitely generated as a ring.) Clearly J satisfies (2.4'). J is not empty since the prime subfields of A, B are isomorphic. Thus (2.5') is satisfied. Suppose $f: A' \to B'$ is in J and c is an element of A. We want to find a matching element d in B. There are two cases. First suppose c is algebraic over A'. Then c is determined up to isomorphism over A' by its minimal polynomial $p(x)$ over A'. Now f carries $p(x)$ to a polynomial $fp(x)$ over B', and B contains a root d of $fp(x)$ since it is algebraically closed. Thus f extends to an isomorphism $g: A'(c) \to B'(d)$. Second, suppose c is transcendental over A'. Since B' is finitely generated and B has infinite transcendence degree, there is an element d of B which is transcendental over B'. Thus again f extends to an isomorphism $g: A'(c) \to B'(d)$. Either way, condition (2.6') is satisfied. By symmetry, so is (2.7'). So J defines a back-and-forth system from A to B (namely the set of pairs $(\bar{a}, f\bar{a})$ with $f \in J$); now quote Lemma 3.2.2.

If $A \subseteq B$ in the example above, then we can say a little more. For every finitely generated subfield C of A, there is a system J as above, such that every map in J pointwise fixes C. In terms of back-and-forth systems, this says that if $\bar{e}$ is a tuple of elements which generate C, then there is a back-and-forth system I from A to B in which every pair has the form $(\bar{e}\bar{a}, \bar{e}\bar{b})$.

Consequences of back-and-forth equivalence

If two structures A and B are back-and-forth equivalent, they are in some sense hard to tell apart. The next two theorems illustrate this. A **position** of length n in a play of the back-and-forth game $\mathrm{EF}_\gamma(A, B)$ is a pair $(\bar{c}, \bar{d})$ of n-tuples, where $\bar{c}$ (resp. $\bar{d}$) lists in order the elements of A (resp. B) chosen in the first n moves. (A **position** is a position of some finite length.) The position is said to be **winning** for one of the players if that player has a strategy which enables him or her to win in $EF_\gamma(A, B)$ whenever the first n moves are $(\bar{c}, \bar{d})$.

It is not hard to see that $(\bar{c}, \bar{d})$ is a winning position for a player if and only if that player has a winning strategy for the game $EF_\gamma((A, \bar{c}), (B, \bar{d}))$. In particular the starting position (of length 0) is winning for player $\exists$ if and only if A and B are back-and-forth equivalent.

The next result says that for countable structures, back-and-forth equivalence is the same thing as isomorphism.

Theorem 3.2.3. *Let L be any signature (not necessarily countable) and let A and B be L-structures.*

(a) *If $A \cong B$ then A is back-and-forth equivalent to B.*

(b) *Suppose A, B are at most countable. If A is back-and-forth equivalent to B then $A \cong B$. In fact, if $\bar{c}$, $\bar{d}$ are tuples from A, B respectively, such that $(\bar{c}, \bar{d})$ is a winning position for player $\exists$ in $EF_\omega(A, B)$, then there is an isomorphism from A to B which takes $\bar{c}$ to $\bar{d}$.*

Proof. (a) is a special case of Lemma 3.2.1(a).

(b) Since the game $EF_\omega(A, B)$ has infinite length, if A and B are at most countable then player $\forall$ can list all the elements of A and of B among his choices. Let player $\exists$ play to win, and let $(\bar{a}, \bar{b})$ be the resulting play. Since player $\exists$ wins, the diagram lemma (Lemma 1.4.2) gives an isomorphism $f: A = \langle \bar{a} \rangle_A \to \langle \bar{b} \rangle_B = B$. The last sentence is proved the same way, but starting the play at $(\bar{c}, \bar{d})$.   □

Example 3: *Dense linear orderings without endpoints.* An old theorem of Cantor states that if A and B are countable dense linear orderings without endpoints (see (2.32) in section 2.2) then $A \cong B$. This follows at once from Theorem 3.2.3(b), when we show that A is back-and-forth equivalent to B. The required back-and-forth system consists of all pairs of tuples $(\bar{a}, \bar{b})$ such that, for some $n < \omega$, $\bar{a} = (a_0, \ldots, a_{n-1})$ is a tuple of elements of A, $\bar{b} = (b_0, \ldots, b_{n-1})$ is a tuple of elements of B, and, for all $i < j < n$, $a_i \gtrsim a_j \Leftrightarrow b_i \gtrsim b_j$.

Example 4: *Atomless boolean algebras.* Let A and B be countable atomless boolean algebras (see (2.30) in section 2.2). Then $A \cong B$. Again we show this by Theorem 3.2.3(b). Let J be the set of all isomorphisms from finite subalgebras of A to finite subalgebras of B. Then (2.4') and (2.5') clearly hold. For (2.6'), suppose $f \in J$ and let $a_0, \ldots, a_{k-1}$ be the atoms of the boolean algebra A' which is the domain of f. Then the isomorphism type of any element c of A over A' is determined once we are told, for each $i < k$, whether $c \wedge a_i$ is 0, a_i or neither. Since B is atomless, there is an element d of B such that for each $i < k$,

(2.9) $d \wedge f(a_i)$ is 0 (resp. $f(a_i)$) $\Leftrightarrow c \wedge a_i$ is 0 (resp. a_i).

So f can be extended to an isomorphism whose domain includes c. Thus J satisfies (2.6′), and by symmetry it satisfies (2.7′) too. We have proved that $A \cong B$.

Incidentally the results of Examples 3 and 4 are as false as they possible could be when we replace 'countable' by an uncountable cardinal κ. For example one can show that there are 2^κ non-isomorphic dense linear orderings and 2^κ non-isomorphic atomless boolean algebras, all of cardinality κ.

Nevertheless back-and-forth equivalence gives us useful information about uncountable structures.

Theorem 3.2.4. *Let A, B be L-structures and $\bar{a}$, $\bar{b}$ n-tuples from A, B respectively. If $(\bar{a}, \bar{b})$ is a winning position for player $\exists$ in $EF_\omega(A, B)$, then $(A, \bar{a}) \equiv_{\infty\omega} (B, \bar{b})$. In particular if A and B are back-and-forth equivalent, then they are $L_{\infty\omega}$-equivalent.*

Proof. We prove that if $\phi(\bar{x})$ is any formula of $L_{\infty\omega}$ and $(\bar{a}, \bar{b})$ is a winning position for player $\exists$, then $A \vDash \phi(\bar{a}) \Leftrightarrow B \vDash \phi(\bar{b})$. The proof is by induction on the construction of ϕ.

If ϕ is atomic, the result follows from the diagram lemma (Lemma 1.4.2) and the definition of winning.

If ϕ is of the form $\neg\psi$, $\bigwedge\Phi$ or $\bigvee\Phi$, then we have the result straightforwardly from the induction hypothesis.

Next let ϕ be $\exists y \, \psi(\bar{x}, y)$. Suppose $A \vDash \phi(\bar{a})$; then there is an element c in A such that $A \vDash \psi(\bar{a}, c)$. Since the position $(\bar{a}, \bar{b})$ is winning for player $\exists$, she has a winning strategy from this position onward; this strategy tells her to choose a certain element d of B if player $\forall$ makes his next move by choosing c. So $(\bar{a}c, \bar{b}d)$ must still be a winning position for player $\exists$. Then the induction hypothesis tells us that $B \vDash \psi(\bar{b}, d)$, so that $B \vDash \exists y \, \psi(\bar{b}, y)$ as required. Likewise in the other direction from B to A.

Finally suppose ϕ is $\forall y \psi$. We reduce this to the previous cases by writing $\neg\exists y\neg$ for $\forall y$.   □

In fact the converse is true too, though it is a little harder to prove: if A and B are $L_{\infty\omega}$-equivalent then they are back-and-forth equivalent. So back-and-forth equivalence is not a good criterion for elementary equivalence, because it proves too much. We shall remedy this fault in the next section.

The following theorem is important, though we have no space to prove it here. A sentence σ_B as in the theorem is called a **Scott sentence** of B. No satisfactory analogue of the theorem is known for uncountable cardinalities.

Theorem 3.2.5 (Scott's isomorphism theorem). *Let L be a countable signature and B a countable L-structure. Then there is a sentence σ_B of $L_{\omega_1\omega}$ such that the models of σ_B are exactly the L-structures which are back-and-forth equivalent to B. In particular B is (up to isomorphism) the only countable model of σ_B.* □

Exercises for section 3.2

1. Prove Lemma 3.2.1(b).

2. (a) Show that, in the game $EF_\gamma(A, B)$, a strategy σ for a player can be written as a family $(\sigma_i : i < \gamma)$ where for each $i < \gamma$, σ_i is a function which picks the player's ith choice $\sigma_i(\bar{x})$ as a function of the sequence $\bar{x}$ of previous choices of the two players. (b) Show that σ can also be written as a family $(\sigma_i' : i < \gamma)$ where for each $i < \gamma$, σ_i' is a function which picks the player's ith choice $\sigma_i'(\bar{y})$ as a function of the sequence $\bar{y}$ of previous choices of the *other* player. (c) How can the functions σ_i be found from the functions σ_i', and vice versa? [To simplify the statements, assume that the domains of A and B are disjoint.]

3. (a) The game $P_\omega(A, B)$ is defined exactly like $EF_\omega(A, B)$ except that player $\forall$ must always choose from structure A and player $\exists$ from structure B. Show that, if A is at most countable, then player $\exists$ has a winning strategy for $P_\omega(A, B)$ if and only if A is embeddable in B. (b) What if player $\forall$ must choose from structure A in even-numbered steps and from structure B in odd-numbered steps (and $\exists$ vice versa)?

4. The game $H_\omega(A, B)$ is defined exactly like $EF_\omega(A, B)$ except that player $\exists$ wins the play $(\bar{a}, \bar{b})$ iff for every atomic formula ϕ, $A \vDash \phi(\bar{a}) \Rightarrow B \vDash \phi(\bar{b})$. Show that, if A and B are at most countable, then player $\exists$ has a winning strategy for $H_\omega(A, B)$ if and only if B is a homomorphic image of A.

5. Suppose A and B are two countable dense linearly ordered sets without endpoints, both partitioned into classes $P_0, \ldots, P_{n-1}$ so that each class P_i occurs densely in both orderings. Show that A is isomorphic to B.

6. (a) If ζ is a linear ordering, let ζ^+ be the ordering which we get by replacing each point of ζ by a pair of points a, b with $a < b$. Show that if ζ and ξ are dense linear orderings without endpoints then ζ^+ and ξ^+ are back-and-forth equivalent. (b) Show

that, if ζ and ξ are dense linear orderings which have first points but no last points, then ζ is back-and-forth equivalent to ξ.

7. Let A, B and C be fields; suppose $A \subseteq B$, $A \subseteq C$, and both B and C are algebraically closed and of infinite transcendence degree over A. Let (B, A) be the structure consisting of B with a 1-ary relation symbol P added so as to pick out A; and likewise with (C, A) and C. Show that (B, A) is back-and-forth equivalent to (C, A).

Suppose L, L' are signatures with no function symbols, and with no symbols in common. We form the signature $L + L'$ as follows: the symbols of $L + L'$ are those of L, those of L' and two new 1-ary relation symbols P and Q. The **disjoint sum** *of an L-structure A and an L'-structure B is the $(L + L')$-structure $A + B$ whose domain is the disjoint union of $\mathrm{dom}(A)$ and $\mathrm{dom}(B)$; the symbols of L are interpreted on $\mathrm{dom}(A)$ exactly as in A, and those of L' are interpreted on $\mathrm{dom}(B)$ as in B; P and Q are interpreted as names of $\mathrm{dom}(A)$ and $\mathrm{dom}(B)$ respectively.*
8. Show that if A, B are respectively back-and-forth equivalent to A', B', then their disjoint sum $A + B$ is back-and-forth equivalent to $A' + B'$.

3.3 Games for elementary equivalence

In section 3.2 we noticed that there are Ehrenfeucht–Fraïssé games $\mathrm{EF}_k(A, B)$ of finite length k. But we did nothing with them. In this section we shall see that after a small piece of cosmetic surgery, these finite games are a powerful tool of first-order model theory. They have hundreds of applications – I shall discuss just two.

First comes the surgery. In place of $\mathrm{EF}_k(A, B)$ we devise a game $\mathrm{EF}_k[A, B]$, which is played exactly like $\mathrm{EF}_k(A, B)$ but with a different criterion for winning. The players between them make k pairs of choices, and at the end of the play, when tuples $\bar{c}$ from A and $\bar{d}$ from B have been chosen, player $\exists$ wins the game $\mathrm{EF}_k[A, B]$ iff

(3.1) for every *unnested* atomic formula ϕ of L, $A \vDash \phi(\bar{c}) \Leftrightarrow B \vDash \phi(\bar{d})$.

If the signature L contains no function symbols or constants, then every formula of L is unnested anyway, so that $\mathrm{EF}_k(A, B)$ and $\mathrm{EF}_k[A, B]$ are exactly the same. So readers who are only interested in linear orderings can read square brackets as round brackets throughout. The games $\mathrm{EF}_k[A, B]$ are called **unnested Ehrenfeucht–Fraïssé games**.

We write $A \approx_k B$ to mean that player $\exists$ has a winning strategy for the game $\mathrm{EF}_k[A, B]$. Then $\approx_k$ is an equivalence relation on the class of L-structures – this is immediate from the argument of Lemma 3.2.1(c).

It will be useful to allow the structures to carry some parameters with them. Thus if $n < \omega$ and $\bar{a}$, $\bar{b}$ are n-tuples of elements of A, B respectively, we write $(A, \bar{a}) \approx_k (B, \bar{b})$ to mean that player $\exists$ has a winning strategy for the game $\text{EF}_k[(A, \bar{a}), (B, \bar{b})]$. The condition for player $\exists$ to win this game, when the play has chosen k-tuples $\bar{c}$, $\bar{d}$ from A, B respectively, is just that

(3.2)   for every unnested atomic formula ϕ of L,
   $A \vDash \phi(\bar{a}, \bar{c}) \Leftrightarrow B \vDash \phi(\bar{b}, \bar{d})$.

This is a restatement of the condition (3.1) with $(A, \bar{a})$ and $(B, \bar{b})$ in place of A and B.

Lemma 3.3.1. *Let A and B be structures of the same signature. Suppose $n, k < \omega$; suppose $\bar{a}$, $\bar{b}$ are n-tuples of elements of A, B respectively. Then the following are equivalent.*

(a) $(A, \bar{a}) \approx_{k+1} (B, \bar{b})$.

(b) *For every element c of A there is an element d of B such that $(A, \bar{a}, c) \approx_k (B, \bar{b}, d)$; and for every element d of B there is an element c of A such that $(A, \bar{a}, c) \approx_k (B, \bar{b}, d)$.*

Proof. First suppose (a) holds. Let c be an element of A. Then player $\exists$ can regard c as player $\forall$'s first choice in a play of $\text{EF}_{k+1}[(A, \bar{a}), (B, \bar{b})]$. Let her use her winning strategy σ to choose d as her reply to c. Now if the two players decide to play the game $\text{EF}_k[(A, \bar{a}, c), (B, \bar{b}, d)]$, player $\exists$ can win by regarding this second game as the last k steps in the play of $\text{EF}_{k+1}[(A, \bar{a}), (B, \bar{b})]$, using σ to choose her moves. This proves the first half of (b), and the second half follows by symmetry.

Conversely suppose (b) holds. Then player $\exists$ can win the game $\text{EF}_{k+1}[(A, \bar{a}), (B, \bar{b})]$ as follows. If player $\forall$ opens by choosing some element c of A, then player $\exists$ chooses d as in (b), and for the rest of the game she follows her winning strategy for $\text{EF}_k[(A, \bar{a}, c), (B, \bar{b}, d)]$. Likewise if player $\forall$ started with an element d of B. $\square$

We turn to the fundamental theorem about the equivalence relations $\approx_k$ between structures of the form $(A, \bar{a})$ with A an L-structure. It will say among other things that for each k there are just finitely many equivalence classes of $\approx_k$, and that each equivalence class is definable by a formula of L. It will also put a bound on the complexity of these defining formulas, in terms of the following notion.

For any formula ϕ of the first-order language L, we define the **quantifier rank** of ϕ, $\text{qr}(\phi)$, by induction on the construction of ϕ, as follows.

(3.3)   If ϕ is atomic then $\text{qr}(\phi) = 0$.

(3.4) $\text{qr}(\neg \psi) = \text{qr}(\psi)$.

(3.5) $\text{qr}(\bigwedge \Phi) = \text{qr}(\bigvee \Phi) = \max \{\text{qr}(\psi): \psi \in \Phi\}$.

(3.6) $\text{qr}(\forall x \, \psi) = \text{qr}(\exists x \, \psi) = \text{qr}(\psi) + 1$.

Thus $\text{qr}(\phi)$ measures the nesting of quantifiers in ϕ.

Theorem 3.3.2 *(Fraïssé–Hintikka theorem). Let L be a first-order language with finite signature. Then we can effectively find for each k, $n < \omega$ a finite set $\Theta_{n,k}$ of unnested formulas $\theta(x_0, \ldots, x_{n-1})$ of quantifier rank at most k, such that*

(a) *for every L-structure A, all k, $n < \omega$ and each n-tuple $\bar{a} = (a_0, \ldots, a_{n-1})$ of elements of A, there is exactly one formula θ in $\Theta_{n,k}$ such that $A \vDash \theta(\bar{a})$.*

(b) *for all k, $n < \omega$ and every pair of L-structures A, B, if $\bar{a}$ and $\bar{b}$ are respectively n-tuples of elements of A and B, then $(A, \bar{a}) \approx_k (B, \bar{b})$ if and only if there is θ in $\Theta_{n,k}$ such that $A \vDash \theta(\bar{a})$ and $B \vDash \theta(\bar{b})$.*

(c) *for every $k < \omega$ and every unnested formula $\phi(\bar{x})$ of L with n free variables $\bar{x}$ and quantifier rank at most k, we can effectively find a disjunction $\theta_0 \vee \ldots \vee \theta_{m-1}$ of formulas $\theta_i(\bar{x})$ in $\Theta_{n,r}$ which is logically equivalent to ϕ.*

Proof. First let me describe the sets $\Theta_{n,k}$. When we know what these sets are, an induction on k quickly gives property (a), and (c) is an exercise. This will leave property (b) to be proved.

We write ϕ^1 for ϕ and ϕ^0 for $\neg \phi$. We write $^m2$ for the set of maps $s: m \to 2$, taking m as $\{0, \ldots, m-1\}$ and 2 as $\{0,1\}$.

We begin with $k = 0$ and a fixed $n < \omega$. There are just finitely many unnested atomic formulas $\phi(x_0, \ldots, x_{n-1})$ of L; list them as $\phi_0, \ldots, \phi_{m-1}$. Take $\Theta_{n,0}$ to be the set of all formulas of the form $\phi_0^{s(0)} \wedge \ldots \wedge \phi_{m-1}^{s(m-1)}$ as s ranges over $^m2$. Thus $\Theta_{n,0}$ lists all the possible unnested quantifier-free types of n-tuples of elements of an L-structure. (In general it includes some impossible types too, bearing in mind formulas with a conjunct $x_0 \neq x_0$. But no matter.)

When $\Theta_{n+1,k}$ has been defined, we list the formulas in it as $\chi_0(x_0, \ldots, x_n)$, $\ldots, \chi_{j-1}(x_0, \ldots, x_n)$. Then we define $\Theta_{n,k+1}$ to be the set of all formulas

$$(3.7) \qquad \bigwedge_{i \in X} \exists x_n \chi_i(x_0, \ldots, x_n) \wedge \forall x_n \bigwedge_{i \in X} \chi_i(x_0, \ldots, x_n)$$

as X ranges over the subsets of j. (Thus each formula in $\Theta_{n,k+1}$ lists the ways in which the n-tuple can be extended to an $(n+1)$-tuple, in terms of the formulas of quantifier rank k satisfied by the $(n+1)$-tuple.)

We prove (b) by induction on k, for all n simultaneously. Let A and B be L-structures and let $\bar{a}$, $\bar{b}$ be n-tuples of elements of A, B respectively.

First suppose $k = 0$. Then by definition of $\approx_0$, $(A, \bar{a}) \approx_0 (B, \bar{b})$ if and only if for every unnested atomic formula ϕ of L, $A \vDash \phi(\bar{a}) \Leftrightarrow B \vDash \phi(\bar{b})$. But this in turn holds if and only if $\bar{a}$ and $\bar{b}$ have the same unnested quantifier-free type in A and B respectively; in other words, if and only if there is some θ in $\Theta_{n,0}$ such that $A \vDash \theta(\bar{a})$ and $B \vDash \theta(\bar{b})$, as required. Clearly θ is unique.

Now we prove the result for $k + 1$, assuming it proved for k. In the notation of (3.7), let X be the set of i such that A has an element c for which $A \vDash \chi_i(\bar{a}, c)$, and for this choice of X let $\theta'(x_0, \ldots, x_{n-1})$ be the formula (3.7), which is in $\Theta_{n,k+1}$. Then certainly $A \vDash \theta'(\bar{a})$. But now use property (a), Lemma 3.3.1 and the induction hypothesis to see what the statement '$(A, \bar{a}) \approx_{k+1} (B, \bar{b})$' means. It means that first, for every $i \in X$ there is an element d of B such that $B \vDash \chi_i(\bar{b}, d)$, and second, for every element d of D there is $i \in X$ such that $B \vDash \chi_i(\bar{b}, d)$. In short, it means exactly that $B \vDash \theta'(\bar{b})$. $\qquad\square$

I shall refer to the formulas in the sets $\Theta_{n,k}$ as **formulas in game-normal form**, or more briefly **game-normal formulas**. By (c) of the theorem, every first-order formula ϕ is logically equivalent to a disjunction of formulas in game-normal form with at most the same free variables as ϕ. If ϕ was unnested, the game-normal formulas can be chosen to be of the same quantifier rank as ϕ; but note that the process of reducing a formula to unnested form (Theorem 2.6.1) will generally raise the quantifier rank.

Corollary 3.3.3. *Let L be a first-order language of finite signature. Then for any two L-structures A and B the following are equivalent.*
(a) $A \equiv B$.
(b) *For every $k < \omega$, $A \approx_k B$.*

Proof. By the theorem, (b) says that A and B agree on all unnested sentences of finite quantifier rank. So (a) certainly implies (b). By Corollary 2.6.2, every first-order sentence is logically equivalent to an unnested sentence of finite quantifier rank; so (b) implies (a) too. $\qquad\square$

Application 1: elimination sets

Ehrenfeucht–Fraïssé games form a useful tool for quantifier elimination. They give us a way of finding elimination sets by thinking about the structures themselves, rather than about the theories of the structures.

Suppose $\mathbf{K}$ is a class of L-structures. For each structure A in $\mathbf{K}$, write $\mathrm{tup}(A)$ for the set of all pairs $(A, \bar{a})$ where $\bar{a}$ is a tuple of elements of A; write $\mathrm{tup}(\mathbf{K})$ for the union of the sets $\mathrm{tup}(A)$ with A in $\mathbf{K}$. By an **(unnested) graded back-and-forth system** for $\mathbf{K}$ we mean a family of equivalence relations $(E_k : k < \omega)$ on $\mathrm{tup}(\mathbf{K})$ with the following properties:

(3.8)   if $\bar{a}$, $\bar{b}$ are in $\mathrm{tup}(A)$, $\mathrm{tup}(B)$ respectively and $\bar{a} E_0 \bar{b}$, then for every unnested atomic formula $\phi(\bar{x})$ of L, $A \vDash \phi(\bar{a})$ iff $B \vDash \phi(\bar{b})$;

(3.9)   if $\bar{a}$, $\bar{b}$ are in tup(A), tup(B) respectively, $\bar{a}\ E_{k+1}\ \bar{b}$ and c is
         any element of A, then there is an element d of B such that
         $\bar{a}c\ E_k\ \bar{b}d$.

Lemma 3.3.4. *Suppose* $(E_k: k < \omega)$ *is a graded back-and-forth system for* **K**.
Then $(A, \bar{a})\ E_k\ (B, \bar{b})$ *implies* $(A, \bar{a}) \approx_k (B, \bar{b})$.

Proof. By (3.9), player $\exists$ can choose so that after the 0-th step in the game
$\mathrm{EF}_k[(A, \bar{a}), (B, \bar{b})]$ we have $\bar{a}c_0\ E_{k-1}\ \bar{b}d_0$, after the 1-th step we have
$\bar{a}c_0c_1\ E_{k-2}\ \bar{b}d_0d_1$, and so on until $\bar{a}\bar{c}\ E_0\ \bar{b}\bar{d}$ after k steps. But then player $\exists$
wins by (3.8). $\square$

Lemma 3.3.5. *Suppose* $(E_k: k < \omega)$ *is a graded back-and-forth system for* **K**.
*Suppose that, for each n and k, E_k has just finitely many equivalence classes
on n-tuples, and each of these classes is definable by a formula* $\chi_{k,n}(\bar{x})$. *Then
the set of all formulas* $\chi_{k,n}$ $(k, n < \omega)$ *forms an elimination set for* **K**.

Proof. We have to show that each formula $\phi(\bar{x})$ of the language L is
equivalent throughout **K** to a boolean combination of formulas $\chi_{k,n}(\bar{x})$
$(k, n < \omega)$. By Corollary 2.6.2 we can suppose that ϕ is unnested, and so by
Theorem 3.3.2(c), ϕ is logically equivalent to a boolean combination of
game-normal formulas $\theta(\bar{x})$. Now by Lemma 3.3.4, each equivalence class
under $\approx_k$ is a union of equivalence classes of E_k. It follows by Theorem
3.3.2(b) that each game-normal formula $\phi(\bar{x})$ is equivalent to a disjunction of
formulas $\chi_{k,n}(\bar{x})$. $\square$

This is all the general theory that we need. Now we turn to an example.

Consider the ordered group of integers. The appropriate language for this
group is a language L whose symbols are $+$, $-$, 0, 1 and $<$. The ordered
group of integers forms an L-structure which we shall write as $\mathbb{Z}$.

Our aim is to find an elimination set for Th($\mathbb{Z}$). Where to look? We cast
around for possible equivalence relations E_k. Roughly speaking, we shall
formalise the notion '$\bar{a}$ can't be distinguished from $\bar{b}$ without mentioning
more than m elements'.

More precisely, suppose $\bar{x}$ is $(x_0, \ldots, x_{n-1})$ and m is a positive integer.
Then by an m-term $t(\bar{x})$ we shall mean a term $\Sigma_{i<m} s_i$ where each s_i is either
0 or 1 or -1 or x_j or $-x_j$ for some $j < n$. Let $\bar{a} = (a_0, \ldots, a_{n-1})$ and
$\bar{b} = (b_0, \ldots, b_{n-1})$ be two n-tuples of elements of $\mathbb{Z}$. We say that $\bar{a}$ is
m-**equivalent** to $\bar{b}$ if for every m-term $t(\bar{x})$ the following hold in $\mathbb{Z}$:

(3.10) $t(\bar{a}) > 0 \Leftrightarrow t(\bar{b}) > 0$;

(3.11) $t(\bar{a})$ is congruent to $t(\bar{b})$ (mod q) (for each integer q, $1 \leqslant q \leqslant m$).

Note that if $\bar{a}$ is m-equivalent to $\bar{b}$, then $\bar{a}$ is m'-equivalent to $\bar{b}$ for all $m' < m$.

Lemma 3.3.6. *Suppose $\bar{a}$ and $\bar{b}$ are n-tuples of elements of $\mathbb{Z}$ which are 3-equivalent. Then for every unnested atomic formula $\phi(\bar{x})$ of L, $\mathbb{Z} \vDash \phi(\bar{a})$ iff $\mathbb{Z} \vDash \phi(\bar{b})$.*

Proof. For example, $\mathbb{Z} \vDash a_0 + a_1 = a_2 \quad \Leftrightarrow \quad \mathbb{Z} \nvDash (a_0 + a_1 - a_2 > 0 \vee - a_0 - a_1 + a_2 > 0) \quad \Leftrightarrow \quad \mathbb{Z} \nvDash (b_0 + b_1 - b_2 > 0 \vee -b_0 - b_1 + b_2 > 0) \quad \Leftrightarrow \quad \mathbb{Z} \vDash b_0 + b_1 = b_2$. $\qquad \square$

Lemma 3.3.7. *Suppose m is a positive integer, and $\bar{a}$ and $\bar{b}$ are n-tuples of elements of $\mathbb{Z}$ which are m^{2m}-equivalent. Then for every element c of $\mathbb{Z}$ there is an element d of $\mathbb{Z}$ such that the tuples $\bar{a}c$, $\bar{b}d$ are m-equivalent.*

Proof. Take an element c, and consider all the true sentences of the form

(3.12) $t(\bar{a}) + ic \equiv j \pmod{q}$
                    where $t(\bar{x})$ is an $(m-1)$-term, $0 < i < m$ and $j < q \leqslant m$.

Since $\bar{a}$ and $\bar{b}$ are m^{2m}-equivalent, $t(\bar{a})$ and $t(\bar{b})$ are certainly congruent modulo $m!$. Let α be the remainder when c is divided by $m!$. Then if d is any element of $\mathbb{Z}$ which is congruent to α modulo $m!$, we have

(3.13) $t(\bar{b}) + id \equiv j \pmod{q}$ whenever $t(\bar{a}) + ic \equiv j \pmod{q}$.

This tells us how to find a d to take care of the conditions (3.11).

Turning to the conditions (3.10), consider the set of all true statements of the forms

(3.14) $t(\bar{a}) + ic > 0, \; t(\bar{a}) + ic \leqslant 0$
                    where $t(\bar{x})$ is an $(m-1)$-term and $0 < i < m$.

After multiplying by suitable integers, we can bring these inequalities to the forms

(3.15) $t(\bar{a}) + m!c > 0, \; t(\bar{a}) + m!c \leqslant 0$
                    where $t(\bar{x})$ is an $m!.(m-1)$-term.

Taking greatest and least values in the obvious way, we can reduce (3.15) to a condition of the form

(3.16) $-t_1(\bar{a}) < m!c \leqslant -t_2(\bar{a})$,

together with a set of inequalities $\Phi(\bar{a})$ which don't mention c. (Possibly we reach a single inequality in (3.16), if $m!c$ is bounded only on one side.) So by (3.16), there is a number x in $\mathbb{Z}$ such that

(3.17) $-t_1(\bar{a}) < x \leqslant -t_2(\bar{a})$, and x is congruent to $m!\alpha \pmod{(m!)^2}$.

Now $-t_1(\bar{a})$ is at most an $m!.(m-1)$-term, and so by assumption it is congruent modulo $(m!)^2$ to $-t_1(\bar{b})$; similarly with $-t_2(\bar{a})$. Hence there is also a number y in $\mathbb{Z}$ such that

(3.18) $\qquad -t_1(\bar{b}) < y \leqslant -t_2(\bar{b})$, and y is congruent to $m!\alpha$ (mod $(m!)^2$).

Put $d = y/m!$. Then d is congruent to α modulo $m!$. We have

(3.19) $\qquad\qquad\qquad -t_1(\bar{d}) < m!d \leqslant -t_2(\bar{d})$

(cf. (3.16)), and the inequalities $\Phi(\bar{b})$ also hold since they use at worst $m! \cdot 2(m-1)$-terms. So tracing backwards along the path from (3.14) to (3.16), we have all the corresponding

(3.20) $\qquad\qquad\qquad t(\bar{a}) + ic > 0 \Leftrightarrow t(\bar{b}) + id > 0$

$\qquad\qquad\qquad\qquad$ where $t(\bar{x})$ is an $(m-1)$-term and $0 < i < m$.

Thus d serves for the lemma. $\qquad\qquad\qquad\qquad\qquad\qquad\qquad\qquad$ $\square$

We define $m_0, m_1, \ldots$ inductively by

(3.21) $\qquad\qquad\qquad m_0 = 3, \; m_{i+1} = m_i^{2m_i}$.

We define the equivalence relations E_k by $(\mathbb{Z}, \bar{a}) \, E_k \, (\mathbb{Z}, \bar{b})$ if $\bar{a}$ is m_k-equivalent to $\bar{b}$. By Lemmas 3.3.6 and 3.3.7, $(E_k: k < \omega)$ is a graded back-and-forth system for $\{\mathbb{Z}\}$. So by Lemma 3.3.5 we have found an elimination set for Th$(\mathbb{Z})$. The formulas in the elimination set are all fairly simple: each of them is either an inequality or a congruence to some fixed modulus.

Theorem 3.3.8. Th$(\mathbb{Z})$ *is decidable.*

Proof. First we show that for any tuple $\bar{a}$ in $\mathbb{Z}$ and any $k < \omega$ we can compute a bound $\delta(\bar{a}, k)$ such that

(3.22) $\quad$ for every c there is d with $|d| < \delta(\bar{a}, k)$ such that $(\mathbb{Z}, \bar{a}c) \, E_k \, (\mathbb{Z}, \bar{a}d)$.

The calculations in the proof of Lemma 3.3.7 show that $\delta(\bar{a}, k)$ can be chosen to be $m^{2m} \cdot \mu$ where m is m_k and μ is $\max\{|a_i|: a_i$ occurs in $\bar{a}\}$.

It follows by induction on k that if $\phi(\bar{x})$ is a formula of L of quantifier rank k, and $\bar{a}$ is a tuple of elements of $\mathbb{Z}$, then we can compute in a bounded number of steps whether or not $\mathbb{Z} \vdash \phi(\bar{a})$. Suppose for example that ϕ is $\exists y \, \psi(\bar{x}, y)$, where ψ has quantifier rank $k-1$. If there is an element c such that $\mathbb{Z} \vDash \psi(\bar{a}, c)$, then there is such an element $c < \delta(\bar{a}, k-1)$. So we only need check the truth of $\mathbb{Z} \vDash \psi(\bar{a}, c)$ for these finitely many c; by induction hypothesis this takes only a finite number of steps. $\qquad\qquad$ $\square$

The proof of Theorem 3.3.8 gives a primitive recursive bound $f(n)$ on the number of steps needed to check the truth of a sentence of length n. The bound $f(n)$ rises very fast with n. Exercising a little more thrift in the proof of Lemma 3.3.7, one can bring the bound down to something of the order of $2^{2^{2^{\kappa n}}}$ for some constant κ. But Fischer and Rabin showed that any decision procedure for $\text{Th}(\mathbb{Z})$ needs at least 2^{2^n} steps to settle the truth of sentences with n symbols (in the limit as n tends to infinity).

Corollary 3.3.9. *The linear orderings $\omega^* + \omega$ and ω both have decidable theories.*

Proof. For $\omega^* + \omega$ we have already proved it: a sentence ϕ of the first-order language of linear orderings is true in $\omega^* + \omega$ if and only if it is true in $\mathbb{Z}$. For ω we note that statements about ω can be translated into statements about the non-negative elements of $\mathbb{Z}$. $\square$

Application 2: replacements preserving $\equiv$

Theorem 3.3.10. *Let G_1, G_2 and H be groups. Assume $G_1 \equiv G_2$. Then $G_1 \times H \equiv G_2 \times H$.*

Proof. By Corollary 3.3.3 it suffices to show that if $k < \omega$ and $G_1 \approx_k G_2$ then $G_1 \times H \approx_k G_2 \times H$. Assume henceforth that $G_1 \approx_k G_2$. Then player $\exists$ has a winning strategy σ for the game $\text{EF}_k[G_1, G_2]$.

Let the two players meet to play the game $\text{EF}_k[G_1 \times H, G_2 \times H]$. This will be one of those many occasions when player $\exists$ will guide her choices by playing another game on the side. The side game will in fact be $\text{EF}_k[G_1, G_2]$. Whenever player $\forall$ offers an element, say the element $a \in G_1 \times H$, player $\exists$ will first split it into a product $a = g \cdot h$ with $g \in G_1$ and $h \in H$. Then she will pretend that player $\forall$ has just chosen g in the side game, and she will use her strategy σ to choose a reply $g' \in G_2$ in the side game. Her public reply to the element a will then be the element $b = g' \cdot h \in G_2 \times H$. Likewise the other way round if player $\forall$ chose from $G_2 \times H$.

At the end of the game let the play be $(g_0 \cdot h_0, \ldots, g_{k-1} \cdot h_{k-1}; g'_0 \cdot h_0, \ldots, g'_{k-1} \cdot h_{k-1})$. Player $\exists$ has won the side game. Now the unnested atomic formulas of the language L of groups are the formulas of the form $x = y$, $1 = y$, $x_0 \cdot x_1 = y$ and $x^{-1} = y$. So for all $i, j, l < k$ we have

(3.23)
$$g_i = g_j \text{ iff } g'_i = g'_j,$$
$$1 = g_i \text{ iff } 1 = g'_i,$$
$$g_i \cdot g_j = g_l \text{ iff } g'_i \cdot g'_j = g'_l,$$
$$g_i^{-1} = g_j \text{ iff } g'^{-1}_i = g'_j.$$

By the definition of cartesian products of groups, this implies that for all $i, j, l < k$ we also have

$$(3.24) \qquad g_i \cdot h_i = g_j \cdot h_j \text{ iff } g_i' \cdot h_i = g_j' \cdot h_j,$$

$$1 = g_i \cdot h_i \text{ iff } 1 = g_i' \cdot h_i,$$

$$g_i \cdot h_i \cdot g_j \cdot h_j = g_l \cdot h_l \text{ iff } g_i' \cdot h_i \cdot g_j' \cdot h_j = g_l' \cdot h_l,$$

$$(g_i \cdot h_i)^{-1} = g_j \cdot h_j \text{ iff } (g_i' \cdot h_i)^{-1} = g_j' \cdot h_j.$$

(Thus for example $g_i \cdot h_i = g_j \cdot h_j \Leftrightarrow (g_i = g_j$ and $h_i = h_j) \Leftrightarrow (g_i' = g_j'$ and $h_i = h_j) \Leftrightarrow g_i' \cdot h_i = g_j' \cdot h_j$.) So player $\exists$ wins the public game too, which proves the theorem. $\square$

The proof of Theorem 3.3.10 uses very few facts about groups. It would work equally well in any case where a part of a structure can be isolated and replaced: for example if an interval in a linear ordering is replaced by an elementarily equivalent linear ordering, the whole resulting ordering is elementarily equivalent to the original one. Also we could take an infinite product of groups and make replacements at all factors simultaneously.

Exercises for section 3.3

1. Prove Theorem 3.3.2(c).

2. Show that in the statement of Theorem 3.3.2 the formulas in $\Theta_{n,k}$ can all be taken to be $\exists_{k+1}$ formulas. Show also that they can all be taken to be $\forall_{k+1}$ formulas. [Prove both by simultaneous induction.]

3. Show that every unnested first-order formula of quantifier rank k is logically equivalent to an unnested first-order formula of quantifier rank k which is in negation normal form.

4. Find a simple set of axioms for $\text{Th}(\mathbb{Z}, +, <)$. [They should say that $\mathbb{Z}$ is an ordered group with a least positive element 1, and for every positive integer n there should be an axiom expressing that for all x, exactly one of $x, x + 1, \ldots, x + n - 1$ is divisible by n. Rework the proof of Theorem 3.3.8, using the class of all models of your axioms in place of $\mathbb{Z}$.]

5. Let L be the first-order language of linear orderings. (a) Show that if $h < 2^k$ then there is a formula $\phi(x, y)$ of L of quantifier rank $\leq k$ which expresses (in any linear ordering) '$x < y$ and there are at least h elements strictly between x and y'. (b) Let A be the ordering of the integers, and write $s(a, b)$ for the number of integers strictly between a and b. Show that if $a_0 < \ldots < a_{n-1}$ and $b_0 < \ldots < b_{n-1}$ in A, then $(A, a_0, \ldots, a_{n-1}) \approx_k (A, b_0, \ldots, b_{n-1})$ iff for all $m < n - 1$ and all $i < 2^k$, $s(a_m, a_{m+1}) = i \Leftrightarrow s(b_m, b_{m+1}) = i$.

6. Show that if G, G', H and H' are groups with $G \leqslant G'$ and $H \leqslant H'$, then $G \times H \leqslant G' \times H'$.

7. Show that there is no formula of first-order logic which expresses '$\langle a, b \rangle$ is in the transitive closure of R', even on finite structures. (For infinite structures it is easy to show there is no such formula.)

8. Let A and B be structures of the same signature. Immerman's **pebble game** on A, B of length k with p pebbles is played as follows. Pebbles π_0, $\ldots$, π_{p-1}, ρ_0, $\ldots$, ρ_{p-1} are given. The game is played like $EF_k[A, B]$, except that at each step, player $\forall$ must place one of the pebbles on his choice (one of the π_i if he chose from A, one of the ρ_i if he chose from B), then player $\exists$ must put the corresponding ρ_i (π_i) on her choice. (At the beginning the pebbles are not on any elements; later in the game the players may have to move pebbles from one element to another.) The condition for player $\exists$ to win is that after every step, if $\bar{a} = (a_0, \ldots, a_{p-1})$ is the sequence of elements of A with pebbles π_0, $\ldots$, π_{p-1} resting on them (where we ignore any pebbles not resting on an element), and likewise $\bar{b} = (b_0, \ldots, b_{p-1})$ the elements of B labelled by ρ_0, $\ldots$, ρ_{p-1}, then for every unnested atomic $\phi(x_0, \ldots, x_{p-1})$, $A \vDash \phi(\bar{a}) \Leftrightarrow B \vDash \phi(\bar{b})$. Show that player $\exists$ has a winning strategy for this game if and only if A and B agree on all first-order sentences which have quantifier rank $\leqslant k$ and use at most p distinct variables.

9. Let A and B be structures of the same signature. Show that A is back-and-forth equivalent to B if and only if player $\exists$ has a winning strategy for the game $EF_\omega[A, B]$.

Further reading

One great merit of the back-and-forth criterion for elementary equivalence is that we can adjust it to other languages by altering the rules of the game. For example it lifts to infinitary languages:

Barwise, J. Back and forth through infinitary logic. In *Studies in model theory*, Morley, M. D. (ed), pp. 5–34. Studies in Mathematics, Mathematical Association of America, 1973.

It also attracts the goodwill of complexity theorists and database theorists by suggesting any number of interesting equivalences between finite structures:

Ebbinghaus, H.-D. and Flum, J. *Finite model theory*. Perspectives in Mathematical Logic. Springer-Verlag, Berlin, 1995.

The following book contains a rich collection of examples of back-and-forth techniques in connection with linear orderings.

Rosenstein, J. G. *Linear orderings*. New York: Academic Press, 1982.

In another direction altogether, Hilary Putnam once claimed that 'Models are not lost noumenal waifs looking for someone to name them; they are constructions within our theory itself, and they have names from birth'. He wrote this in a paper discussing possible philosophical consequences of the downward Löwenheim—Skolem theorem, but it is hard to square his remark with the upward version of this theorem, Corollary 5.1.4 below. Putnam's arguments are in:

Putnam, H. Models and reality. In *Philosophy of mathematics, selected readings*, Benacerraf, P. and Putnam, H. (eds), pp. 421–444. Second edition. Cambridge: Cambridge University Press, 1983.

4

Interpretations

She turnd hersell into an eel,
To swim into yon burn,
And he became a speckled trout,
To gie the eel a turn.

Then she became a silken plaid,
And stretched upon a bed,
And he became a green covering,
And gaind her maidenhead.

Scots ballad from F. J. Child, The English and Scottish Popular Ballads.

The twenty-three year old Felix Klein in his famous Erlanger Programm (1872) proposed to classify geometries by their automorphisms. He hit on something fundamental here: in a sense, *structure is whatever is preserved by automorphisms*. One consequence – if slogans can have consequences – is that a model-theoretic structure implicitly carries with it all the features which are set-theoretically definable in terms of it, since these features are preserved under all automorphisms of the structure.

There is a rival model-theoretic slogan: *structure is whatever is definable*. Surprisingly, this slogan points in the same direction as the previous one. For example, if we have a field K, we can define the projective plane over K. But precisely because the projective plane is definable from K, any automorphism of K will induce an automorphism of the plane too. Either way, the plane comes with the field; in some abstract sense it is the field, but looked at from an unusual point of view.

A merit of the second slogan is that it gives us a means of controlling the host of 'implicit' features of a structure A. If these features are definable in terms of A, they must be definable by some kind of sentences. So we can consider those features definable by sentences of a certain form, or in a certain language. Two examples are the relativised reducts of A and the structures interpretable in A. Much recent research has invoked these notions. It is not hard to see why they should be important for methodology: they allow us to see properties of a structure that might otherwise be hard to uncover. The homological invariants of a topological space are important for a similar reason.

4.1 Automorphisms

Let A be an L-structure. Every automorphism of A is a permutation of $\text{dom}(A)$. By Theorem 1.2.1(c), the collection of all automorphisms of A is a group under composition. We write $\text{Aut}(A)$ for this group, regarded as a permutation group on $\text{dom}(A)$, and we call it the **automorphism group** of A.

$\text{Aut}(A)$ as a permutation group

For any set Ω, the group of all permutations of Ω is called the **symmetric group** on Ω, in symbols $\text{Sym}(\Omega)$. Several important properties of a structure A are really properties of its automorphism group as a subgroup of $\text{Sym}(\text{dom }A)$. In the next few definitions, suppose G is a subgroup of $\text{Sym}(\Omega)$.

First, if X is a subset of Ω, then the **pointwise stabiliser** of X in G is the set $\{g \in G : g(a) = a$ for all $a \in X\}$. This set forms a subgroup of G, and we write it $G_{(X)}$ (or $G_{(\bar{a})}$ where $\bar{a}$ is a sequence listing the elements of X). The **setwise stabiliser** of X in G, $G_{\{X\}}$, is the set $\{g \in G : g(X) = X\}$, which is also a subgroup of G. In fact we have $G_{(X)} \subseteq G_{\{X\}} \subseteq G$.

If a is an element of Ω, the **orbit** of a under G is the set $\{g(a) : g \in G\}$. The orbits of all elements of Ω under G form a partition of Ω. If the orbit of every element (or the orbit of one element – it comes to the same thing) is the whole of Ω, we say that G is **transitive on** Ω.

We say that a structure A is **transitive** if $\text{Aut}(A)$ is transitive on $\text{dom}(A)$. The opposite case is where A has no automorphisms except the identity 1_A; in this case we say that A is **rigid**. Here are two examples.

Example 1: *Ordinals*. Let the structure A be an ordinal $(\alpha, <)$, so that $<$ well-orders the elements of A. Then A is rigid. For suppose f is an automorphism of A which is not the identity. Then there is some element a such that $f(a) \neq a$; replacing f by f^{-1} if necessary, we can suppose that $f(a) < a$. Since f is a homomorphism, $f^2(a) = f(f(a)) < f(a)$, and so by induction $f^{n+1}(a) < f^n(a)$ for each $n < \omega$. Then $a > f(a) > f^2(a) > \ldots$, contradicting that $<$ is a well-ordering.

Example 2: *Affine space*. Let D be the direct sum of countably many cyclic groups of order 2. (Or equivalently, let D be a countable-dimensional vector space over the two-element field $\mathbb{F}_2$.) On D we define a relation

(1.1) $R(x, y, z, w) \Leftrightarrow x + y = z + w.$

The structure A consists of the set D with the relation R. If d is any element of D, there is a permutation e_d of the set D, defined by $e_d(a) = a + d$. This permutation is clearly an automorphism of A taking 0 to d – which shows that A is a transitive structure. Also if we fix d, we can make D into an abelian group with d as the identity, by defining an addition operation $+_d$ in terms of R:

$$(1.2) \qquad x +_d y = z \Leftrightarrow R(x, y, z, d).$$

So A is what remains of the group D when we forget which element is 0. It is known to geometers as the **countable-dimensional affine space over** $\mathbb{F}_2$.

Returning to our group G of permutations of Ω, we write Ω^n for the set of all ordered n-tuples of elements of Ω (where n is a positive integer). Then G automatically acts as a set of permutations of Ω^n too, putting $g(a_0, \ldots, a_{n-1}) = (ga_0, \ldots, ga_{n-1})$. So we can talk about the **orbits** of G on Ω^n. When n is greater than 1 and Ω has more than one element, then G is certainly not transitive on Ω^n. But model theory has a special regard for the following possibility, which is not far off transitivity.

We say that G is **oligomorphic** (on Ω) if for every positive integer n, the number of orbits of G on Ω^n is finite. We say that a structure A is **oligomorphic** if $\operatorname{Aut}(A)$ is oligomorphic on $\operatorname{dom}(A)$. In section 6.3 below we shall see that for countable structures, oligomorphic is the same thing as ω-categorical; this will give us dozens of examples. But for the moment, consider the ordered set $A = (\mathbb{Q}, <)$ of rational numbers. If $\bar{a}$ and $\bar{b}$ are any two n-tuples whose elements are in the same relative order in $\mathbb{Q}$, then there is an automorphism of A which takes $\bar{a}$ to $\bar{b}$. The number of possible $<$-orders of the elements of an n-tuple $(a_0, \ldots, a_{n-1})$ is at most, say, $(2n - 1)!$ (thus a_1 is either $< a_0$ or $= a_0$ or $> a_0$; then there are at most five cases for a_2; etc). So A is oligomorphic.

$\operatorname{Aut}(A)$ as a topological group

Suppose G is a group of permutations of a set Ω. If we know that G is $\operatorname{Aut}(A)$ for some structure A with domain Ω, what does this tell us about G? The answer needs some topology.

Let H be a subgroup of G. We say that H is **closed** in G if the following holds:

(1.3)   suppose $g \in G$ and for every tuple $\bar{a}$ of elements of Ω there is h in H such that $g\bar{a} = h\bar{a}$; then $g \in H$.

We say that the group G is **closed** if it is closed in the symmetric group $\operatorname{Sym}(\Omega)$. Note that if G is closed and H is closed in G, then H is closed; this is immediate from the definitions.

Theorem 4.1.1. *Let Ω be a set; let G be a subgroup of $\mathrm{Sym}(\Omega)$ and H a subgroup of G. Then the following are equivalent.*
(a) *H is closed in G.*
(b) *There is a structure A with $\mathrm{dom}(A) = \Omega$ such that $H = G \cap \mathrm{Aut}(A)$.*
In particular a subgroup H of $\mathrm{Sym}(\Omega)$ is of form $\mathrm{Aut}(B)$ for some structure B with domain Ω if and only if H is closed

Proof. (a) $\Rightarrow$ (b). For each $n < \omega$ and each orbit $\triangle$ of H on Ω^n, choose an n-ary relation symbol $R_\triangle$. Take L to be the signature consisting of all these relation symbols, and make Ω into an L-structure A by putting $R_\triangle^A = \triangle$. Every permutation in H takes $R_\triangle$ to $R_\triangle$, so that $H \subseteq G \cap \mathrm{Aut}(A)$. For the converse, suppose g is an automorphism of A and $g \in G$. For each n-tuple $\bar{a}$, if $\bar{a}$ lies in $\triangle = R_\triangle^A$ then so does $g\bar{a}$, and so $g\bar{a} = h\bar{a}$ for some h in H. Since H is closed in G, it follows that g is in H.

(b) $\Rightarrow$ (a). Assuming (b), we show that H is closed in G. Let g be an element of G such that for each finite subset W of Ω there is $h \in H$ with $g|W = h|W$. Let $\phi(\bar{x})$ be an atomic formula of the signature of A, and $\bar{a}$ a tuple of elements of A. Then choose W above so that it contains $\bar{a}$. We have $A \vDash \phi(\bar{a}) \Leftrightarrow A \vDash \phi(h\bar{a})$ (since $h \in \mathrm{Aut}(A)$) $\Leftrightarrow A \vDash \phi(g\bar{a})$. Thus g is an automorphism of A. $\qquad\square$

When H is closed, the structure A constructed in the proof of (a) $\Rightarrow$ (b) above is called the **canonical structure** for H. By the proof, A can be chosen to be an L-structure with $|L| \leqslant |\Omega| + \omega$.

The word 'closed' suggests a topology, and here it is. We say that a subset S of $\mathrm{Sym}(\Omega)$ is **basic open** if there are tuples $\bar{a}$ and $\bar{b}$ in Ω such that $S = \{g \in \mathrm{Sym}(\Omega): g\bar{a} = \bar{b}\}$ (write this set as $S(\bar{a}, \bar{b})$). In particular $\mathrm{Sym}(\Omega)_{(\bar{a})}$ is a basic open set. An **open subset** of $\mathrm{Sym}(\Omega)$ is a union of basic open subsets. If $\Omega = \mathrm{dom}(A)$, we define a (**basic**) **open subset** of $\mathrm{Aut}(A)$ to be the intersection of $\mathrm{Aut}(A)$ with some (basic) open subset of $\mathrm{Sym}(\Omega)$.

Lemma 4.1.2. *Let A be a structure and write G for $\mathrm{Aut}(A)$.*
  (a) *The definitions above define a topology on G; it is the topology induced by that on $\mathrm{Sym}(\Omega)$. Under this topology, G is a topological group, i.e. multiplication and inverse in G are continuous operations.*
  (b) *A subgroup of G is open if and only if it contains the pointwise stabiliser of some finite set of elements of A.*
  (c) *A subset F of G is closed under this topology if and only if it is closed in the sense of (1.3) above (with F for H).*
  (d) *A subgroup H of G is dense in G if and only if H and G have the same orbits on $(\mathrm{dom}\, A)^n$ for each positive integer n.*

Proof. (a) A permutation g takes $\bar{a}_1$ to $\bar{b}_1$ and $\bar{a}_2$ to $\bar{b}_2$ if and only if it takes $\bar{a}_1\bar{a}_2$ to $\bar{b}_1\bar{b}_2$; so the intersection of two basic open sets is again basic open. The first sentence of (a) follows at once by general topology. For the second sentence, $g \in S(\bar{a}, \bar{b})$ if and only if $g^{-1} \in S(\bar{b}, \bar{a})$, which proves the continuity of inverse. Finally if $gh \in S(\bar{a}, \bar{b})$, write $\bar{c}$ for $h\bar{a}$; then $g \in S(\bar{c}, \bar{b})$, $h \in S(\bar{a}, \bar{c})$ and $S(\bar{c}, \bar{b}) \cdot S(\bar{a}, \bar{c}) \subseteq S(\bar{a}, \bar{b})$; so multiplication is continuous.

(b) For each tuple $\bar{a}$ the pointwise stabiliser $G_{(\bar{a})}$ is $G \cap S(\bar{a}, \bar{a})$, which is open. Every subgroup containing $G_{(\bar{a})}$ is a union of cosets of $G_{(\bar{a})}$, hence it is open too. In the other direction, suppose H is an open subgroup containing a non-empty basic open set $G \cap S(\bar{a}, \bar{b})$. Then H contains $G_{(\bar{a})}$, since every element of $G_{(\bar{a})}$ can be written as gh with $g \in G \cap S(\bar{b}, \bar{a}) \subseteq H$ and $h \in G \cap S(\bar{a}, \bar{b}) \subseteq H$.

(c) A set $F \subseteq G$ is closed in the topology if and only if for every g in $\mathrm{Aut}(A)$, if each basic neighbourhood of g meets F then g lies in F. This is exactly (1.3).

(d) A subgroup H of G is dense if and only if for every g in $\mathrm{Aut}(A)$, each basic neighbourhood of g meets H. $\qquad\square$

As we pass from A to $\mathrm{Aut}(A)$ as permutation group, then to $\mathrm{Aut}(A)$ as topological group and finally to $\mathrm{Aut}(A)$ as abstract group, we keep throwing away information. How much of this information can be recovered? In some cases, precious little of it – consider Example 1 above. On the whole, the larger the automorphism group of a structure, the better the chances of reconstructing the structure from the automorphism group. In a series of papers, Mati Rubin has shown that an impressive number of structures are essentially determined by their abstract automorphism groups.

Automorphisms of countable structures

If A is a countable structure, we can build up automorphisms of A from finite approximations. From this observation it's a short step to the next theorem. If G is a group and H is a subgroup of G, we write $(G:H)$ for the index of H in G.

Theorem 4.1.3. *Let G be a closed group of permutations of ω and H a closed subgroup of G. Then the following are equivalent.*
(a) *H is open in G.*
(b) *$(G : H) \leqslant \omega$.*
(c) *$(G : H) < 2^{\omega}$.*

Proof. (a) $\Rightarrow$ (b). Suppose (a) holds. Then there is some tuple $\bar{a}$ of elements of ω such that the stabiliser $G_{(\bar{a})}$ of $\bar{a}$ lies in H. Suppose now that g, j are two elements of G such that $g\bar{a} = j\bar{a}$; then $j^{-1}g \in G_{(\bar{a})} \subseteq H$ and so the cosets gH, jH are equal. Since there are only countably many possibilities for $g\bar{a}$, the index $(G{:}H)$ must be at most countable.

(b) $\Rightarrow$ (c) is trivial.

(c) $\Rightarrow$ (a). We suppose that H is not open in G, and we construct continuum many left cosets of H in G.

We define by induction sequences $(\bar{a}_i : i < \omega)$, $(\bar{b}_i : i < \omega)$ of tuples of elements of ω and a sequence $(g_i : i < \omega)$ of elements of G such that the following hold for all i.

(1.4) $\bar{b}_0 = \langle \rangle$; $\bar{b}_{i+1}$ is a concatenation of all the sequences

$$(k_0 \ldots k_i)(\bar{a}_0{}^{\frown} \ldots {}^{\frown}\bar{a}_i)$$

where each k_j is in $\{1, g_0, \ldots, g_i\}$;

(1.5) $g_i\bar{b}_i = \bar{b}_i$;

(1.6)   there is no $h \in H$ such that $h\bar{a}_i = g_i\bar{a}_i$;

(1.7) i is an item in $\bar{a}_i$.

When $\bar{b}_i$ has been chosen, we have by assumption that $G_{(\bar{b}_i)} \not\subseteq H$, and so there is some automorphism $g_i \in G$ which fixes $\bar{b}_i$ (giving (1.5)) and is not in H. Since H is closed in G, it follows that there is a tuple $\bar{a}_i$ such that $h\bar{a}_i \neq g_i\bar{a}_i$ for all h in H. This ensures (1.6); adding i to $\bar{a}_i$ if necessary, we have (1.7) too.

Now for any subset S of $\omega\backslash\{0\}$, define $g_i^S = g_i$ if $i \in S$, and $= 1$ if $i \notin S$. Put $f_i^S = g_i^S \ldots g_0^S$. For each $j > i$ we have $f_j^S\bar{a}_i = f_i^S\bar{a}_i$, by (1.4) and (1.5). So by (1.7) we can define a map $g_S : \omega \to \omega$ by:

(1.8)   for each $i < \omega$, $g_S(i) = f_j^S(i)$ for all $j \geq i$.

Since the maps f_i^S are automorphisms, g_S is injective. But also g_S is surjective; for consider any $i \in \omega$ and put $j = (f_i^S)^{-1}(i)$. If $j \leq i$ then $g_S(j) = f_i^S(f_i^S)^{-1}(i) = i$. If $(f_i^S)^{-1}(i) = j > i$, then $g_S(j) = f_j^S(f_i^S)^{-1}(i) = g_j^S \ldots g_{i+1}^S(i) = i$ by (1.4), (1.5) and (1.7). So g_S is a permutation of ω. Since the f_i^S are in the closed group G and, for each tuple $\bar{a}$ in ω, g_S agrees on $\bar{a}$ with some f_i^S, it follows that g_S is in G.

There are 2^ω distinct subsets S of $\omega\backslash\{0\}$. It remains only to show that the corresponding permutations g_S lie in different right cosets of H.

Suppose $S \neq T$. Then there is some least $i > 0$ which is, say, in S but not in T. By (1.6) there is no element of H which agrees with g_i on $\bar{a}_i$. Put $f = f_{i-1}^S = f_i^T$. Now consider $f^{-1}\bar{a}_i$, and choose some $j \geq i$ such that all the items in $f^{-1}\bar{a}_i$ are $\leq j$. We have, for all h in H,

(1.9) $g_S(f^{-1}\bar{a}_i) = f_j^S f^{-1}(\bar{a}_i) = g_j^S \ldots g_{i+1}^S g_i(\bar{a}_i) = g_i(\bar{a}_i) \neq h(\bar{a}_i)$

$= hg_j^T \ldots g_{i+1}^T(\bar{a}_i) = hf_j^T f^{-1}(\bar{a}_i) = (hg_T)(f^{-1}\bar{a}_i)$.

So $g_S \notin Hg_T$, which finishes the proof. $\qquad\qquad\square$

At first sight this theorem is not very model-theoretic. We can translate it into model theory as follows.

Let A be a countable L^+-structure, suppose $L^- \subseteq L^+$ and let B be the L^--reduct $A|L^-$ of A. Then $H = \mathrm{Aut}(A)$ is a subgroup of $G = \mathrm{Aut}(B)$. Let g be any element of G, and consider the structure gA; gA is exactly like A except that for each symbol S of L^+, $S^{gA} = g(S^A)$. In particular the domain of gA is $\mathrm{dom}(A)$, and $g(S^A) = S^A$ for each symbol S in L^-, so that the reduct $(gA)|L^-$ is exactly B again.

Suppose now that k is another element of G. When is gA equal to kA? The answer is: when $g(S^A) = k(S^A)$ for each symbol S, or in other words, when $k^{-1}g$ is an automorphism of A – or, in other words again, when the cosets gH and kH in G are equal. This shows that *the index of* $\mathrm{Aut}(A)$ *in* $\mathrm{Aut}(B)$ *is equal to the number of different ways in which the symbols of* $L^+\backslash L^-$ *can be interpreted in* B *so as to give a structure isomorphic to* A.

Theorem 4.1.4 (Kueker–Reyes theorem). *Let* L^- *and* L^+ *be signatures with* $L^- \subseteq L^+$. *Let* A *be a countable* L^+-*structure and let* B *be the reduct* $A|L^-$. *Put* $G = \mathrm{Aut}(B)$. *Then the following are equivalent.*

(a) *There is a tuple* $\bar{a}$ *of elements of* A *such that* $G_{(\bar{a})} \subseteq \mathrm{Aut}(A)$.

(b) *There are at most countably many distinct expansions of* B *which are isomorphic to* A.

(c) *The number of distinct expansions of* B *which are isomorphic to* A *is less than* 2^ω.

(d) *There is a tuple* $\bar{a}$ *of elements of* A *such that for each atomic formula* $\phi(x_0, \ldots, x_{n-1})$ *of* L^+ *there is a formula* $\psi(x_0, \ldots, x_{n-1}, \bar{y})$ *of* $L^-_{\omega_1\omega}$ *such that* $A \vDash \forall \bar{x}(\phi(\bar{x}) \leftrightarrow \psi(\bar{x}, \bar{a}))$.

Proof. Our translation of Theorem 4.1.3 gives the equivalence of (a), (b) and (c) at once. It remains to show that (a) is equivalent to (d).

From (d) to (a) is clear. For the converse, suppose $G_{(\bar{a})} \subseteq \mathrm{Aut}(A)$, and let $\phi(x_0, \ldots, x_{n-1})$ be an atomic formula of L^+. Without loss we can suppose that ϕ is unnested, and for simplicity let us assume too that ϕ is $R(x_0, \ldots, x_{n-1})$ where R is some n-ary relation symbol. For each n-tuple $\bar{c}$ in $\phi(A^n)$ let $\sigma_{\bar{c}}(\bar{a}, \bar{c})$ be the Scott sentence of the structure $(B, \bar{a}, \bar{c})$ (see section 3.2). Now, if $\bar{c}$ is in $\phi(A^n)$ and $\bar{d}$ is an n-tuple such that $A \vDash \sigma_{\bar{c}}(\bar{a}, \bar{d})$, then $(B, \bar{a}, \bar{c}) \cong (B, \bar{a}, \bar{d})$, so $(B, \bar{a}, \bar{c}, R^A) \cong (B, \bar{a}, \bar{d}, R^A)$ by (a) and hence $A \vDash \phi(\bar{d})$. It follows that $A \vDash \forall \bar{x}(\phi(\bar{x}) \leftrightarrow \bigvee_{\bar{c} \in \phi(A^n)} \sigma_{\bar{c}}(\bar{a}, \bar{x}))$. $\qquad\square$

Corollary 4.1.5. *Let* A *be a countable structure. Then the following are equivalent.*

(a) $|\mathrm{Aut}(A)| \leqslant \omega$.

(b) $|\mathrm{Aut}(A)| < 2^\omega$.

(c) *There is a tuple* $\bar{a}$ *in* A *such that* $(A, \bar{a})$ *is rigid.*

Proof. The implications (c) $\Rightarrow$ (a) $\Rightarrow$ (b) are immediate. The implication (b) $\Rightarrow$ (c) follows from the theorem by adding a constant for each element of A. Alternatively, use (c) $\Rightarrow$ (a) from Theorem 4.1.3 with $G = \text{Aut}(A)$, $H = \{1\}$.   □

Exercises for section 4.1

1. Show that for every abstract group G there is a structure with domain G whose automorphism group is isomorphic to G. [Let X be a set of generators of G, and for each $x \in X$ introduce a function $f_x \colon g \to g \cdot x$ on the set G. Consider the structure consisting of the set G and the functions f_x. This structure is essentially the **Cayley graph** of the group G.]

2. Show that if the structure B is an expansion of A, then there is a continuous embedding of $\text{Aut}(B)$ into $\text{Aut}(A)$. (b) Show that if B is a definitional expansion, then this embedding is an isomorphism.

3. Show that if G is a group of permutations of a set Ω, $\bar{a}$ is a tuple of elements of Ω and h is a permutation of Ω, then $G_{(h\bar{a})} = h(G_{(\bar{a})})h^{-1}$.

4. Show that if G is an oligomorphic group of permutations of a set Ω, and X is a finite subset of Ω, then $G_{(X)}$ is also oligomorphic.

5. Suppose G is a subgroup of $\text{Sym}(\Omega)$. (a) Show that the topology on G is Hausdorff. (b) Show that the basic open sets are exactly the right cosets of basic open subgroups; show that they are also exactly the left cosets of basic open subgroups.

6. Show that every open subgroup of $\text{Aut}(A)$ is closed. Give an example to show that the converse fails.

7. Show that a subgroup of $\text{Aut}(A)$ is open if and only if it has non-empty interior.

8. Show that if A is an infinite structure then $\text{Aut}(A)$ has a dense subgroup of cardinality at most $\text{card}(A)$.

9. Suppose K, H and G are subgroups of $\text{Sym}(\Omega)$ with K a dense subgroup of H and H a dense subgroup of G. Show that K is a dense subgroup of G.

10. Show that if A is a countable structure which is $L_{\omega_1\omega}$-equivalent to some uncountable structure, then A has 2^ω automorphisms.

11. Show that if A is a countable structure and every orbit of $\text{Aut}(A)$ on elements of A is finite, then $|\text{Aut}(A)|$ is either finite or 2^ω.

12. Let G be a closed subgroup of $\mathrm{Sym}(\omega)$ and H any subgroup of G. (a) Show that the closure of H in G is a subgroup of G. (b) Show that if $(G:H) < 2^\omega$ then there is some tuple $\bar{a}$ such that $H_{(\bar{a})}$ is a dense subgroup of $G_{(\bar{a})}$.

4.2 Relativisation

Often in algebra one considers a pair of structures, for example a field and its algebraic closure, or a group that acts on a set. Model theory is not very good at handling pairs of structures. Instead a model theorist will usually try to represent the two structures as parts of some larger structure.

This may be a complicated matter. Let us start with the simplest case, where one structure can be picked out inside another one by a single 1-ary relation symbol.

Consider two signatures L and L' with $L \subseteq L'$. Let C be an L'-structure and B a substructure of the reduct $C|L$. Then we can make the pair of structures C, B into a single structure as follows. Take a new 1-ary relation symbol P, and write L^+ for L' with P added. Expand C to an L^+-structure A by putting $P^A = \mathrm{dom}(B)$. Then we can recover C and B from A by

(2.1) $C = A|L'$,

  $B = $ the substructure of $A|L$ whose domain is P^A.

Thus C is a reduct of A. We call B a **relativised reduct** of A, meaning that to get B from A we have to 'relativise' the domain to a definable subset of $\mathrm{dom}(A)$ as well as removing some symbols.

From now on, we forget about C. The setting is that L and L^+ are signatures with $L \subseteq L^+$, and P is a 1-ary relation symbol in $L^+ \backslash L$.

Let A be an L^+-structure. Lemma 1.2.2 gave necessary and sufficient conditions for P^A to be the domain of a substructure of $A|L$. When these conditions are satisfied, the substructure is uniquely determined. We write it A_P, and we call it the P-**part** of A. Otherwise A_P is not defined. (From Lemma 1.2.2 one can write these necessary and sufficient conditions as a set of first-order sentences that A must satisfy. We call them the **admissibility conditions** for relativisation to P.) Of course A_P depends on the language L as well as A and P.

The next theorem says that facts about A_P can be translated systematically into facts about A.

Theorem 4.2.1 (*Relativisation theorem*). *Let L and L^+ be signatures such that $L \subseteq L^+$, and P a 1-ary relation symbol in $L^+ \backslash L$. Then for every formula $\phi(\bar{x})$ of $L_{\infty\omega}$ there is a formula $\phi^P(\bar{x})$ of $L^+_{\infty\omega}$ such that the following holds:*

> *if A is an L^+-structure such that A_P is defined, and $\bar{a}$ is a sequence of elements from A_P, then*

(2.2) $A_P \vDash \phi(\bar{a})$ *if and only if* $A \vDash \phi^P(\bar{a})$.

Proof. We define ϕ^P by induction on the complexity of ϕ:

(2.3) ϕ^P is ϕ when ϕ is atomic;

(2.4) $\left(\bigwedge_{i \in I} \psi_i \right)^P$ is $\bigwedge_{i \in I} (\psi_i^P)$, and likewise with $\bigvee$ for $\bigwedge$;

(2.5) $(\neg \phi)^P$ is $\neg (\psi^P)$;

(2.6) $(\forall y \ \psi(\bar{x}, y))^P$ is $\forall y (Py \rightarrow \psi^P(\bar{x}, y))$, and
            $(\exists y \ \psi(\bar{x}, y))^P$ is $\exists y (Py \wedge \psi^P(\bar{x}, y))$.

Then (2.2) follows at once by induction on the complexity of ϕ.   □

The formula ϕ^P in this theorem is called the **relativisation** of ϕ to P. Note that if ϕ is first-order then so is ϕ^P. In fact the passage from ϕ to ϕ^P preserves the form of ϕ rather faithfully.

Corollary 4.2.2. *Let L and L^+ be signatures with $L \subseteq L^+$ and P a 1-ary relation symbol in $L^+ \backslash L$. If A and B are L^+-structures such that $A \leqslant B$ and A_P is defined, then B_P is defined and $A_P \leqslant B_P$.*

Proof. Exercise.   □

Example 1: *Linear groups.* Suppose G is a group of $n \times n$ matrices over a field F. Then we can make G and F into a single structure A as follows. The signature of A has 1-ary relation symbols *group* and *field*, 3-ary relation symbols *add* and *mult*, and n^2 2-ary relation symbols *coeff*$_{ij}$ $(1 \leqslant i,j \leqslant n)$. The sets *group*^{A} and *field*^{A} consist of the elements of G and F respectively. The relations *add*^{A} and *mult*^{A} express addition and multiplication in F. For each matrix $g \in G$, the ijth entry in g is the unique element f such that *coeff*$_{ij}(g, f)$ holds. There's no need to put in a symbol for multiplication in G, because it can be defined in terms of the field operations, using the symbols *coeff*$_{ij}$ (see Exercise 4). Note that in this example there are no function or constant symbols, and so B_P and B_Q are automatically defined for any structure B of the same signature as A.

Sometimes a structure B is picked out inside a structure A, not by a 1-ary relation symbol P but by a formula $\theta(x)$. When θ is in the first-order language of A, then again we call B a **relativised reduct** of A. The case considered above, where $\theta(x)$ is Px, becomes a special case. If θ also contains parameters from A, we call B a **relativised reduct with parameters**. One can adapt the relativisation theorem straightforwardly by putting θ in place of P everywhere.

Example 2: *ω as a relativised reduct.* Logicians are often pleased when they find the natural numbers as a part of some other structure. It gives them access to techniques of several kinds – for example recursion theory or nonstandard methods. One simple example is where A is a transitive model of Zermelo–Fraenkel set theory and $\theta(x)$ is the formula '$x \in \omega$'. Then the ordering $<$ on ω coincides with $\in$, and we can write set-theoretic formulas that define $+$ and $\cdot$. Note that in this example ω satisfies a rather strong form of the Peano axioms, as follows:

(2.7)   0 is not of the form $x + 1$; if $x, y \in \omega$ and $x + 1 = y + 1$ then $x = y$;

(2.8)   for every formula $\phi(x)$ of the first-order language of A, possibly with parameters from A, if $\phi(0)$ and $\forall x(x \in \omega \wedge \phi(x) \rightarrow \phi(x + 1))$ both hold in A then $\forall x(x \in \omega \rightarrow \phi(x))$ holds in A.

(2.8) is the induction axiom schema for subsets of ω which are first-order definable (with parameters) in A. Of course this includes the subsets of ω which are first-order definable in the structure $\langle \omega, < \rangle$ itself, by the relativisation theorem. But it may contain a great many more subsets besides – perhaps all the subsets of ω. That depends on our choice of A.

Example 3: *Relativised reducts of the rationals as an ordered set.* Let A be the following structure: the domain of A is the set $\mathbb{Q}$ of rational numbers, and the relations of A are all those which are $\varnothing$-definable from the usual ordering $<$ of the rationals. What are the relativised reducts of A (without parameters)? First note that $\mathrm{Aut}(A)$ is exactly $\mathrm{Aut}(\mathbb{Q}, <)$ since A is a definitional expansion of $(\mathbb{Q}, <)$. Next, $\mathrm{Aut}(A)$ is transitive on $\mathbb{Q}$, and it follows that any subset of $\mathbb{Q}$ which is definable without parameters is either empty or the whole of $\mathbb{Q}$. So we can forget the relativisation. Thirdly, if B is any reduct of A then $\mathrm{Aut}(A) \subseteq \mathrm{Aut}(B) \subseteq \mathrm{Sym}(\mathbb{Q})$, and $\mathrm{Aut}(B)$ is closed in $\mathrm{Sym}(\mathbb{Q})$ by Theorem 4.1.1. And finally, $\mathrm{Aut}(\mathbb{Q}, <)$ is oligomorphic and its orbits on n-tuples are all $\varnothing$-definable; so every orbit of $\mathrm{Aut}(B)$ on n-tuples is a union of finitely many orbits of $\mathrm{Aut}(\mathbb{Q}, <)$ and hence is defined by some relation of A. It follows that, up to definitional equivalence, the relativised reducts of A correspond exactly to the closed groups lying between $\mathrm{Aut}(A)$ and $\mathrm{Sym}(\mathbb{Q})$.

It can be shown that apart from $\mathrm{Aut}(A)$ and $\mathrm{Sym}(\mathbb{Q})$, there are just three such groups. The first is the group of all permutations of A which either preserve the order or reverse it. The second is the group of all permutations which preserve the cyclic relation '$x < y < z$ or $y < z < x$ or $z < x < y$'; this corresponds to taking an initial segment of $\mathbb{Q}$ and moving it to the end. The third is the group generated by these other two: it consists of those permutations which preserve the relation 'exactly one of x, y lies between z and w'.

One often finds some further devices attached to relativisation. For example there are the **relativised quantifiers** $(\forall x \in P)$ and $(\exists x \in P)$. These are defined thus:

(2.9) $(\forall x \in P)\phi$ means $\forall x(Px \rightarrow \phi)$; $(\exists x \in P)\phi$ means $\exists x(Px \wedge \phi)$.

Some logicians introduce variables x_P which range over those elements which satisfy Px; for example in set theory one has variables α, β etc. that range over the class of ordinals. Some writers use **sortal signatures**, whose structures carry partial functions; the functions are defined only when the arguments come from some given sets, known as **sorts**. I shall not pursue this further.

Pseudo-elementary classes

Many classes of mathematical structures are defined in terms of some feature that can be added to them.

Example 4: *Orderable groups.* An **ordered group** is a group G which carries a linear ordering $<$ such that if g, h and k are any elements of G, then

(2.10) $g < h$ implies $k \cdot g < k \cdot h$ and $g \cdot k < h \cdot k$.

A group is **orderable** if a linear ordering can be added so as to make it into an ordered group. Clearly an orderable group can't have elements $\neq 1$ of finite order. But this is not a sufficient condition for orderability (unless the group happens to be abelian; see Exercise 5.5.10).

For the remainder of this section, let L be a first-order language. A **pseudo-elementary class** (for short, a **PC class**) of L-structures is a class of structures of the form $\{A|L: A \vDash \phi\}$ for some sentence ϕ in a first-order language $L^+ \supseteq L$. A **PC$_\triangle$ class** of L-structures is a class of the form $\{A|L: A \vDash U\}$ for some theory U in a first-order language $L^+ \supseteq L$.

For example the class of orderable groups is a PC class. But we get a more intriguing example by turning Example 4 on its head:

Example 5: *Ordered abelian groups.* Let L be the first-order language of linear orderings (with symbol $<$) and let U be the theory of ordered abelian groups. Then the class $\mathbf{K} = \{A|L: A \vDash U\}$ is the class of all linear orderings which are orderings of abelian groups. This is a PC class, since U can be written as a finite theory and hence as a single sentence. By a result of Mal'tsev, the countable order-types which are in $\mathbf{K}$ are precisely those of the form ζ^α or $\zeta^\alpha \cdot \eta$ where ζ is the order-type of the integers, η the order-type of the rationals, α is an ordinal $< \omega_1$, and ζ^α is the order-type defined as follows. Write $^{(\alpha)}\mathbb{Z}$ for the set of all sequences $(n_i: i < \alpha)$ where n_i is an integer for each $i < \alpha$, and $n_i \neq 0$ for only finitely many i. If $m = (m_i: i < \alpha)$

and $n = (n_i: i < \alpha)$ are two distinct elements of $^{(\alpha)}\mathbb{Z}$, write $m < n$ iff $m_i < n_i$ for the greatest i at which $m_i \neq n_i$. Then ζ^α is the order-type of $(^{(\alpha)}\mathbb{Z}, <)$. It follows that $\mathbf{K}$ is not first-order axiomatisable. For example the linear ordering ζ is elementarily equivalent to $\zeta(1 + \eta)$ (e.g. by Exercise 2.7.6(a)), but the first is in $\mathbf{K}$ and the second is not.

It also follows that $\mathbf{K}$ contains, up to isomorphism, just ω_1 linear orderings of cardinality ω. This is interesting, because an old conjecture of Vaught states that the number of countable models of a countable first-order theory (up to isomorphism) is always either 2^ω or $\leqslant \omega$. Morley showed that if $\mathbf{J}$ is a class of the form $\{A|L: A \vDash U\}$ for some countable first-order theory U, then the number of countable structures in $\mathbf{J}$, up to isomorphism, is either $\leqslant \omega_1$ or $= 2^\omega$. Since there are plenty of examples for cardinalities $\leqslant \omega$ and $= 2^\omega$, the example $\mathbf{K}$ above shows that Morley's result is best possible.

One can generalise these notions, using relativised reducts A_P. We define a $\mathbf{PC}'_\triangle$ **class** of L-structures to be a class of the form $\{A_P: A \vDash U$ and A_P is defined$\}$ for some theory U in a language $L^+ \supseteq L \cup \{P\}$. By the admissibility conditions, every $\mathbf{PC}'_\triangle$ class can be written as $\{A_P: A \vDash U'\}$ for some theory U' in L^+.

A natural example of a $\mathbf{PC}'_\triangle$ class is the class of multiplicative groups of fields. Here U is the theory of fields together with a symbol P which picks out the non-zero elements, and L has only the symbol for multiplication. One can show that this class is not first-order axiomatisable.

It turns out that $\mathbf{PC}'_\triangle$ is not really a generalisation of $\mathbf{PC}_\triangle$; it's exactly the same thing.

Theorem 4.2.3. *The $\mathbf{PC}'_\triangle$ classes are exactly the $\mathbf{PC}_\triangle$ classes. More precisely, let $\mathbf{K}$ be a class of L-structures.*

(a) *If $\mathbf{K}$ is a $\mathbf{PC}'_\triangle$ class $\{A_P: A \vDash U$ and A_P is defined$\}$ for some theory U in a first-order language L^+, then $\mathbf{K}$ is also a $\mathbf{PC}_\triangle$ class $\{A|L: A \vDash U^*\}$ for some theory U^* in a first-order language L^* with $|L^*| \leqslant |L^+|$.*

(b) *If $\mathbf{K}$ is a $\mathbf{PC}'$ class and all structures in $\mathbf{K}$ are infinite, then $\mathbf{K}$ is a $\mathbf{PC}$ class.*

Proof. To make a $\mathbf{PC}_\triangle$ class into a $\mathbf{PC}'_\triangle$ class, add the symbol P with the axiom $\forall x\, Px$.

The full proof of the other direction in (a) is surprisingly subtle, and I omit it. But if $\mathbf{K}$ is a $\mathbf{PC}'_\triangle$ class in which every structure has cardinality $\geqslant |L^+|$, then we can show that $\mathbf{K}$ is a $\mathbf{PC}_\triangle$ class by the following handy argument. The same argument proves (b).

Let $\mathbf{K}$ be $\{A_P: A \vDash U$ and A_P is defined$\}$, and suppose that each structure in $\mathbf{K}$ has cardinality $\geqslant |L^+|$. Then by the downward Löwenheim–Skolem theorem, each structure in **K** is of the form $B = A_P$ for some model A of U with $|A| = |B|$. For each symbol S in the signature of L^+, introduce a copy S^*. Add a 1-ary function symbol F. If A and B are as above, take an arbitrary bijection $f: \mathrm{dom}(A) \to \mathrm{dom}(B)$, and interpret each symbol S^* on $\mathrm{dom}(B)$ as the image of S^A under f. Interpret F as the restriction of f to $\mathrm{dom}(B)$. Using the symbols of L together with the symbols S^* and F, we can write down a theory U^* which expresses that the interpretations of the symbols S^* make $\mathrm{dom}(B)$ into a model D of U, and F maps B isomorphically onto the P^*-part. Then $\mathbf{K} = \{D | L: D \vDash U^*\}$, and so **K** is a $\mathrm{PC}_\triangle$ class. $\square$

Exercises for section 4.2

1. Show that if A is a relativised reduct of B and B is a relativised reduct of C, then A is a relativised reduct of C.

2. Describe the admissibility formulas for relativisation to P.

3. Prove Corollary 4.2.2.

4. In Example 1, write out a formula $\psi(x, y, z)$ which expresses that the matrix z is the product of the matrices x and y.

5. Show that the structure $\langle \omega, + \rangle$ is a relativised reduct of the ring of integers. [Sums of squares.]

6. Show that if B is a relativised reduct of A, then there is an induced continuous homomorphism $h: \mathrm{Aut}(A) \to \mathrm{Aut}(B)$.

7. Show that the downward Löwenheim–Skolem theorem holds for $\mathrm{PC}_\triangle$ classes in the following sense: if $L \subseteq L^+$, U is a theory in L^+ and **K** is the class of all L-reducts of models of U, then for every structure A in **K** and every set X of elements of A, there is an elementary substructure of A of cardinality $\leqslant |X| + |L^+|$ which contains all the elements of X.

8. In Example 5, show that each ordering ζ^α is isomorphic to the reverse ordering $(\zeta^\alpha)^*$.

9. Show that the class of multiplicative groups of real-closed fields is first-order axiomatisable.

10. Show that in Theorem 4.2.3(b) the condition that all structures in **K** are infinite can't be dropped. [For a PC class **K** the set $\{|A|: A \in \mathbf{K}, |A| < \omega\}$ is primitive recursive.]

4.3 Interpreting one structure in another

Let K and L be signatures, A a K-structure and B an L-structure, and n a positive integer. An (n-dimensional) **interpretation** Γ **of B in** A is defined to consist of three items,

(3.1)   a formula $\partial_\Gamma(x_0, \ldots, x_{n-1})$ of signature K,

(3.2)   for each unnested atomic formula $\phi(y_0, \ldots, y_{m-1})$ of L, a
         formula $\phi_\Gamma(\bar{x}_0, \ldots, \bar{x}_{m-1})$ of signature K in which the $\bar{x}_i$ are
         disjoint n-tuples of distinct variables,

(3.3)   a surjective map $f_\Gamma \colon \partial_\Gamma(A^n) \to \text{dom}\,(B)$,

such that for all unnested atomic formulas ϕ of L and all $\bar{a}_i \in \partial_\Gamma(A^n)$,

(3.4) $B \vDash \phi(f_\Gamma\bar{a}_0, \ldots, f_\Gamma\bar{a}_{m-1}) \Leftrightarrow A \vDash \phi_\Gamma(\bar{a}_0, \ldots, \bar{a}_{m-1})$.

The formula ∂_Γ is the **domain formula** of Γ; the formulas ∂_Γ and ϕ_Γ (for all unnested atomic ϕ) are the **defining formulas** of Γ. The map f_Γ is the **coordinate map** of Γ. It assigns to each element $f_\Gamma\bar{a}$ of B the 'coordinates' $\bar{a}$ in A; in general one element may have several different tuples of coordinates.

Unless anything is said to the contrary, we assume that the defining formulas of Γ are all first-order. For example we say that B is **interpretable** in A if there is an interpretation of B in A with all its defining formulas first-order. We say that B is **interpretable in** A **with parameters** if there is a sequence $\bar{a}$ of elements of A such that B is interpretable in $(A, \bar{a})$.

We shall write $=_\Gamma$ for ϕ_Γ when ϕ is the formula $y_0 = y_1$. Wherever possible we shall abbreviate $(\bar{a}_0, \ldots, \bar{a}_{m-1})$ and $(f\bar{a}_0, \ldots, f\bar{a}_{m-1})$ to $\bar{a}$ and $f\bar{a}$ respectively.

Example 1: *Relativised reductions*. Using the notation of section 4.2, suppose B is the relativised reduct A_P. Then there is a one-dimensional interpretation Γ of B in A as follows.

$\partial_\Gamma(x) := Px$.

$\phi_\Gamma := \phi(\bar{x})$, for each unnested atomic formula $\phi(\bar{y})$.

The coordinate map $f_\Gamma \colon P^A \to \text{dom}\,(A)$ is simply the inclusion map. We call the interpretation Γ a **relativised reduction**.

Example 2: *Rationals and integers*. The familiar interpretation of the rationals in the integers is a two-dimensional interpretation Γ, as follows.

$\partial_\Gamma(x_0, x_1) := x_1 \neq 0$,

$=_\Gamma(x_{00}, x_{01}; x_{10}, x_{11}) := x_{00} \cdot x_{11} = x_{01} \cdot x_{10}$,

$plus_\Gamma(x_{00}, x_{01}; x_{10}, x_{11}; x_{20}, x_{21})$
$:= x_{21} \cdot (x_{00} \cdot x_{11} + x_{01} \cdot x_{10}) = x_{01} \cdot x_{11} \cdot x_{20}$.

$$times_\Gamma(x_{00}, x_{01}; x_{10}, x_{11}; x_{20}, x_{21})$$
$$:= x_{00} \cdot x_{10} \cdot x_{21} = x_{01} \cdot x_{11} \cdot x_{20}.$$

The coordinate map takes each pair (m, n) with $n \neq 0$ to the rational number m/n. The formulas ψ_Γ for the remaining unnested atomic formulas ψ express addition and multiplication of rationals in terms of addition and multiplication of integers, just as in the algebra texts.

Example 3: Algebraic extensions. Let A be a field, $p(X)$ an irreducible polynomial of degree n over A and ξ a root of $p(X)$ in some field extending A. Then there is an n-dimensional interpretation Γ of $A[\xi]$ in A. If $\bar{a} = (a_0, \ldots, a_{n-1})$ is an n-tuple of elements of A, write $q_{\bar{a}}(X)$ for the polynomial $X^n + a_{n-1}X^{n-1} + \ldots + a_1X + a_0$. Then $\partial_\Gamma(A^n)$ is the whole of A^n and $f_\Gamma(\bar{a})$ is $q_{\bar{a}}(\xi)$. The formula $=_\Gamma(\bar{a}, \bar{b})$ will say that $p(X)$ divides $(q_{\bar{a}}(X) - q_{\bar{b}}(X))$; I leave it to the reader to verify that $=_\Gamma$ can be written as a p.p. formula. The formulas $(y_0 + y_1 = y_2)_\Gamma$ and $(y_0 \cdot y_1 = y_2)_\Gamma$ follow the usual definitions of addition and multiplication of polynomials, and again these give us p.p. formulas. In fact ∂_Γ is quantifier-free and ϕ_Γ is p.p. for every unnested atomic ϕ. Just as in Example 2, we are looking at a familiar algebraic object from a slightly peculiar angle.

If Γ is an interpretation of an L-structure B in a K-structure A, then there are certain sentences of signature K which must be true in A just because Γ is an interpretation, regardless of what A and B are. These sentences say
(i) $=_\Gamma$ defines an equivalence relation on $\partial_\Gamma(A^n)$,
(ii)   for each unnested atomic formula ϕ of L, if $A \vDash \phi_\Gamma(\bar{a}_0, \ldots, \bar{a}_{n-1})$ with $\bar{a}_0, \ldots, \bar{a}_{n-1}$ in $\partial_\Gamma(A^n)$, then also $A \vDash \phi_\Gamma(\bar{b}_0, \ldots, \bar{b}_{n-1})$ when each $\bar{b}_i$ is an element of $\partial_\Gamma(A^n)$ which is $=_\Gamma$-equivalent to $\bar{a}_i$,
(iii) if $\phi(y_0)$ is a formula of L of form $c = y_0$, then there is $\bar{a}$ in $\partial_\Gamma(A^n)$ such that for all $\bar{b}$ in $\partial_\Gamma(A^n)$, $A \vDash \phi_\Gamma(\bar{b})$ if and only if $\bar{b}$ is $=_\Gamma$-equivalent to $\bar{a}$,
(iv)   a clause like (iii) for each function symbol.
These first-order sentences are called the **admissibility conditions** of Γ. They generalise the admissibility conditions for a relativised reduct. Note that they depend only on parts (3.1) and (3.2) of Γ, and not on the coordinate map.

Suppose Γ interprets B in A. Then in a sense, A knows everything there is to know about B, and so we can answer questions about B by reducing them to questions about A. The next theorem develops this idea. Bearing in mind Example 1, this theorem is best seen as a refinement of the relativisation theorem (Theorem 4.2.1).

Theorem 4.3.1 (Reduction theorem). *Let A be a K-structure, B an L-structure and Γ an n-dimensional interpretation of B in A. Then for every formula $\phi(\bar{y})$*

of the language $L_{\infty\omega}$ there is a formula $\phi_\Gamma(\bar{x})$ of the language $K_{\infty\omega}$ such that for all $\bar{a}$ from $\partial_\Gamma(A^n)$,

$$(3.5) \qquad\qquad B \vDash \phi(f_\Gamma\bar{a}) \Leftrightarrow A \vDash \phi_\Gamma(\bar{a}).$$

(Recall our notational conventions. Since Γ is n-dimensional, $\bar{a}$ will be a tuple of n-tuples.)

Proof. By Corollary 2.6.2, every formula of $L_{\infty\omega}$ is equivalent to a formula of $L_{\infty\omega}$ in which all atomic subformulas are unnested. So we can prove the theorem by induction on the complexity of formulas, and clause (3.4) in the definition of interpretations already takes care of the atomic formulas. For compound formulas we define

$$(3.6) \quad (\neg\phi)_\Gamma = \neg(\phi_\Gamma),$$

$$(3.7) \quad (\bigwedge_{i \in I} \phi_i)_\Gamma = \bigwedge_{i \in I}(\phi_i)_\Gamma, \quad \text{and likewise with } \bigvee \text{ for } \bigwedge,$$

$$(3.8) \quad (\forall y\, \phi)_\Gamma = \forall x_0 \ldots x_{n-1}(\partial_\Gamma(x_0, \ldots, x_{n-1}) \rightarrow \phi_\Gamma),$$

$$(3.9) \quad (\exists y\, \phi)_\Gamma = \exists x_0 \ldots x_{n-1}(\partial_\Gamma(x_0, \ldots, x_{n-1}) \wedge \phi_\Gamma). \qquad \square$$

This fundamental but trivial theorem calls for several remarks.

Remark 1. The map $\phi \mapsto \phi_\Gamma$ of the theorem depends only on parts (3.1) and (3.2) of Γ, and not at all on the coordinate map f_Γ. We shall exploit this fact through the rest of the chapter. Parts (3.1) and (3.2) of the definition of Γ form an **interpretation of L in K**, and the map $\phi \mapsto \phi_\Gamma$ of the reduction theorem is the **reduction map** of this interpretation.

Remark 2. We have been rather careless about variables. For example if ϕ in the theorem is $\phi(z)$, what are the variables of ϕ_Γ? I avoid answering this question. For purposes of the reduction theorem, all formulas of $L_{\infty\omega}$ are of form $\phi(y_0, y_1, \ldots)$ and all formulas of $K_{\infty\omega}$ are correspondingly of form $\psi(x_{00}, \ldots, x_{0,n-1}; x_{10}, \ldots, x_{1,n-1}; \ldots)$. Also I assume tacitly that if ϕ and θ are the same unnested atomic formula of $L_{\infty\omega}$ up to permutation of variables, then ϕ_Γ and θ_Γ are the same formula up to a corresponding permutation of their free variables.

Remark 3. If ∂_Γ and the formulas ϕ_Γ (for unnested atomic ϕ) are $\exists_1^+$ first-order formulas, then for every $\exists_1^+$ first-order formula ψ of L, ψ_Γ is also $\exists_1^+$ first-order. This is true even when ψ contains nested atomic formulas, since the removal of nesting introduces existential quantifiers at worst (see Theorem 2.6.1). More generally but more vaguely, if we have any reasonable measure of the complexity of formulas, then ϕ_Γ will be only a bounded amount more complex than ϕ.

Remark 4. If L is a recursive language (see section 2.3 above) and the map $\phi \mapsto \phi_\Gamma$ in (3.2) is recursive, then we call Γ a **recursive interpretation**. For a recursive interpretation the reduction map (restricted to first-order formulas) is recursive too. With inessential changes in the proof of Theorem 4.3.1 we can arrange that the reduction map is also 1–1 on first-order formulas. Recursion theorists will deduce that if Γ is a recursive interpretation of B in A, then $\mathrm{Th}(B)$ is 1–1 reducible to $\mathrm{Th}(A)$.

The associated functor

Theorem 4.3.2. *Let Γ be an n-dimensional interpretation of a signature L in a signature K, and let $\mathrm{Admis}(\Gamma)$ be the set of admissibility conditions of Γ. Then for every K-structure A which is a model of $\mathrm{Admis}(\Gamma)$, there are an L-structure B and a map $f \colon \partial_\Gamma(A^n) \to \mathrm{dom}(B)$ such that*
(a) *Γ with f forms an interpretation of B in A, and*
(b) *if g and C are such that Γ and g form an interpretation of C in A, then there is an isomorphism $i \colon B \to C$ such that $i(f\bar{a}) = g(\bar{a})$ for all $\bar{a} \in \partial_\Gamma(A^n)$.*

Proof. Let A be a model of Γ. Then we build an L-structure B as follows. Define a relation $\sim$ on $\partial_\Gamma(A^n)$ by

$$(3.10) \qquad \bar{a} \sim \bar{a}' \text{ iff } A \models =_\Gamma(\bar{a}, \bar{a}').$$

By (i) of the admissibility conditions, $\sim$ is an equivalence relation. Write $\bar{a}^\sim$ for the equivalence class of $\bar{a}$. The domain of B will be the set of all equivalence classes $\bar{a}^\sim$ with $\bar{a}$ in $\partial_\Gamma(A^n)$. (Readers who don't allow empty structures should add $\exists \bar{x}\, \partial_\Gamma(\bar{x})$ to the admissibility conditions, here and henceforth.)

For every relation symbol R of L, we define the relation R^B by

$$(3.11) \quad (\bar{a}_0{}^\sim, \ldots, \bar{a}_{m-1}{}^\sim) \in R^B \text{ iff } A \models \phi_\Gamma(\bar{a}_0, \ldots, \bar{a}_{m-1})$$
$$\text{where } \phi(y_0, \ldots, y_{m-1}) \text{ is } R y_0 \ldots y_{m-1}.$$

By (ii) of the admissibility conditions, this is a sound definition. The definitions of c^B and F^B are similar, relying on (iii) and (iv) of the admissibility conditions. This defines the L-structure B. We define $f \colon \partial_\Gamma(A^n) \to \mathrm{dom}(B)$ by $f\bar{a} = \bar{a}^\sim$. Then f is surjective, and B has been defined so as to make (3.4) true. Hence Γ and f are an interpretation of B in A. This proves (a).

To prove (b), suppose Γ and g are an interpretation of C in A. For each tuple $\bar{a} \in \partial_\Gamma(A^n)$, define $i(f\bar{a})$ to be $g\bar{a}$. We claim this is a sound definition of an isomorphism $i \colon B \to C$. If $f\bar{a} = f\bar{a}'$ then $A \models =_\Gamma(\bar{a}, \bar{a}')$ and hence $g\bar{a} = g\bar{a}'$ by (3.4); thus the definition of i is sound. A similar argument in the other direction shows that i is injective; moreover i is surjective since g is surjective by (3.3). Then (3.4) for f and g shows that i is an embedding. This proves the claim, and with it the theorem. $\square$

We write ΓA for the structure B of the theorem. The reduction theorem (Theorem 4.3.1) applies to ΓA as follows:

(3.12) for all formulas $\phi(\bar{y})$ of L, all K-structures A satisfying the admissibility conditions of Γ, and all tuples $\bar{a} \in \partial_\Gamma(A^n)$,
$$\Gamma A \vDash \phi(\bar{a}^\sim) \Leftrightarrow A \vDash \phi_\Gamma(\bar{a}).$$

Let Γ be as in the theorem, let A and A' be models of the admissibility conditions of Γ, and let $e: A \to A'$ be an elementary embedding. Then for every tuple $\bar{a} \in \partial_\Gamma(A^n)$, $e\bar{a}$ is in $\partial_\Gamma(A'^n)$. Moreover if $\bar{c}$ is another tuple in $\partial_\Gamma(A^n)$ and $A \vDash =_\Gamma(\bar{a}, \bar{c})$, then $A' \vDash =_\Gamma(e\bar{a}, e\bar{c})$. It follows that there is a well-defined map Γe from $\mathrm{dom}(\Gamma A)$ to $\mathrm{dom}(\Gamma A')$, defined by $(\Gamma e)(\bar{a}^\sim) = (e\bar{a})^\sim$. One easily verifies that $\Gamma(1_A)$ is the map $1_{\Gamma A}$. Also if $e_1: A \to A'$ and $e_2: A' \to A''$ are elementary embeddings, then $\Gamma(e_2 e_1) = (\Gamma e_2)(\Gamma e_1)$.

Furthermore, Γe is an elementary embedding of ΓA into $\Gamma A'$. For let $\bar{a}$ be a sequence of tuples from $\partial_\Gamma(A^n)$ and ϕ a formula of L. Then we have

(3.13) $\Gamma A \vDash \phi(\bar{a}^\sim) \Leftrightarrow A \vDash \phi_\Gamma(\bar{a}) \Rightarrow A' \vDash \phi_\Gamma(e\bar{a}) \Leftrightarrow \Gamma A' \vDash \phi((\Gamma e)\bar{a}^\sim)$.

Here the left-hand equivalence is by the reduction theorem ((3.12) above), the central implication is because e is elementary, and the right-hand equivalence is by the reduction theorem and the definition of Γe.

In fact the definition of Γe makes sense whenever A, A' are models of the admissibility conditions of Γ and $e: A \to A'$ is any homomorphism which preserves the formulas ∂_Γ and $=_\Gamma$. If e also preserves all the formulas ϕ_Γ for unnested atomic formulas ϕ of L, then (3.13) is good for these formulas ϕ too, and so Γe is a homomorphism from ΓA to $\Gamma A'$. To sum up, we have the following.

Theorem 4.3.3. *Let Γ be an interpretation of a signature L in a signature K, with admissibility conditions* $\mathrm{Admis}(\Gamma)$.

(a) *Γ induces a functor, written* $\mathrm{Func}(\Gamma)$, *from the category of models of* $\mathrm{Admis}(\Gamma)$ *and elementary embeddings, to the category of L-structures and elementary embeddings.*

(b) *If the formulas ∂_Γ and ϕ_Γ (for unnested atomic ϕ) are $\exists_1^+$ formulas, then we can extend the functor* $\mathrm{Func}(\Gamma)$ *in (a), replacing 'elementary embeddings' by 'homomorphisms'.* $\square$

We call the functor $\mathrm{Func}(\Gamma)$ of Theorem 4.3.3, in either the (a) or the (b) version, the **associated functor** of the interpretation Γ. Usually we shall write it just Γ; there is little danger of confusing the interpretation with the functor.

Suppose Γ is the associated functor of an interpretation of L in K. Then whenever ΓA is defined, we have a group homomorphism $\alpha \mapsto \Gamma\alpha$ from $\mathrm{Aut}(A)$ to $\mathrm{Aut}(\Gamma A)$. What can be said about this homomorphism?

Theorem 4.3.4. *Let Γ be an interpretation of L in K, and let A be an L-structure such that ΓA is defined. Then the induced homomorphism h: $\mathrm{Aut}(A) \to \mathrm{Aut}(B)$ is continuous.*

Proof. It suffices to show that if F is a basic open subgroup of $\mathrm{Aut}(B)$, then there is an open subgroup E of $\mathrm{Aut}(A)$ such that $h(E) \subseteq F$. Let F be $\mathrm{Aut}(B)_{(\bar{b})}$ for some tuple $\bar{b}$ of elements of B. Let X be a finite set of elements of A such that each element in $\bar{b}$ is of form $f_\Gamma(\bar{a})$ for some tuple $\bar{a}$ of elements of X. Then by the definition of h, $h(\mathrm{Aut}(A)_{(X)}) \subseteq \mathrm{Aut}(B)_{(\bar{b})}$. $\square$

Exercises for section 4.3

1. Show that if Γ is an n-dimensional interpretation of L in K, then for every K-structure A for which ΓA is defined, $|\Gamma A| \leqslant |A|^n$.

2. Let A, B and C be structures. Show that if B is interpretable in A and C is interpretable in B, then C is interpretable in A.

3. Write down an interpretation Γ such that for every abelian group A, ΓA is the group $A/5A$. By applying Γ to the inclusion $\mathbb{Z} \to \mathbb{Q}$, show that Theorem 4.3.3(b) fails if we replace $\exists_1^+$ by $\exists_1$ and 'homomorphisms' by 'embeddings'.

4. Let A be an L-structure with at least two elements. Show that the disjoint sum $A + A$ (see Exercise 3.2.8) is interpretable in A.

5. Let G be a group and A a normal abelian subgroup of G such that G/A is finite. Show that G is interpretable in A with parameters. [The cosets of A can be written as $h_1 A + \ldots + h_n A$. Interpret in $A + \ldots + A$ (n summands).]

*Suppose the languages K and L have no relation symbols. A **polynomial interpretation** of L in K is a map Δ which assigns to each constant c of L a closed term c_Δ of K, and to each m-ary function symbol F of L a term $F_\Delta(x_0, \ldots, x_{m-1})$ of K.*
6. Show how a polynomial interpretation Δ of L in K induces an interpretation Γ of L in K, in which for every equation ϕ, ϕ_Γ is also an equation. Show that for every K-structure A, ΓA exists and is a reduct of a definitional extension of A. (We write ΔA for ΓA.)

7. Write down a polynomial interpretation Δ such that for every ring A, ΔA is the Lie ring of A.

8. Let A be a field and n a positive integer. (a) Let R be the set of all n-tuples $\bar{a} = (a_0, \ldots, a_{n-1})$ of elements of A such that the polynomial $X^n + a_{n-1}X^{n-1} + \ldots + a_1 X + a_0$ is irreducible over A, and if α is a root of this polynomial then the field $A[\alpha]$ is Galois over A. Show that R is $\varnothing$-definable over A. [Split 'Galois' into 'normal' and 'separable'. Note that every element of $A[\alpha]$ can be regarded as an equivalence class of n-tuples of elements of A, by an equivalence relation which is definable in terms of $\bar{a}$.] (b) Let G be a finite group. Let R_G be the set of all n-tuples as above, such that the Galois group $A[\alpha]/A$ is isomorphic to G. Show that R_G is also $\varnothing$-definable over A.

4.4 Imaginary elements

Suppose a structure B is interpretable in a structure A. Then we can think of the elements of B as 'implicit' elements of A. The reduction theorem supports this intuition – it tells us that we can talk about the elements of B by making statements about elements of A. Probably the most familiar example is the complex numbers: we can think of them as pairs of real numbers, so that the complex numbers are 'implicit' in the field of real numbers.

This aspect of interpretations was the motivation for the definitions which follow. But for the moment we shall forget about interpretations and start again from scratch. In the proofs of two of the theorems in this section, we shall establish some uniformities by using the compactness theorem for first-order logic. Since that theorem is not proved until the next section, first-time readers may prefer to skip these proofs. None of the notions defined in this section depend on the compactness theorem.

Let L be a first-order language and A an L-structure. An **equivalence formula** of A is a formula $\phi(\bar{x}, \bar{y})$ of L, without parameters, such that the relation $\{(\bar{a}, \bar{b}): A \vDash \phi(\bar{a}, \bar{b})\}$ is a non-empty equivalence relation E_ϕ. We write ∂_ϕ for the set $\{\bar{a}: A \vDash \phi(\bar{a}, \bar{a})\}$. We write $\bar{a}/\phi$ for the E_ϕ-equivalence class of the tuple $\bar{a} \in \partial_\phi$. Items of the form $\bar{a}/\phi$, where ϕ is an equivalence formula and $\bar{a}$ a tuple, are known as **imaginary elements** of A.

Note that in every L-structure A there is a natural correspondence between the elements a of A and the imaginary elements $a/(x = y)$. (Every real element is imaginary!) Likewise every tuple $(a_0, \ldots, a_{n-1})$ of elements of A can be identified with a single element $\bar{a}/\theta$ where $\theta(\bar{x}, \bar{y})$ is the formula $\bigwedge_{i<n} x_i = y_i$. Note also that, if $\phi(x_0, \ldots, x_{n-1})$ is any formula of L, there are imaginary elements corresponding to the set $\phi(A^n)$ and its complement, thanks to the equivalence formula $\phi(\bar{x}) \leftrightarrow \phi(\bar{y})$.

Saharon Shelah suggested a way of regarding imaginary elements of a structure as genuine elements. Let A be an L-structure. Shelah defined another structure A^{eq} (pronounced 'A E Q') which contains A as a definable part. The definition of A^{eq} is as follows.

Let L be a signature and A an L-structure. Let $\theta(\bar{x}, \bar{y})$ be an equivalence formula of A. We write I_θ for the set of all classes $\bar{a}/\theta$ with $\bar{a} \in \partial_\theta$, and $f_\theta: \partial_\theta \to I_\theta$ for the map $\bar{a} \mapsto \bar{a}/\theta$. The elements of A^{eq} are the equivalence classes $\bar{a}/\theta$ with θ an equivalence formula of A and $\bar{a} \in \partial_\theta$. We identify each element a of A with the equivalence class $a/(x = y)$. For each equivalence formula θ there are a 1-ary relation symbol P_θ and a relation symbol R_θ, interpreted in A^{eq} so that P_θ names I_θ and R_θ names (the graph of) the function f_θ. Also the symbols of L are interpreted in A^{eq} just as in A, on the elements in $I_{(x=y)}$. Two small refinements are called for here. First, since $I_{(x=y)}$ is not the whole domain of A^{eq}, the function symbols of L will have to be replaced by relation symbols; see Exercise 2.6.1. In talking about A^{eq} we can keep the function symbols as a shorthand. And second, it is tidier (and makes the proof of Theorem 4.4.1 smoother) if we ensure that the different sets I_θ are disjoint. One can do that by the usual set-theoretic device of replacing each element $\bar{a}/\theta$ by the ordered pair $(\bar{a}/\theta, \theta)$.

This defines the structure A^{eq}. We write L^{eq} for its signature. The sets I_θ are called the **sorts** of A^{eq}.

Theorem 4.4.1. *Let A be a structure of signature L.*

(a) $\text{dom}(A)$ *is $\varnothing$-definable in A^{eq}, and so are all the relations R^A, functions F^A and constants c^A of A.*

(b) *For every element a of A^{eq} there are a formula $\phi(\bar{x}, y)$ of L^{eq} and a tuple $\bar{b}$ in A such that a is the unique element of A^{eq} satisfying $\phi(\bar{b}, y)$.*

(c) *For every first-order formula $\phi(\bar{x})$ of L^{eq} there is a first-order formula $\phi^\downarrow(\bar{x})$ of L such that, for every tuple $\bar{a}$ of elements of A, $A^{\text{eq}} \vDash \phi(\bar{a})$ $\Leftrightarrow A \vDash \phi^\downarrow(\bar{a})$.*

(d) *If $e: A \to B$ is an elementary embedding, then e extends to an elementary embedding $e^{\text{eq}}: A^{\text{eq}} \to B^{\text{eq}}$.*

(e) *If X is a set of elements of A, then $\text{acl}_A(X) = \text{acl}_{A^{\text{eq}}}(X) \cap \text{dom}(A)$.*

Proof. Left to the reader.   □

Example 1: *Projective planes.* Suppose A is a three-dimensional vector space over a finite field, and let L be the first-order language of A. Then we can write a formula $\theta(x, y)$ of L which expresses 'vectors x and y are non-zero and are linearly dependent on each other'. The formula θ is an equivalence formula of A, and the sort I_θ is the set of points of the projective plane P associated with A. Likewise we can write a formula $\eta(x_1, x_2, y_1, y_2)$ which expresses 'x_1 and x_2 are linearly independent; so are y_1 and y_2; and x_1 and x_2 together generate the same plane as y_1 and y_2'. Again η is an equivalence formula of A, and the sort I_η is the set of lines of P. Using the relations R_θ and R_η we easily define the incidence relation of P. So statements about the projective plane P can be translated rather directly into statements about A^{eq}.

Perhaps it was overkill to define the whole of A^{eq} at once. In most applications we are only concerned with a finite number of the sorts I_θ. A **finite slice** of A^{eq} is a structure defined exactly the same way as A^{eq}, but using only finitely many of the equivalence formulas θ (always including the equivalence formula $x = y$).

There is a close relationship between interpretations and A^{eq}. In fact some writers invoke the following theorem as an excuse for working with A^{eq} instead of interpretations; it's purely a matter of taste.

Theorem 4.4.2. *Let A be a K-structure with more than one element, and B an L-structure. Then B is interpretable in A if and only if B is isomorphic to a relativised reduct of a definitional expansion of a finite slice of A^{eq}.*

Proof. Suppose first that there is an n-dimensional interpretation Γ of B in A. Let θ be the formula $\partial_\Gamma(\bar{x}_1) \wedge \partial_\Gamma(\bar{x}_2) \wedge =_\Gamma(\bar{x}_1, \bar{x}_2)$. Then θ is an equivalence formula, so that I_θ is a sort of A^{eq}. The imaginary elements $\bar{a}/\theta$ are in one-one correspondence with the elements $f_\Gamma(\bar{a})$ ($\bar{a} \in \partial_\Gamma(A^n)$) of B. I leave it to the reader to check that the relations of B correspond to $\varnothing$-definable relations on A^{eq}, using the symbol R_θ.

In the other direction, suppose B is a relativised reduct A^*_P where A^* is a definitional extension of a finite slice of A^{eq}. By time-sharing (Exercise 2.1.6, using two or more elements of A) we can suppose that the domain of B lies inside some one set I_θ where $\theta(\bar{x}_1, \bar{x}_2)$ is an n-ary equivalence formula. Then the required interpretation Γ has

$$\partial_\Gamma(\bar{x}) := P^*(\bar{x}),$$
$$=_\Gamma := \theta^*$$

where $P^*(\bar{x})$ and θ^* are formulas of K^{eq} which are equivalent to $P(\bar{x})$ and θ; and likewise with the remaining symbols of Γ. $\qquad\square$

Finite cover property

The following notion plays an important role in the study of uncountably categorical structures. I mention it here because one of its more memorable formulations is in terms of A^{eq}.

Let T be a complete theory in a first-order language L. We say that T **doesn't have the finite cover property** (for short, doesn't have the f.c.p.) if for every formula $\theta(\bar{x}, \bar{y}, \bar{z})$ of L with $\bar{x}$, $\bar{y}$ of the same length n, there is a formula $\phi(\bar{z})$ of L with the following property:

(4.1)   in any model A of T, if $\bar{a}$ is a tuple of elements and $\theta(\bar{x}, \bar{y}, \bar{a})$ defines an equivalence relation E on the set of n-tuples of elements of A, then E has infinitely many classes if and only if $A \vDash \phi(\bar{a})$.

Warning. This is the definition commonly used in practice, but it is not quite equivalent to the original definition by Keisler, which is given in Exercise 3 below. Our formulation is due to Shelah, who shows that it's equivalent to Keisler's for a large and important class of complete theories known as 'stable theories' (see section 9.4 below). According to Keisler's definition an unstable complete theory always has the f.c.p. (cf. Exercise 9.4.10).

There is another way of phrasing the f.c.p. If L is a first-order language and A is an L-structure, we say that **infinity is definable in** A if for every formula $\phi(\bar{x}, \bar{y})$ of L there is a formula $\sigma(\bar{y})$ of L such that for each tuple $\bar{a}$ in A,

$$(4.2) \qquad A \vDash \sigma(\bar{a}) \Leftrightarrow \phi(A^n, \bar{a}) \text{ is infinite.}$$

For example, if A is an algebraically closed field and $\phi(x, \bar{y})$ is an equation $p(x, \bar{y}) = 0$, where p is a polynomial with integer coefficients, then for every tuple $\bar{a}$ from A, $\phi(A, \bar{a})$ is infinite if and only if $p(x, \bar{a})$ is the zero polynomial; we can write this as a condition $\sigma(\bar{a})$ on $\bar{a}$.

Unpacking the definition of A^{eq}, we see that *T fails to have the f.c.p. if and only if for every model A of T, infinity is definable in A^{eq}*.

The definability of infinity *for all models* of a complete first-order theory T is equivalent to something simpler. (This is our first application of the compactness theorem for first-order logic.)

Theorem 4.4.3. *Let L be a first-order language and T a complete theory in L. Then the following are equivalent:*
(a) Infinity is definable in every model of T.
(b) If $\phi(\bar{x}, \bar{y})$ is any formula of L, then there is some $k < \omega$ such that for every model A of T and each tuple $\bar{a}$ in A, either $|\phi(A^n, \bar{a})| \leqslant k$ or $\phi(A^n, \bar{a})$ is infinite.

Proof. (a) $\Rightarrow$ (b): Suppose A is a model of T and $\sigma(\bar{x})$, $\phi(\bar{x}, \bar{y})$ are formulas of L such that (4.2) holds. For contradiction assume that (b) is false, i.e. that for every $k < \omega$ there is a tuple $\bar{a}_k$ in A such that $k < |\phi(A^n, \bar{a}_k)| < \omega$. Then take an m-tuple $\bar{d}$ of distinct new constants, and consider the following theory T':

$$(4.3) \qquad T \cup \{\neg \sigma(\bar{d})\} \cup \{\exists_{\geqslant k} \bar{x} \phi(\bar{x}, \bar{d}) : k < \omega\}.$$

If T_0 is any finite subset of T', then T_0 contains only finitely many sentences $\exists_{\geqslant h} \bar{x} \phi(\bar{x}, \bar{d})$. Let k be a finite number greater than all these h; then we can construct a model of T_0 by taking A and interpreting $\bar{d}$ as names for the elements $\bar{a}_k$. Hence every finite subset of T' has a model. So it follows by the compactness theorem for first-order logic (Theorem 5.1.1) that T' has a model B. Then (4.2) fails when we put $B|L$ for A and $\bar{d}^B$ for $\bar{a}$. This contradicts (a). Hence our assumption that (b) fails was false.

(b) $\Rightarrow$ (a): If (b) holds then we can write down a formula for $\sigma(\bar{y})$, namely $\exists_{>k}\bar{x}\,\phi(\bar{x}, \bar{y})$. $\square$

Note that the compactness theorem gives us a uniformity: the formula σ can be chosen to depend only on ϕ and not on the choice of model A.

Eliminating imaginaries

Following Bruno Poizat, we say that an L-structure A has **elimination of imaginaries** if for every equivalence formula $\theta(\bar{x}, \bar{y})$ of A and each tuple $\bar{a}$ in A there is a formula $\phi(\bar{x}, \bar{z})$ of L such that the equivalence class $\bar{a}/\theta$ of $\bar{a}$ can be written as $\phi(A^n, \bar{b})$ for some *unique* tuple $\bar{b}$ from A. We allow $\bar{z}$ and $\bar{b}$ to be empty here, in which case $\bar{a}/\theta$ is $\varnothing$-definable.

More usefully, we say that A has **uniform elimination of imaginaries** if the same holds, except that ϕ depends only on θ and not on $\bar{a}$. Another way of saying this is that for every equivalence formula $\theta(\bar{x}, \bar{y})$ of A there is a function F which is definable without parameters, taking tuples as values, such that for all $\bar{a}_1$ and $\bar{a}_2$ in A, $\bar{a}_1$ is θ-equivalent to $\bar{a}_2$ if and only if $F(\bar{a}_1) = F(\bar{a}_2)$.

The uniformity is helpful, because it creates bijections between sets of tuples and sets of definable classes in A, thus:

Theorem 4.4.4. *Let A be an L-structure. Then the following are equivalent.*
(a) *A has uniform elimination of imaginaries.*
(b) *For every formula $\psi(\bar{x}, \bar{y})$ of L, with $\bar{x}$ of length n, there is a formula $\chi(\bar{x}, \bar{z})$ of L such that for each tuple $\bar{a}$ in A there is a unique tuple $\bar{b}$ in A such that $\psi(A^n, \bar{a}) = \chi(A^n, \bar{b})$.*

Proof. (a) $\Rightarrow$ (b). Write $\theta(\bar{y}, \bar{y}')$ for the formula

$$(4.4) \qquad\qquad \forall \bar{x}(\psi(\bar{x}, \bar{y}) \leftrightarrow \psi(\bar{x}, \bar{y}')).$$

Then θ is an equivalence formula, so by (a) there is a formula $\phi(\bar{y}, \bar{z})$ of L such that for each $\bar{b}$ in A, $\theta(A^m, \bar{b}) = \phi(A^m, \bar{c})$ for some unique tuple $\bar{c}$. Let $\chi(\bar{x}, \bar{z})$ be the formula $\exists \bar{y}\,(\psi(\bar{x}, \bar{y}) \wedge \phi(\bar{y}, \bar{z}))$.
(b) $\Rightarrow$ (a) is immediate. $\square$

We say that a theory T in L **has elimination of imaginaries** if all the models of T have elimination of imaginaries. Likewise T **has uniform elimination of imaginaries** if every model of T has uniform elimination of imaginaries, in such a way that ϕ in the definition above depends only on θ and not on the choice of model. But in fact we can say something tidier in this case.

Theorem 4.4.5. *Let T be a theory in L. The following are equivalent.*
(a) *T has uniform elimination of imaginaries.*
(b) *If $\theta(\bar{x}, \bar{y})$ is any formula of L such that $T \vdash$ 'θ defines an equivalence relation', then there is a formula $\phi(\bar{x}, \bar{z})$ of L such that $T \vdash \forall \bar{y} \exists_{=1} \bar{z} \forall \bar{x} (\theta(\bar{x}, \bar{y}) \leftrightarrow \phi(\bar{x}, \bar{z}))$.*
(c) *For every formula $\theta(\bar{x}, \bar{y})$ of L there is a formula $\phi(\bar{x}, \bar{z})$ of L such that $T \vdash$ ('θ defines an equivalence relation' $\rightarrow \quad \forall \bar{y} \exists_{=1} \bar{z} \forall \bar{x} (\theta(\bar{x}, \bar{y}) \leftrightarrow \phi(\bar{x}, \bar{z})))$.*

Proof. (c) $\Rightarrow$ (a) $\Rightarrow$ (b) are immediate. Assuming (b) we get (c) as follows: when $\bar{x}$ and $\bar{y}$ both have the same length n, write $\theta^*(\bar{x}, \bar{y})$ for the formula

(4.5)   ('θ defines an equivalence relation' $\rightarrow \theta$).

Then in any L-structure A, θ^* defines an equivalence relation E; but if θ is not an equivalence formula of A then E is the trivial equivalence relation on $(\mathrm{dom}\, A)^n$ with just one class. Choose ϕ by putting θ^* for θ in (b). Then ϕ works for (c). $\qquad \square$

   The next result (which again rests on the compactness theorem) tells us that if T is a first-order theory satisfying a fairly mild condition, then elimination of imaginaries for T is always uniform. This explains why model theorists are apt to leave out the word 'uniform' in this context – the uniformity comes for free.

Theorem 4.4.6. *Let T be a theory in a first-order language L, and suppose that there are closed terms s, t of L such that $T \vdash s \neq t$. Then the following are equivalent:*
(a) *T has elimination of imaginaries.*
(b) *T has uniform elimination of imaginaries.*

Proof. We only need prove (a) $\Rightarrow$ (b). Let $\theta(\bar{x}, \bar{y})$ be a formula of L, and write Eq_θ for the sentence of L which expresses that θ is an equivalence formula. Take a tuple $\bar{c}$ of distinct new constants, and let T' be the theory

(4.6) $T \cup \{\mathrm{Eq}_\theta\} \cup$

{'There is not a unique $\bar{z}$ such that $\forall \bar{x} (\phi(\bar{x}, \bar{z}) \leftrightarrow \theta(\bar{x}, \bar{c}))$': ϕ a formula of L}.

If T' had a model B, then $B|L$ would be a model of T without elimination of imaginaries. So T' has no model, and hence by the compactness theorem for first-order logic (Theorem 5.1.1) there is a finite subset T_0 of T' which also has no models. Let $\phi_i(\bar{x}, \bar{z}_i)$ $(i < k)$ be the finitely many formulas which appear in T_0 in the role of ϕ. Then

(4.7) $$T \cup \{Eq_\theta\} \vdash$$

$$\bigvee_{i<k} \text{'There is a unique } \bar{z}_i \text{ such that } \forall \bar{x}(\phi_i(\bar{x}, \bar{z}_i) \leftrightarrow \theta(\bar{x}, \bar{c}))\text{'.}$$

Since the constants $\bar{c}$ are distinct and appear nowhere in $T \cup Eq_\theta$, the lemma on constants (Lemma 2.3.2) allows us to replace the second line of (4.7) by

(4.8) $\quad \forall \bar{y} \bigvee_{i<k} \text{'There is a unique } \bar{z}_i \text{ such that } \forall \bar{x}(\phi_i(\bar{x}, \bar{z}_i) \leftrightarrow \theta(\bar{x}, \bar{y}))\text{'.}$

Using Exercise 2.1.6 (time-sharing with formulas) and the two closed terms s, t, we can build a formula

(4.9) $$\psi(\bar{x}, \bar{z}_0, \ldots, \bar{z}_{k-1}, \bar{w})$$

which turns into formulas equivalent (modulo the theory $\{s \neq t\}$) to $\phi_i(\bar{x}, \bar{z}_i)$ when we put appropriate combinations of the terms s, t in place of the variables $\bar{w}$. With a little ingenuity (which I leave to the reader) we can code up the formulas ϕ_i using the formula ψ, so that we get

(4.10) $$T \cup \{Eq_\theta\} \vdash$$
$$\forall \bar{x} \text{'There is a unique } \bar{z} \text{ such that } \forall \bar{y}(\phi(\bar{x}, \bar{z}) \leftrightarrow \theta(\bar{x}, \bar{y}))\text{'.}$$

Doing the same for each formula θ, T has a uniform elimination of imaginaries. $\qquad\square$

Fields with elimination of imaginaries

We shall show that the theory of real-closed fields and the theory algebraically closed fields of a fixed characteristic both have elimination of imaginaries. Since each of these theories entails $0 \neq 1$, the elimination of imaginaries is uniform.

Recall that Theorems 2.7.2 and 2.7.3 stated quantifier elimination results for these theories. We shall use two corollaries of quantifier elimination for algebraically closed fields. By Example 2 in section 7.4, the theory of algebraically closed fields of a fixed characteristic is complete. By Example 1 in section 9.2, if A is an algebraically closed field and X is a set of elements of A which is definable by a first-order formula with parameters from A, then either X or $(\text{dom } A) \backslash X$ is a finite set; this says that A is a **minimal structure**.

Theorem 4.4.7. *Let T be the theory of real-closed fields, in the first-order language L. Then T has elimination of imaginaries.*

Proof. Let A be a model of T, $\theta(\bar{x}, \bar{y})$ an equivalence formula for T and $\bar{b}$ a tuple from A. Then $\theta(\bar{x}, \bar{b})$ defines the θ-equivalence class $\bar{b}/\theta$ of $\bar{b}$. We can suppose that $\bar{x} = (x_0, \ldots, x_{n-1})$.

We claim that for each $i < n$ there are elements $a_0, \ldots, a_{i-1}$ and a formula $\psi_i(x_0, \ldots, x_{i-1}, \bar{y})$ of L such that

(4.11)   for every $\bar{d} \in \bar{b}/\theta$, $(a_0, \ldots, a_{i-1})$ is the only i-tuple
of elements of A satisfying $\psi_i(x_0, \ldots, x_{i-1}, \bar{d})$ in A,

(4.12)   there are $a'_i, \ldots, a'_{n-1}$ such that $(a_0, \ldots, a_{i-1}, a'_i, \ldots, a'_{n-1}) \in \bar{b}/\theta$.

Clause (4.11) says that the tuple $(a_0, \ldots, a_{i-1})$ is definable from $\bar{b}/\theta$ in A^{eq}.

We prove the claim by induction on i. When $i = 0$ it is trivially true. Assuming it is true for i, we prove it for $i + 1$ as follows.

Consider the non-empty set W of all elements a of A such that for

$$A \vDash \exists x_{i+1} \ldots x_{n-1} \theta(a_0, \ldots, a_{i-1}, a, x_{i+1}, \ldots, x_{n-1}, \bar{b}).$$

By the quantifier elimination theorem for real-closed fields, W is a finite union of singletons and open intervals whose endpoints are either $-\infty$, ∞ or elements definable from $\bar{b}$. Since W is not altered if we replace $\bar{b}$ by any other tuple from $\bar{b}/\theta$, these endpoints are in fact definable from $\bar{b}/\theta$. We look at the possibilities. If W is the whole of A, we can take a_i to be 0; then ψ_{i+1} can be ψ_i together with a new conjunct '$x_i = 0$'. If not, but W contains the set of elements $< d$, where d is an element definable from $\bar{b}/\theta$, then we can choose a_i to be $d - 1$, with a suitable ψ_{i+1}; similarly if W contains a final segment of A. Likewise, if W contains the open interval (d_1, d_2) where d_1 and d_2 are both definable from $\bar{b}/\theta$, then we can take a_i to be $(d_1 + d_2)/2$. There remains the case where W contains no non-empty open interval. Then W must be a finite set of elements definable from $\bar{b}/\theta$, and so we can take a_i to be the first of these elements. In each case it is easy to write down a suitable formula ψ_{i+1}. This proves the claim.

By the claim for $i = n$, $(a_0, \ldots, a_{n-1})$ is the unique n-tuple $\bar{a}$ in A such that $\bar{b}/\theta$ is defined by the formula $\theta(\bar{x}, \bar{a}) \wedge \psi_n(\bar{a}, \bar{a})$. $\square$

Theorem 4.4.8. *Let T be the theory of algebraically closed fields of some fixed characteristic, in the first-order language L. Then T has elimination of imaginaries.*

Proof. (The following model-theoretic argument is due to Anand Pillay. One can also give geometric arguments.) Let A be a model of T, $\theta(\bar{x}, \bar{y})$ an equivalence formula for T and $\bar{b}$ a tuple from A. Then $\theta(\bar{x}, \bar{b})$ defines the θ-equivalence class $\bar{b}/\theta$ of $\bar{b}$. We can suppose that $\bar{x} = (x_0, \ldots, x_{n-1})$.

We claim that for each $i < n$ there is a formula $\psi_i(x_0, \ldots, x_{i-1}, \bar{y})$ of L such that the set $\psi_i(A^i, \bar{d})$ is the same for each $\bar{d} \in \bar{b}/\theta$, and moreover the set U_i of i-tuples $(a_0, \ldots, a_{i-1}) \in \psi_i(A^i, \bar{b})$ such that

(4.13)   there are $a'_i, \ldots, a'_{n-1}$ such that $(a_0, \ldots, a_{i-1}, a'_i, \ldots,$
$a'_{n-1}) \in \bar{b}/\theta$

is finite and non-empty. We prove the claim by induction on i. When $i = 0$ it is trivially true. Assuming it is true for i, we prove it for $i + 1$ as follows.

Let $(a_0, \ldots, a_{i-1})$ be some tuple in U_i, and consider the non-empty set W of all elements a of A such that

$$A \models \exists x_{i+1} \ldots x_{n-1} \theta(a_0, \ldots, a_{i-1}, a, x_{i+1}, \ldots, x_{n-1}, \bar{b}).$$

Since W is a definable subset of A and A is a minimal structure, there are two possibilities: either W is finite, or $(\text{dom } A)\backslash W$ is finite. In the first case suppose W has exactly k elements. Then take ψ_{i+1} to be the formula

$$\psi_i \wedge \exists_{=k} x_i \exists x_{i+1} \ldots x_{n-1} \theta(\bar{x}, \bar{y}) \wedge \exists x_{i+1} \ldots x_{n-1} \theta(\bar{x}, \bar{y}).$$

In the second case, since the field of algebraic elements of A is infinite, W must meet it, say in some element a. Then a satisfies some nontrivial polynomial equation, say $p(x) = 0$. We take ψ_{i+1} to be the formula

$$\psi_i \wedge (p(x_i) = 0) \wedge \exists x_{i+1} \ldots x_{n-1} \theta(\bar{x}, \bar{y}).$$

This proves the claim.

When $i = n$, the claim gives us a formula $\psi(\bar{x}, \bar{y})$ (namely $\psi_n(\bar{x}, \bar{y}) \wedge \theta(\bar{x}, \bar{y})$) such that just finitely many tuples $\bar{a}$ (viz. those in U_n) satisfy $\psi(\bar{x}, \bar{b})$ in A, and they all lie in $\bar{b}/\theta$. We need a trick to turn this finite set U_n into a single tuple; the tuple must be determined by the set, independent of any ordering of the set. Here we use algebra. For each tuple $\bar{a} = (a_0, \ldots, a_{n-1})$ in U_n we take $q_{\bar{a}}(X)$ to be the polynomial

$$a_0 + a_1 X + \ldots + a_{n-1} X^{n-1} + X^n.$$

Then we write $q(X)$ for the polynomial

$$\prod_{\bar{a} \in X_n} q_{\bar{a}}.$$

By the unique factorisation theorem for polynomials over a field, the set of polynomials $q_{\bar{a}}$, and hence the set U_n, can be recovered from q. So for our tuple representing $\bar{b}/\theta$ we can take the sequence of coefficients of q. $\square$

Exercises for section 4.4

1. Show that if E is an equivalence relation on tuples in a set I_θ in A^{eq}, and E is definable in A^{eq} without parameters, then E is equivalent in a natural way to an equivalence relation on tuples of elements of A which is definable in A without parameters. (So there is nothing to be gained by passing to $(A^{\text{eq}})^{\text{eq}}$).

2. Let A be an L-structure and B a finite slice of A^{eq}. (a) Show that the restriction map $g \mapsto g | \text{dom}(A)$ defines an isomorphism from $\text{Aut}(B)$ to $\text{Aut}(A)$. (b) Show that if $\text{Aut}(A)$ is oligomorphic, then so is $\text{Aut}(B)$.

3. Let L be a first-order language and T a complete theory in L. Show that, if T has the finite cover property (in Shelah's sense), then there is a formula $\phi(x, \bar{y})$ of L such that, for arbitrarily large finite n, T implies that there are $\bar{a}_0, \ldots, \bar{a}_{n-1}$ for which $\neg \exists x \bigwedge_{i \in n} \phi(x, \bar{a}_i)$ holds, but $\exists x \bigwedge_{i \in W} \phi(x, \bar{a}_i)$ holds for each proper subset W of n. *This conclusion is Keisler's form of the finite cover property.* [The easy part is to deduce it with $\bar{x}$ in place of x. Then use induction on the length of $\bar{x}$.]

4. Show that if the L-structure A has uniform elimination of imaginaries, and $\bar{a}$ is a sequence of elements of A, then $(A, \bar{a})$ also has uniform elimination of imaginaries.

5. Show that ZFC (Zermelo–Fraenkel set theory with Choice) has elimination of imaginaries. [For each equivalence class, choose the set of all elements of least rank in the class.] (b) Show that first-order Peano arithmetic has uniform elimination of imaginaries.

6. Show that if V is a vector space of dimension at least 2 over a finite field of at least three elements, then V doesn't have elimination of imaginaries.

Further reading

There is a much fuller treatment of interpretations in my *Model theory*. Interpretations first appeared in model theory as a method for proving the undecidability of some first-order theories. Here as in many other areas, Tarski laid the foundations. But his notion of interpretations was less general than ours, which can be traced to Anatolii Mal'tsev:

Mal'cev, A. I., *The metamathematics of algebraic systems. Collected papers: 1936–1967*, trans. B. F. Wells III. Amsterdam: North-Holland 1971.

One can read about some other kinds of interpretation in:

Rubin, M., *The reconstruction of trees from their automorphism groups*. Providence RI: American Mathematical Society.

From the mid 1980s much research went into characterising the groups which are interpretable in various kinds of field – it became clear that this was important for model-theoretic approaches to algebraic geometry. At present the best results are when the group is 'definable in' (i.e. a relativised reduct of) the field. A good representative of this area is:

Pillay, A., Differential algebraic groups and the number of countable differentially closed fields. In *Model theory of fields*, Marker, D., Messmer, M. and Pillay, A., Lecture Notes in Logic, pp. 114–134. Berlin: Springer-Verlag, 1996.

The idea of eliminating imaginaries is due to Bruno Poizat (though there are geometric antecedents due to Weil and Chow). He discusses it in his privately published textbook, which we can hope will soon be available in English:

Poizat, B., *Cours de théorie des modèles*. Villeurbanne: Nur al-Mantiq wal Ma'rifah, 1985.

For an optimist, Theorem 4.4.8 says that in doing algebraic geometry with an algebraically closed field K, we don't need to go beyond the field itself, because other structures (such as projective groups over the field) can already be represented within the field. The pessimist's reading of Theorem 4.4.8 is different; it says that interpretations give the model theorist no extra leverage in algebraic geometry, because they lead to nothing new. Unfortunately the pessimist is onto something here. Even when model theory makes deep inroads into questions of algebraic geometry (as with Hrushovski's work on the Mordell–Lang conjecture – see below), there is a serious communication problem, because model-theoretic methods seem totally unrelated to the extra apparatus (Lie algebras, sheaves, cohomology, K-theory etc.) which is the stock-in-trade of modern geometers. The fact that I can't refer the reader to any literature discussing this problem is itself an indication of the work that needs to be done here.

A good route into Ehud Hrushovski's model-theoretic proof of the function field case of the Mordell-Lang conjecture (from diophantine geometry) is the following book:

Bouscaren, E. (ed.). Model Theory and Algebraic Geometry, An introduction to E. Hrushovski's proof of the geometric Mordell-Lang-conjecture. Lecture Notes in Mathematics 1696. Berlin, Springer-Verlag, 1998.

One of the central lemmas used in Hrushovski's argument is well worth studying for its own sake; it gives a model-theoretic axiomatisation of the Zariski topology:

Hrushovski, E. and Zilber, B., Zariski geometries. *Journal of American Mathematical Society* **9** (1996), pp. 1–56.

5

The first-order case: compactness

A given species of bird would show the same ability of grasping . . . numbers
. . . but the ability differs with the species. Thus with pigeons it may be five
or six according to experimental conditions, with jackdaws it is six and with
ravens and parrots, seven.
O. Koehler, The ability of birds to 'count', *Bull. Animal Behaviour* **9** (1950) 41–5.

Ravens, so we read, can only count up to seven. They can't tell the difference
between two numbers greater than or equal to eight. First-order logic is much
the same as ravens, except that the cutoff point is rather higher: it's ω instead
of 8.

This chapter is wholly devoted to the model theory of first-order languages. First-order model theory has always been the heart of model theory.
The main reason for this is that first-order logic, for all its expressive power,
is too weak to distinguish between one large number and another. The result
is that there are a number of constructions which give models of a first-order
theory, or turn a given model into a new one. In this chapter we study two
such constructions. The first is a combination of the compactness theorem
with (Robinson) diagrams. The second is amalgamation; it can be seen as an
application of the first. The idea of amalgamation is very powerful, and I
have used it whenever I can. Subtly different kinds of amalgamation produce
preservation theorems and definability theorems.

The last section of the chapter proves Ramsey's theorem as a direct
consequence of the fact that first-order logic can't recognise ω.

5.1 Compactness for first-order logic

If any theorem is fundamental in first-order model theory, it must surely be
the compactness theorem.

Theorem 5.1.1 *(Compactness theorem for first-order logic). Let T be a first-
order theory. If every finite subset of T has a model then T has a model.*

Proof. Let L be a first-order language and T a theory in L. Assume first that every finite subset of T has a non-empty model. We shall show that T can be extended to a Hintikka set T^+ (see section 2.3 above) in a larger first-order language L^+. By Theorem 2.3.3 it follows that some L^+-structure A is a model of T^+, so the reduct $A^+|L$ will be a model of T.

Write κ for the cardinality of L. Let c_i $(i < \kappa)$ be distinct constants not in L; we call these constants **witnesses**. Let L^+ be the first-order language got by adding the witnesses c_i to the signature of L. Then L^+ has κ sentences, and we can list them as ϕ_i $(i < \kappa)$. We shall define an increasing chain $(T_i : i \le \kappa)$ of theories in L^+, so that the following hold. (All models are assumed to be L^+-structures.)

(1.1)   For each $i \le \kappa$, every finite subset of T_i has a model.

(1.2)   For each $i < \kappa$, the number of witnesses c_k which are used in

T_i but not in $\bigcup_{j<i} T_j$ is finite.

The definition is by induction on i. We put $T_0 = T$ and at limit ordinals δ we take $T_\delta = \bigcup_{i<\delta} T_i$. Clearly these definitions respect (1.1) and (1.2); (1.1) is true at T_0 because of our assumption that every finite subset of T has a non-empty model.

For successor ordinals $i + 1$ we first define

$$(1.3) \quad T'_{i+1} = \begin{cases} T_i \cup \{\phi_i\} & \text{if every finite subset of this set has a model,} \\ T_i & \text{otherwise.} \end{cases}$$

If $\phi_i \in T'_{i+1}$ and ϕ_i has the form $\exists x\, \psi$ for some formula $\psi(x)$, then we choose the earliest witness c_j which is not used in T'_{i+1} (by (1.2) there is such a witness), and we put

$$(1.4) \quad\quad\quad\quad\quad T_{i+1} = T'_{i+1} \cup \{\psi(c_j)\}.$$

If $\phi_i \notin T'_{i+1}$ or ϕ_i is not of the form $\exists x\, \psi$, we put $T_{i+1} = T'_{i+1}$. These definitions clearly ensure (1.2), but we must show that (1.1) remains true when (1.4) holds. Let U be a finite subset of T'_{i+1} and let A be any L^+-structure which is a model of $U \cup \{\exists x\, \psi\}$. Then there is an element a of A such that $A \vDash \psi(a)$. Take such an element a, and let B be the L^+-structure which is exactly like A except that $c_j^B = a$. Since the witness c_j never occurs in U, B is still a model of U, and since c_j never occurs in $\psi(x)$, $B \vDash \psi(a)$ and so $B \vDash \psi(c_j)$. Condition (1.1) is secured.

We claim that T_κ is a Hintikka set for L^+. By Theorem 2.3.4 it suffices to prove three things.

(a) Every finite subset of T_κ has a model. This holds by (1.1).

(b) For every sentence ϕ of L^+, either ϕ or $\neg\phi$ is in T_κ. To prove this, suppose ϕ is ϕ_i and $\neg\phi$ is ϕ_j. If $\phi \notin T_\kappa$ then $\phi_i \notin T_{i+1}$, and by (1.3) this means that there is a finite subset U of T_i such that $U \cup \{\phi\}$ has no model. By the same argument, if $\neg\phi \notin T_\kappa$ then there is a finite subset U' of T_j such that $U' \cup \{\neg\phi\}$ has no model. Now $U \cup U'$ is a finite subset of T_κ, so it has a model A. Either $A \vDash \phi$ or $A \vDash \neg\phi$, and we have a contradiction either way. Thus at least one of ϕ, $\neg\phi$ is in T_κ.

(c) For every sentence $\exists x \, \psi(x)$ in T_κ there is a closed term t of L^+ such that $\psi(t) \in T_\kappa$. For this, suppose $\exists x \, \psi(x)$ is ϕ_i. Since $\phi_i \in T_\kappa$, (1.4) applies, and so T_{i+1} contains a sentence $\psi(c_j)$ where c_j is a witness. Then $\psi(c_j)$ is in T_κ.

Thus T_κ is a Hintikka set T^+ for L^+ and it contains T; so T has a model. This completes the argument, except in the freak case where some finite subset of T has only the empty model. In this case the empty L-structure must be a model of all T. □

Corollaries of compactness

The first two corollaries are immediate. In fact Corollary 5.1.2 is just another way of stating the theorem. (See Exercise 1.)

Corollary 5.1.2. *If T is a first-order theory, ψ a first-order sentence and $T \vdash \psi$, then $U \vdash \psi$ for some finite subset U of T.*

Proof. Suppose to the contrary that $U \nvdash \psi$ for every finite subset U of T. Then every finite subset of $T \cup \{\neg\psi\}$ has a model, so by the compactness theorem $T \cup \{\neg\psi\}$ has a model. Therefore $T \nvdash \psi$. □

The next corollary is one of our few encounters with formal inference rules. If we overlook the fact that computers have finite memories, a set is recursively enumerable (r.e. for short) if and only if it can be listed by a computer. (Recursive languages were defined in section 2.3.)

Corollary 5.1.3. *Suppose L is a recursive first-order language, and T is a recursively enumerable theory in L. Then the set of consequences of T in L is also recursively enumerable.*

Proof. Using one's favourite proof calculus, one can recursively enumerate all the consequences in L of a finite set of sentences. Since T is r.e., we can recursively enumerate its finite subsets; Corollary 5.1.2 says that every consequence of T is a consequence of one of these finite subsets. □

Since first-order logic can't distinguish between one infinite cardinal and another, there should be no surprise that every infinite structure has arbitrarily large elementary extensions. The compactness theorem will give us this result very quickly.

I follow custom in calling Corollary 5.1.4 the upward Löwenheim–Skolem theorem. But in fact Skolem didn't even believe it, because he didn't believe in the existence of uncountable sets.

Corollary 5.1.4 *(Upward Löwenheim–Skolem theorem).* *Let L be a first-order language of cardinality $\leq \lambda$ and A an infinite L-structure of cardinality $\leq \lambda$. Then A has an elementary extension of cardinality λ.*

Proof. Supplying names for all the elements of A, consider the elementary diagram of A, eldiag(A) for short (see section 2.5 above). Let c_i ($i < \lambda$) be λ new constants, and let T be the theory

(1.5) $\text{eldiag}(A) \cup \{c_i \neq c_j : i < j < \lambda\}.$

We claim that every finite subset of T has a model. For suppose U is a finite subset of T. Then for some $n < \omega$, just n of the new constants c_i occur in U. We can find a model of T by taking n distinct elements of A and letting each of the new constants in U stand for one of these elements; this is possible because A is infinite.

It follows by the compactness theorem that T has a model B. Since B is a model of eldiag(A), there is an elementary embedding $e: A \to B|L$ (by the elementary diagram lemma, Lemma 2.5.3). Replacing elements of the image of e by the corresponding elements of A, we make $B|L$ an elementary extension of A. Since $B \vDash T$, we have $c_i^B \neq c_j^B$ whenever $i < j < \lambda$, and hence $B|L$ has at least λ elements. To bring the cardinality of $B|L$ down to exactly λ, we invoke the downward Löwenheim–Skolem theorem (Corollary 3.1.4). $\square$

Compactness in infinitary languages?

The compactness theorem fails for infinitary languages. For example, let c_i ($i < \omega$) be distinct constants, and consider the theory

(1.6) $c_0 \neq c_1, \; c_0 \neq c_2, \; c_0 \neq c_2, \ldots,$

$$\bigvee_{0 < i < \omega} c_0 = c_i.$$

This theory has no model, but every proper subset of it has a model.

Exercises for section 5.1

1. Show that each of the following is equivalent to the compactness theorem for first-order logic. (a) For every theory T and sentence ϕ of a first-order language, if $T \vdash \phi$ then for some finite $U \subseteq T$, $U \vdash \phi$. (b) For every theory T and sentence ϕ of a first-order language, if T is equivalent to the theory $\{\phi\}$ then T is equivalent to some finite subset of T. (c) For every first-order theory T, every tuple $\bar{x}$ of distinct variables and all sets $\Phi(\bar{x})$, $\Psi(\bar{x})$ of first-order formulas, if $T \vdash \forall \bar{x}(\bigwedge \Phi \leftrightarrow \bigvee \Psi)$ then there are finite sets $\Phi' \subseteq \Phi$ and $\Psi' \subseteq \Psi$ such that $T \vdash \forall \bar{x}(\bigwedge \Phi' \leftrightarrow \bigvee \Psi')$.

2. (a) Let L be a first-order language, T a theory in L and Φ a set of sentences of L. Suppose that for all models A, B of T, if $A \vDash \phi \Leftrightarrow B \vDash \phi$ for each $\phi \in \Phi$, then $A \equiv B$. Show that every sentence ψ of L is equivalent modulo T to a boolean combination ψ^* of sentences in Φ. (b) Show moreover that if L and T are recursive then ψ^* can be effectively computed from ψ.

3. (Craig's trick) In a recursive first-order language L let T be an r.e. theory. Show that T is equivalent to a recursive theory T^*. [Write ϕ^n for $\phi \wedge \ldots \wedge \phi$ (n times). Try $T^* = \{\phi^n : n$ is the Gödel number of a computation putting ϕ into $T\}$.]

4. Let L be a first-order language, δ a limit ordinal (for example ω) and $(T_i : i < \delta)$ an increasing chain of theories in L, such that for every $i < \delta$ there is a model of T_i which is not a model of T_{i+1}. Show that $\bigcup_{i<\delta} T_i$ is not equivalent to a sentence of L.

5. Show that none of the following classes is first-order definable (i.e. by a single sentence; see section 2.2). (a) The class of infinite sets. (b) The class of torsion-free abelian groups. (c) The class of algebraically closed fields.

6. Let L be a first-order language and T a theory in L. (a) Suppose T has models of arbitrarily high cardinalities; show that T has an infinite model. (b) Let $\phi(x)$ be a formula of L such that for every $n < \omega$, T has a model A with $|\phi(A)| \geq n$. Show that T has a model B for which $\phi(B)$ is infinite.

7. Let L be the first-order language of fields and ϕ a sentence in L. Show that if ϕ is true in every field of characteristic 0, then there is a positive integer m such that ϕ is true in every field of characteristic $\geq m$.

*Recall that we say a theory T is λ-**categorical** if T has, up to isomorphism, exactly one model of cardinality λ.*

8. (a) Let L be a first-order language, T a theory in L and λ a cardinal $\geq |L|$. Show that if T is λ-categorical then T is complete. (b) Use (a) to give quick proofs of the completeness of (i) the theory of (non-empty) dense linear orderings without endpoints and (ii) the theory of algebraically closed fields of a fixed characteristic. Find two other nice examples.

9. Suppose L is a first-order language, A is an L-structure and $\phi(x)$ is a formula of L such that $\phi(A)$ is infinite. Show that for every cardinal $\lambda \geqslant \max(|A|, |L|)$, there is an elementary extension B of A in which $|\phi(B)| = \lambda$.

10. Suppose L is a first-order language, A is an L-structure and λ is a cardinal $\geqslant \max(|A|, |L|)$. Show that A has an elementary extension B of cardinality λ such that for every formula $\phi(x, \bar{y})$ of L and every tuple $\bar{b}$ of elements of B, $\phi(B, \bar{b})$ has cardinality either $= \lambda$ or $< \omega$. [This is a baby version of the constructions of saturated elementary extensions in section 8.2 below. Iterate the construction of the previous exercise to form an elementary chain of length λ, using the Tarski–Vaught theorem on elementary chains, Theorem 2.5.2.]

11. Show that if L is a first-order language of cardinality λ and A is an L-structure of cardinality $\mu \geqslant \lambda$, then A has an elementary extension B of cardinality μ with $|\text{dom}(B)\backslash\text{dom}(A)| = \mu$. If $\nu < \mu$, give an example to show that A need not have an elementary extension B of cardinality μ with $|\text{dom}(B)\backslash\text{dom}(A)| = \nu$.

12. Let L be a first-order language and $\phi(x, y)$ a formula of L. Show that if A is an L-structure in which $\phi(A^2)$ is a well-ordering of an infinite set, then there is an L-structure B elementarily equivalent to A, in which $\phi(B^2)$ is a linear ordering relation which is not well-ordered. [Adapt the proof of Corollary 5.1.4.]

The following result in algebraic geometry was first proved by methods of logic.
13. Let L be the first-order language of fields. (a) Show that for every positive integer n there is an $\forall_2$ sentence ϕ_n of L which expresses (in any field K) 'for any $k, m \leqslant n$, if $\bar{X} = (X_0, \ldots, X_{m-1})$ is an m-tuple of indeterminates and $p_0, \ldots, p_{k-1}, q_0, \ldots,$ q_{m-1} are any polynomials $\in K[\bar{X}]$ of degree $\leqslant n$ such that $q_0, \ldots, q_{m-1}$ define an injective mapping from V to V, where $V = \{\bar{a} \in K^m: p_0(\bar{a}) = \ldots = p_{k-1}(\bar{a}) = 0\}$, then this mapping is surjective'. (b) Show that each sentence ϕ_n is true in every finite field, and hence in any union of a chain of finite fields. [See Theorem 2.4.4.] (c) Show that each sentence ϕ_n is true in every algebraically closed field of prime characteristic. (d) Show that each sentence ϕ_n is true in every algebraically closed field. (e) Deduce that if V is an algebraic variety and F is an injective morphism from V to V, then F is surjective.

14. Let L be a recursive first-order language. Show that if T is a recursively enumerable theory in L, and there are only finitely many inequivalent complete theories $\supseteq T$ in L, then T is decidable.

15. Let L be a recursive first-order language and T a theory in L. Show that if T is decidable then there is a decidable complete theory $T' \supseteq T$ in L.

16. Let L be a recursive first-order language and T a consistent recursively enumerable theory in L. Show that T is decidable if and only if there are models A_i $(i < \omega)$ of T such that $T = \text{Th}\{A_i: i < \omega\}$, and an algorithm which determines, for any $i < \omega$ and any sentence ϕ of L, whether or not $A_i \models \phi$. [For left to right use Exercise 15.]

5.2 Types

The upward Löwenheim–Skolem theorem (Corollary 5.1.4) was a little too crude for comfort. It told us that an infinite structure A can be built up into a larger structure, but it said nothing at all about what the new parts of the structure look like. We want to ask for example whether this piece of A can be kept small while that piece is expanded, whether automorphisms of A extend to automorphisms of the elementary extension of A, and so forth. Questions like these are often hard to answer, and they will keep us occupied right through to Chapter 9.

The main notion used for analysing such questions is the notion of a **type**. We met types briefly in section 2.3. Now it's time to examine them more systematically. One can think of types as a common generalisation of two well-known mathematical notions: the notion of a minimal polynomial in field theory, and the notion of an orbit in permutation group theory.

Let L be a first-order language and A an L-structure. Let X be a set of elements of A and $\bar{b}$ a tuple of elements of A. Let $\bar{a}$ be a sequence listing the elements of X. The **complete type of $\bar{b}$ over X** (with respect to A, in the variables $\bar{x}$) is defined to be the set of all formulas $\psi(\bar{x}, \bar{a})$ such that $\psi(\bar{x}, \bar{y})$ is in L and $A \vDash \psi(\bar{b}, \bar{a})$. More loosely, the complete type of $\bar{b}$ over X is everything we can say about $\bar{b}$ in terms of X. (Although $\bar{a}$ will in general be infinite, each formula $\psi(\bar{x}, \bar{y})$ of L has only finitely many free variables, so that only a finite part of X is mentioned in $\psi(\bar{x}, \bar{a})$.)

We write the complete type of $\bar{b}$ over X with respect to A as $\mathrm{tp}_A(\bar{b}/X)$, or $\mathrm{tp}_A(\bar{b}/\bar{a})$ where $\bar{a}$ lists the elements of X. The elements of X are called the **parameters** of the complete type. Complete types are written p, q, r etc.; one writes $p(\bar{x})$ if one wants to show that the variables of the type are $\bar{x}$. We write $\mathrm{tp}_A(\bar{b})$ for $\mathrm{tp}_A(\bar{b}/\varnothing)$, the type of $\bar{b}$ over the empty set of parameters. Note that if B is an elementary extension of A, then $\mathrm{tp}_B(\bar{b}/X) = \mathrm{tp}_A(\bar{b}/X)$.

Let $p(\bar{x})$ be a set of formulas of L with parameters from X. We shall say that $p(\bar{x})$ is a **complete type over X** (with respect to A, in the variables $\bar{x}$) if it is the complete type of some tuple $\bar{b}$ over X with respect to some elementary extension of A. Putting it loosely again, a complete type over X is everything we can say in terms of X about some possible tuple $\bar{b}$ of elements of A – maybe the tuple $\bar{b}$ is really there in A, or maybe it only exists in an elementary extension of A.

A **type** over X (with respect to A, in the variables $\bar{x}$) is a subset of a complete type over X. We shall write Φ, Ψ, $\Phi(\bar{x})$ etc. for types. A type is called an **n-type** ($n < \omega$) if it has just n free variables. (Some writers insist that these variables should be $x_0, \ldots, x_{n-1}$ from some stock of variables $x_0, x_1, \ldots$. I shall not be so strict.)

We say that a type $\Phi(\bar{x})$ over X is **realised** by a tuple $\bar{b}$ in A if $\Phi \subseteq \mathrm{tp}_A(\bar{b}/X)$. If Φ is not realised by any tuple in A, we say that A **omits** Φ.

The next result, a corollary of the compactness theorem, tells us where to look for types. We say that a set $\Phi(\bar{x})$ of formulas of L with parameters in A is **finitely realised** in A if for every finite subset Ψ of Φ, $A \vDash \exists \bar{x} \bigwedge \Psi$.

Theorem 5.2.1. *Let L be a first-order language, A an L-structure, X a set of elements of A and $\Phi(x_0, \ldots, x_{n-1})$ a set of formulas of L with parameters from X. Then, writing $\bar{x}$ for $(x_0, \ldots, x_{n-1})$,*
(a) $\Phi(\bar{x})$ *is a type over X with respect to A if and only if Φ is finitely realised in A,*
(b) $\Phi(\bar{x})$ *is a complete type over X with respect to A if and only if $\Phi(\bar{x})$ is a set of formulas of L with parameters from X, which is maximal with the property that it is finitely realised in A.*
In particular if Φ is finitely realised in A, then it can be extended to a complete type over X with respect to A.

Proof. (a) Suppose first that Φ is a type over X with respect to A. Then there are an elementary extension B of A and an n-tuple $\bar{b}$ in B such that $B \vDash \bigwedge \Phi(\bar{b})$. So if Ψ is a finite subset of Φ, then $B \vDash \bigwedge \Psi(\bar{b})$ and hence $B \vDash \exists \bar{x} \bigwedge \Psi(\bar{x})$. Since $A \preccurlyeq B$ and the sentence is first-order, this implies $A \vDash \exists \bar{x} \bigwedge \Psi(\bar{x})$.

Conversely suppose Φ is finitely realised in A. Form eldiag(A) and take an n-tuple of distinct new constants $\bar{c} = (c_0, \ldots, c_{n-1})$. Let T be the theory

$$(2.1) \qquad \qquad \text{eldiag}(A) \cup \Phi(\bar{c}).$$

We claim that every finite subset of T has a model. For let U be a finite subset of T, and let Ψ be the set of formulas $\psi(\bar{x})$ of Φ such that $\psi(\bar{c}) \in U$. By assumption $A \vDash \exists \bar{x} \bigwedge \Psi$, and hence there are elements $\bar{a}$ in A such that $A \vDash \bigwedge \Psi(\bar{a})$. By interpreting the constants $\bar{c}$ as names of the elements $\bar{a}$, we make A into a model of U. This proves the claim.

Now by the compactness theorem (Theorem 5.1.1), T has a model C. Since $C \vDash$ eldiag(A), the elementary diagram lemma (Lemma 2.5.3) gives us an elementary embedding $e: A \to C|L$, and by making the usual replacements we can assume that in fact $A \preccurlyeq C|L$. Let $\bar{b}$ be the tuple $\bar{c}^C$. Then $C \vDash \bigwedge \Phi(\bar{b})$ since $C \vDash T$. It follows that $\bar{b}$ satisfies $\Phi(\bar{x})$ in some elementary extension of A, and so Φ is a type over X with respect to A.

(b) If Φ is a complete type over X then Φ contains either ϕ or $\neg \phi$, for each formula $\phi(\bar{x})$ of L with parameters from X. This implies that Φ is a maximal type over X with respect to A. On the other hand if Φ is a maximal type over X with respect to A, then for some tuple $\bar{b}$ in some elementary extension B of A, $B \vDash \bigwedge \Phi(\bar{b})$, so that Φ is included in the complete type of $\bar{b}$ over X. By maximality it must equal this complete type. $\qquad \square$

By this theorem, if X is the empty set of parameters, then the question whether Φ is a type over X with respect to A depends only on $\mathrm{Th}(A)$. Types over the empty set with respect to A are also known as the **types of** $\mathrm{Th}(A)$. More generally let T be any theory in a first-order language. Then a **type of** T is a set $\Phi(\bar{x})$ of formulas of L such that $T \cup \{\exists \bar{x} \bigwedge \Psi\}$ is consistent for every finite subset $\Psi(\bar{x})$ of Φ; a **complete type** of T is a maximal type of T. If T happens to be a complete theory, then we can replace '$T \cup \{\exists \bar{x} \bigwedge \Psi\}$ is consistent' by the equivalent statement '$T \vdash \exists \bar{x} \bigwedge \Psi$'. All this agrees with the definitions in section 2.3.

Suppose A is an L-structure, X is a set of elements of A and n is a positive integer. We write $S_n(X; A)$ for the set of complete n-types over X with respect to A. When A is some fixed structure, we write simply $S_n(X)$. (Some authors write $S^n(X)$.) When T is a complete theory, we write $S_n(T)$ for the set of complete n-types of T.

The sets $S_n(X; A)$ are known as the **Stone spaces** of A. The name refers to the fact that they can be topologised to form the Stone dual spaces of certain boolean algebras of formulas. Like all Stone spaces of boolean algebras, they are compact; this is where the name 'compactness theorem' comes from.

Exercises for section 5.2

1. Let A be an L-structure and B an extension of A. Show that (a)–(c) are equivalent. (a) B is an elementary extension of A. (b) For every tuple $\bar{a}$ of elements of A, $\mathrm{tp}_A(\bar{a}) = \mathrm{tp}_B(\bar{a})$. (c) For every set X of elements of A and every $n < \omega$, $S_n(X; A) = S_n(X; B)$.

2. Let A be an L-structure, X a set of elements of A, $\bar{a}$ a tuple of elements of A and e an automorphism of A which fixes X pointwise. Show that $\bar{a}$ and $e\bar{a}$ have the same complete type over X with respect to A.

For keeping track of the complete types of a theory, quantifier elimination is invaluable. Here are three examples:

3. Let A be the structure $(\mathbb{Q}, <)$ where $\mathbb{Q}$ is the set of rational numbers and $<$ is the usual ordering. Describe the complete 1-types over $\mathrm{dom}(A)$.

4. Let A be an algebraically closed field and C a subfield of A; to save notation I write C also for $\mathrm{dom}(C)$. (a) Show that two elements a, b of A have the same complete type over C if and only if they have the same minimal polynomial over C. (b) Show that for all n, $|S_n(C; A)| = \omega + |C|$.

5. Let A be a vector space over a field k and X a set of elements of A. Describe $S_n(X; A)$ for each $n < \omega$, and show that $|S_n(X; A)| \leq |k| + |X| + \omega$.

6. Let $\mathbb{Z}$ be the ring of integers, and let $\phi(x)$ be a formula in the language of $\mathbb{Z}$ which expresses 'x is a prime number'. (a) Show that $\mathbb{Z}$ has an elementary extension in which some element not in $\mathbb{Z}$ satisfies ϕ. [There is a type containing all the formulas $x > 0$, $x > 1$, ... and ϕ.] *Elements not in $\mathbb{Z}$ which satisfy ϕ are known as* **nonstandard primes**, *as opposed to the* **standard primes** *(i.e. the usual ones).* (b) Show that $\mathbb{Z}$ has an elementary extension in which some non-zero element is divisible by every standard prime.

7. Suppose B is an abelian group and X a subgroup of B. The group B is said to be an **essential extension** of X if for every non-trivial subgroup A of B, $A \cap X \neq \{0\}$. Show that B is an essential extension of X if and only if B omits the type $\{nx \neq a: n \in \mathbb{Z}, a \in X \setminus \{0\}\}$ (which is a 1-type over X with respect to B).

The next three exercises illustrate how we can use the realisation of types to show that certain things aren't first-order expressible.

8. Let L be the language of groups. Show that there is no formula $\phi(x, y)$ of L such that if G is a group and g, h are elements of G then g has the same order as h if and only if $G \vDash \phi(g, h)$. [If a group G contains elements g, h of arbitrarily high finite orders, such that $G \vDash \neg \phi(g, h)$, then compactness gives us an elementary extension of G with two elements g', h' of infinite order such that $G \vDash \neg \phi(g', h')$.]

A group G is **of bounded exponent** *if there is some $n < \omega$ such that $g^n = 1$ for every element g of G.*

9. Let G be a group. Show that the following are equivalent. (a) G is not of bounded exponent. (b) G has an elementary extension in which some element has infinite order.

10. Show that 'simple group' is not first-order expressible, by finding groups G, H such that $G \equiv H$ and G is simple but H is not simple. [Let G be the simple group consisting of all permutations of ω which move just finitely many elements and are even permutations. Using relativisation (see section 4.2), take A to be a structure which contains G, ω and the action of G on ω, and let B be an elementary extension of A which contains an extension H of G in which some element moves infinitely many points. The subgroup of elements moving just finitely many points is normal.]

11. Let L be a first-order language with just finitely many function and constant symbols and no relation symbols, and let V be a set of variables. Assume that the function symbols of L include two distinct 2-ary function symbols F and G. Let A be the term algebra of L with basis V, as in section 1.3. Show that A has an elementary extension in which there is (for example) an element of the form $F(x, F(x, G(x, F(x, F(x, G(x, \ldots) \ldots)$. *For Prolog programmers: what is the relation between this fact and the absence of the 'occur check' in the unification algorithm?*

5.3 Elementary amalgamation

An amalgamation theorem is a theorem of the following shape. We are given two models B, C of some theory T, and a structure A (not necessarily a model of T) which is embedded into both B and C. The theorem states that there is a third model D of T such that both B and C are embeddable into D by embeddings which agree on A:

(3.1)

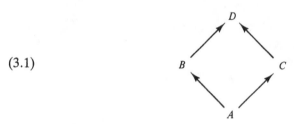

The embeddings may be required to preserve certain formulas.

It was Roland Fraïssé who first called attention to the diagram (3.1). Everything that has happened in model theory during the last thirty years has confirmed how important this diagram is. To see why, imagine some structure M and a small part A of M, and ask 'How does M sit around A?' For the answer, we need to know how A can be extended within M to structures B, C etc. But then we also need to know how any two of these extensions B, C of A are related to each other inside M: what formulas relate the elements of B to the elements of C? An amalgam (3.1) of B and C over A answers this last question.

There are two ways of using these ideas. One is to *build up* a structure M by taking smaller structures, extending them and then amalgamating the extensions. (Fraïssé did exactly this; see section 6.1 below, and section 8.2 on λ-saturated structures for a more general but slightly disguised version of the same idea.) The second way of using amalgamation is not to construct but to *classify*. We classify all the ways of extending the bottom structure A, and then we classify the ways of amalgamating these extensions. In favourable cases this leads to a structural classification of all the models of a theory. Stability theory follows this path; the reader will need to look at another book for details.

One amalgamation theorem is father to the rest. This is the elementary amalgamation theorem, Theorem 5.3.1 below. The other amalgamation results of first-order model theory differ from it in various ways: the maps are not necessarily elementary embeddings, they change the language, the amalgam is required to be strong, and so forth. This section will study the versions where all the maps are elementary. Sections 5.4 and 5.5 will tackle the remainder, so far as they are general results of model theory. Some particular classes of structures have their own particular amalgamation theorems, as we shall see in section 6.1.

Theorem 5.3.1 (*Elementary amalgamation theorem*). *Let L be a first-order language. Let B and C be L-structures and $\bar{a}$, $\bar{c}$ sequences of elements of B, C respectively, such that $(B, \bar{a}) \equiv (C, \bar{c})$. Then there exist an elementary extension D of B and an elementary embedding $g\colon C \to D$ such that $g\bar{c} = \bar{a}$. In a picture,*

(3.2)

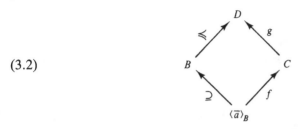

where $f\colon \langle \bar{a} \rangle \to C$ is the unique embedding which takes $\bar{a}$ to $\bar{c}$ (see the diagram lemma, Lemma 1.4.2).

Proof. Replacing C by an isomorphic copy if necessary, we can assume that $\bar{a} = \bar{c}$, and otherwise B and C have no elements in common. Consider the theory $T = \text{eldiag}(B) \cup \text{eldiag}(C)$, where each element names itself.

We claim that T has a model. By the compactness theorem (Theorem 5.1.1), it suffices to show that every finite subset of T has a model. Let T_0 be a finite subset of T. Then T_0 contains just finitely many sentences from eldiag(C). Let their conjunction be $\phi(\bar{a}, \bar{d})$ where $\phi(\bar{x}, \bar{y})$ is a formula of L and $\bar{d}$ consists of pairwise distinct elements which are in C but not in $\bar{a}$. (Of course only finitely many variables in $\bar{x}$ occur free in ϕ.) If T_0 has no model then $\text{eldiag}(B) \vdash \neg \phi(\bar{a}, \bar{d})$. By the lemma on constants (Lemma 2.3.2), since the elements $\bar{d}$ are distinct and they are not in B, $\text{eldiag}(B) \vdash \forall \bar{y} \, \neg \phi(\bar{a}, \bar{y})$. But then $(B, \bar{a}) \vDash \forall \bar{y} \, \neg \phi(\bar{a}, \bar{y})$, and so $(C, \bar{c}) \vDash \forall \bar{y} \, \neg \phi(\bar{c}, \bar{y})$ by the theorem assumption. This contradicts that $\phi(\bar{a}, \bar{d})$ is in eldiag(C). The claim is proved.

Let D^+ be a model of T and let D be the reduct $D^+|L$. By the elementary diagram lemma (Lemma 2.5.3), since $D^+ \vDash \text{eldiag}(B)$, we can assume that D is an elementary extension of B and $b^{D^+} = b$ for all elements b of B. Define $g(d) = d^{D^+}$ for each element d of C. Then by the elementary diagram lemma again, since $D^+ \vDash \text{eldiag}(C)$, g is an elementary embedding of C into D. Finally $g\bar{c} = \bar{a}^{D^+} = \bar{a}$. $\qquad\square$

Note that $\bar{a}$ can be empty in Theorem 5.3.1. In this case the theorem says that any two elementarily equivalent structures can be elementarily embedded together into some structure.

Note also that the theorem can be rephrased as follows. If $(B, \bar{a}) \equiv (C, \bar{c})$ and $\bar{d}$ is any sequence of elements of C, then there is an elementary extension B' of B containing elements $\bar{b}$ so that $(B', \bar{a}, \bar{b}) \equiv (C, \bar{c}, \bar{d})$. Chapter 8 below is an extended meditation on this fact.

One of the most important consequences of the elementary amalgamation theorem is that if A is any structure, we can simultaneously realise all the complete types with respect to A in one and the same elementary extension of A, as follows.

Corollary 5.3.2. *Let L be a first-order language and A an L-structure. Then there is an elementary extension B of A such that every type over $\mathrm{dom}(A)$ with respect to A is realised in B.*

Proof. It suffices to realise all the maximal types over $\mathrm{dom}(A)$ with respect to A. Let these be p_i $(i < \lambda)$ with λ a cardinal. For each $i < \lambda$ let $\bar{a}_i$ be a tuple in an elementary extension A_i of A, such that p_i is $\mathrm{tp}_{A_i}(\bar{a}_i / \mathrm{dom}\, A)$. Define an elementary chain $(B_i : i \leqslant \lambda)$ by induction as follows. B_0 is A, and for each limit ordinal $\delta \leqslant \lambda$, $B_\delta = \bigcup_{i < \delta} B_i$ (which is an elementary extension of each B_i by Theorem 2.5.2). When B_i has been defined and $i < \lambda$, choose B_{i+1} to be an elementary extension of B_i such that there is an elementary embedding $e_i : A_i \to B_{i+1}$ which is the identity on A; this is possible by Theorem 5.3.1. Put $B = B_\lambda$. Then for each $i < \lambda$, $e_i(\bar{a}_i)$ is a tuple in B_λ which realises p_i.   □

In section 8.2 below we shall refine this corollary in several ways.

Heir–coheir amalgams

Consider the case of Theorem 5.3.1 where $\bar{a}$ lists the elements of an elementary substructure A of B. In this case the theorem tells us that if A, B and C are L-structures and $A \leqslant B$ and $A \leqslant C$, then there are an elementary extension D of B and an elementary embedding $g : C \to D$ such that, putting $C' = gC$, the following diagram of elementary inclusions commutes:

(3.3)

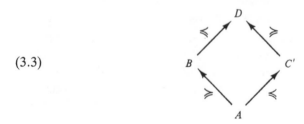

But actually we can say more in this case. Call the commutative diagram (3.3) of L-structures and elementary inclusions an **heir–coheir amalgam** if

(3.4)   for every first-order formula $\psi(\bar{x}, \bar{y})$ of L and all tuples $\bar{b}$, $\bar{c}$ from B, C' respectively, if $D \vDash \psi(\bar{b}, \bar{c})$ then there is $\bar{a}$ in A such that $B \vDash \psi(\bar{b}, \bar{a})$.

We say also that (3.3) is an **heir–coheir amalgam of B and C over A**; in fact it is an **heir–coheir amalgam of B'' and C'' over A** whenever B'' and C'' are elementary extensions of A such that there are isomorphisms $i: B'' \to B$ and $j: C'' \to C'$ which are the identity on A.

Example 1: _Vector spaces._ Suppose A is an infinite vector space over a field K, and B and C are vector spaces with A as subspace. Put $B = B_1 \oplus A$ and $C = C_1 \oplus A$. We can amalgamate B and C over A by putting $D = B_1 \oplus C_1 \oplus A$. Suppose some equation $\sum_{i<m}\lambda_i b_i = \sum_{j<n}\mu_j c_j$ holds in D, where the b_i are in B and the c_j are in C. Let $\pi: D \to B_1 \oplus A$ be the projection along C_1. Then $\sum_{i<m}\lambda_i\pi(b_i) = \sum_{j<n}\mu_j\pi(c_j)$. But $\pi(b_i) = b_i$ and $\pi(c_j)$ lies in A. Thus (3.4) holds when ψ is the formula $\sum_{i<m}\lambda_i x_i = \sum_{j<n}\mu_j y_j$. In fact since A is infinite, one can show that (3.4) holds whenever ψ is quantifier-free; but don't check this by linear algebra – it will fall out almost trivially from the next theorem. By quantifier elimination (see Exercise 2.7.9) it follows that D forms an heir–coheir amalgam of B and C over A. This example is typical: heir–coheir amalgams are 'as free as possible'.

The next theorem says that heir–coheir amalgams always exist when B and C are elementary extensions of A.

Theorem 5.3.3. _Let A, B and C be L-structures such that $A \preccurlyeq B$ and $A \preccurlyeq C$. Then there exist an elementary extension D of B and an elementary embedding $g: C \to D$ such that (3.3) (with $C' = gC$) is an heir–coheir amalgam._

Proof. Much like the proof of Theorem 5.3.1. As before, we assume that $(\mathrm{dom}\ B) \cap (\mathrm{dom}\ C) = \mathrm{dom}(A)$, so that constants behave properly in diagrams. Then we take T to be the theory

(3.5) $\mathrm{eldiag}(B) \cup \mathrm{eldiag}(C) \cup$
          $\{\neg\psi(\bar{b}, \bar{c}): \psi$ is a first-order formula of L and $\bar{b}$ is a tuple in
          B such that $B \vDash \neg\psi(\bar{b}, \bar{a})$ for all $\bar{a}$ in $A\}$.

If T has no model, then by the compactness theorem there are a tuple $\bar{a}$ from A, a tuple $\bar{d}$ of distinct elements in C but not in A, a tuple $\bar{b}$ of elements of B, a sentence $\theta(\bar{a}, \bar{d})$ in $\mathrm{eldiag}(C)$ and sentences $\psi_i(\bar{b}, \bar{a}, \bar{d})$ $(i < k)$ such that $B \vDash \neg\psi_i(\bar{b}, \bar{a}', \bar{a}'')$ for all $\bar{a}'$, $\bar{a}''$ in A, and

(3.6) $\mathrm{eldiag}(B) \vdash \theta(\bar{a}, \bar{d}) \to \psi_0(\bar{b}, \bar{a}, \bar{d}) \vee \ldots \vee \psi_{k-1}(\bar{b}, \bar{a}, \bar{d})$.

Quantifying out the constants $\bar{d}$ by the lemma on constants, we have

(3.7) $B \vDash \forall\bar{y}(\theta(\bar{a}, \bar{y}) \to \psi_0(\bar{b}, \bar{a}, \bar{y}) \vee \ldots \vee \psi_{k-1}(\bar{b}, \bar{a}, \bar{y}))$.

But also $C \vDash \exists\bar{y}\ \theta(\bar{a}, \bar{y})$, so $A \vDash \exists\bar{y}\ \theta(\bar{a}, \bar{y})$ and hence there is a tuple $\bar{a}''$ in A such that $A \vDash \theta(\bar{a}, \bar{a}'')$ and so $B \vDash \theta(\bar{a}, \bar{a}'')$. Applying (3.7), we infer that $B \vDash \psi_i(\bar{b}, \bar{a}, \bar{a}'')$ for some $i < k$; contradiction. The rest of the proof is as before. $\qquad\square$

If (3.3) is an heir–coheir amalgam, then the overlap of B and C' in D is precisely A. For suppose $b = g(c)$ for some b in B and some c in C'. Then by the heir–coheir property, $b = a$ for some a in A. Amalgams with this minimum-overlap property are said to be **strong**. In this terminology we have just shown that first-order logic has the **strong elementary amalgamation property**.

Example 1 *continued.* Here is a more abstract proof that D is an heir–coheir amalgam of B and C over A. Since A is infinite, $A \preccurlyeq B$ and $A \preccurlyeq C$ by quantifier elimination (Exercise 2.7.9). So by the theorem, some vector space D' forms an heir–coheir amalgam of B and C over A. Identifying B and C with their images in D', we can suppose that B and C generate D'; for if D'' is the subspace of D' generated by B and C, then $D'' \preccurlyeq D'$ by quantifier elimination again. Now D' is a strong amalgam of B and C over A. But this means precisely that $D' = B_1 \oplus C_1 \oplus A$, so that D' is D. Thus D is an heir–coheir amalgam of B and C over A.

When do two extensions of a structure have a strong amalgam?

Example 2: *Algebraically closed fields.* Suppose that in (3.1), both B and C are the field of complex numbers, A is the field of reals and D is some algebraically closed field which amalgamates B and C over A. Let i, $-$i be the square roots of -1 regarded as elements of B, and j, $-$j the same elements thought of as in C. Then in D, i must be identified with either j or $-$j, and so the amalgam is not strong.

The moral of this example is that if $\langle a \rangle_B$ in Theorem 5.3.1 is not algebraically closed in B, then in general there is no hope of making the amalgam strong in that theorem. To turn this remark from a moral into a theorem, we need to formalise the notion of 'algebraically closed'.

Strong amalgams and algebraicity

Let B be an L-structure and X a set of elements of B. We say that an element b of B is **algebraic over** X if there are a first-order formula $\phi(x, \bar{y})$ of L and a tuple $\bar{a}$ in X such that $B \models \phi(b, \bar{a}) \wedge \exists_{\leqslant n} x\, \phi(x, \bar{a})$ for some finite n. We write $\mathrm{acl}_B(X)$ for the set of all elements of B which are algebraic over X. If $\bar{a}$ lists the elements of X, we also write $\mathrm{acl}_B(\bar{a})$ for $\mathrm{acl}_B(X)$. We say that X is **algebraically closed** if $X = \mathrm{acl}_B(X)$.

The following facts are an exercise (Exercise 6):

(3.8) $X \subseteq \mathrm{acl}_B(X)$.

(3.9) $Y \subseteq \mathrm{acl}_B(X)$ implies $\mathrm{acl}_B(Y) \subseteq \mathrm{acl}_B(X)$.

(3.10)   if $B \leqslant C$ then $\mathrm{acl}_B(X) = \mathrm{acl}_C(X)$.

By (3.10) we can often write $\mathrm{acl}(X)$ for $\mathrm{acl}_B(X)$ without danger of confusion.

We say that a tuple $\bar{b}$ is **algebraic over** X if every element in $\bar{b}$ is algebraic over X. We say that a type $\Phi(\bar{x})$ over a set X with respect to B is **algebraic** if every tuple realising it is algebraic over X.

Theorem 5.3.5 below will tell us that we can make the amalgam strong in Theorem 5.3.1 whenever the bottom structure $\langle \bar{a} \rangle_B$ is algebraically closed in B (or in C, by symmetry).

Lemma 5.3.4. *Let B be an L-structure, X a set of elements of B listed as $\bar{a}$, and b an element of B. Suppose $b \notin \mathrm{acl}(X)$.*

(a) There is an elementary extension A of B with an element $c \notin \mathrm{dom}(B)$ such that $(B, \bar{a}, b) \equiv (A, \bar{a}, c)$.

(b) There is an elementary extension D of B with an elementary substructure C containing X such that $b \notin \mathrm{dom}(C)$.

Proof. (a) Let c be a new constant and let $p(x)$ be the complete type of b over X. It suffices to show that the theory

(3.11) $\mathrm{eldiag}(B) \cup p(c) \cup \{c \neq d : d \in \mathrm{dom}(B)\}$

has a model. But if it has none, then by the compactness theorem and the lemma on constants, there are finitely many elements $d_0, \ldots, d_{n-1}$ of B and a formula $\phi(x)$ of $p(x)$ (noting that $p(x)$ is closed under $\wedge$), such that

(3.12) $\mathrm{eldiag}(B) \vdash \forall x(\phi(x) \rightarrow x = d_0 \vee \ldots \vee x = d_{n-1})$.

Hence $B \vDash \phi(b) \wedge \exists_{\leqslant n} x\, \phi(x)$, so that $b \in \mathrm{acl}(X)$; contradiction.

(b) Take A and c as in part (a). Since $(A, \bar{a}, b) \equiv (A, \bar{a}, c)$, Theorem 5.3.1 gives us an elementary extension D of A and an elementary embedding $g: A \rightarrow D$ such that $g\bar{a} = \bar{a}$ and $gb = c$. Then D is an elementary extension of gB and $gb = c \notin \mathrm{dom}(B)$. So we get the lemma by taking gB, B for B, C respectively. $\square$

Theorem 5.3.5 (Strong elementary amalgamation over algebraically closed sets). *Let B and C be L-structures and $\bar{a}$ a sequence of elements in both B and C such that $(B, \bar{a}) \equiv (C, \bar{a})$. Then there exist an elementary extension D of B and an elementary embedding $g: C \rightarrow D$ such that $g\bar{a} = \bar{a}$ and $(\mathrm{dom}\, B) \cap g(\mathrm{dom}\, C) = \mathrm{acl}_B(\bar{a})$.*

Proof. Start by repeating the proof of Theorem 5.3.1, but adding to T all the sentences '$b \neq c$' where b is in B but not in $\mathrm{acl}_B(\bar{a})$ and c is in C but not in $\mathrm{acl}_C(\bar{a})$. Write T^+ for this enlarged theory. Suppose D and g are defined using T^+ in place of T. Then $g\bar{a} = \bar{a}$, and it easily follows that g maps $\mathrm{acl}_C(\bar{a})$ onto $\mathrm{acl}_B(\bar{a})$. Thus we have $\mathrm{acl}_B(\bar{a}) \subseteq (\mathrm{dom}\, B) \cap g(\mathrm{dom}\, C)$, and the sentences '$b \neq c$' guarantee the opposite inclusion. It remains only to show that T^+ has a model.

Assume for contradiction that T^+ has no model. Then by the compactness theorem there are finite subsets Y of $\mathrm{dom}(B) \backslash \mathrm{acl}_B(\bar{a})$ and Z of $\mathrm{dom}(C) \backslash \mathrm{acl}_C(\bar{a})$, such that for every elementary extension D of B and elementary embedding $g \colon C \to D$ with $g\bar{a} = \bar{a}$, $Y \cap g(Z) \neq \varnothing$. Choose D and g to make $Y \cap g(Z)$ as small as possible. To save notation we can assume that g is the identity, so that $C \preccurlyeq D$.

Since $Y \cap Z \neq \varnothing$, there is some $b \in Y \cap Z$. By the lemma, there is an elementary extension D' of D with an elementary substructure C' containing $\bar{a}$ such that $b \notin \mathrm{dom}(C')$. Applying Theorem 5.3.3 to the elementary embedding $C' \preccurlyeq D'$ (the same embedding twice over), we find an elementary extension E of D' and an elementary embedding $e \colon D' \to E$ which is the identity on C', such that $(\mathrm{dom}\, D') \cap e(\mathrm{dom}\, D') = \mathrm{dom}(C')$. Now $Y \cap e(Z) \subseteq Y \cap Z$; for if $d \in Y \cap e(Z)$ then d is in C' and hence $ed = d$. But b is in $(Y \cap Z) \backslash (Y \cap e(Z))$; for since b is in D' but not in C', $b \notin e(\mathrm{dom}\, D')$ and hence $b \notin e(Z)$. Thus e contradicts the choice of $Y \cap g(Z)$ as minimal.$\square$

Exercises for section 5.3

1. Let T be the theory of dense linear orderings without endpoints. Describe the heir–coheir amalgams of models of T. [In D, any interval (b_1, b_2) with b_1, b_2 in B which contains elements of C must also contain elements of A. If b in B is $>$ all A, then b is $>$ all C too; and likewise with $<$ for $>$.]

2. Let T be the theory of the linear ordering of the integers. Describe the heir–coheir amalgams of models of T.

*Let A, B and D be L-structures with $A \preccurlyeq B \preccurlyeq D$, and let $\bar{d}$ be a tuple in D. Put $p = \mathrm{tp}_D(\bar{d}/A)$ and $p^+ = \mathrm{tp}_D(\bar{d}/B)$. We say that p^+ is an **heir** of p if for every formula $\phi(\bar{x}, \bar{y})$ of L with parameters in A, and every tuple $\bar{b}$ in B such that $\phi(\bar{x}, \bar{b}) \in p^+$, there is $\bar{a}$ in A such that $\phi(\bar{x}, \bar{a}) \in p$. We say that p^+ is a **coheir** of p if for every formula $\phi(\bar{x}, \bar{y})$ of L and tuple $\bar{b}$ in B such that $\phi(\bar{x}, \bar{b}) \in p^+$, there is $\bar{a}$ in A such that $D \vDash \phi(\bar{a}, \bar{b})$.*

3. Show that the following are equivalent, given the amalgam (3.3) above. (a) The amalgam is heir–coheir. (b) For every tuple $\bar{b}$ in B, $\mathrm{tp}_D(\bar{b}/C)$ is an heir of $\mathrm{tp}_D(\bar{b}/A)$. (c) For every tuple $\bar{c}$ in C', $\mathrm{tp}_D(\bar{c}/B)$ is a coheir of $\mathrm{tp}_D(\bar{c}, A)$. *The type notions introduce a false asymmetry between the two sides of the amalgam. But they do allow us to think about one tuple at a time, and this can be an advantage.*

4. Let $\Phi(\bar{x})$ be a type over a set X with respect to a structure A. Show that the following are equivalent. (a) Φ is algebraic. (b) Φ contains a formula ϕ such that $A \vDash \exists_{\leqslant n} \bar{x}\, \phi(\bar{x})$ for some finite n. (c) In every elementary extension of A, at most finitely many tuples realise Φ.

5. Let L be a first-order language and A an L-structure. Suppose X is a set of elements of A and $\bar{a}$ is a tuple of elements of A, none of which are algebraic over A. Show that some elementary extension B of A contains infinitely many pairwise disjoint tuples $\bar{a}_i$ ($i < \omega$) which all realise $\mathrm{tp}_A(\bar{a}/X)$. [Iterate Theorem 5.3.5.]

6. Let B be an L-structure, C an elementary extension of B and X, Y sets of elements of B. (a) Prove (3.8), (3.9) and (3.10). (b) Deduce that $\mathrm{acl}_B\mathrm{acl}_B(X) = \mathrm{acl}_B(X)$.

7. Let A be an L-structure and X a set of elements of A. Show that there is an elementary extension B of A with a descending sequence $(C_i : i < \omega)$ of elementary substructures such that $\mathrm{acl}_A(X) = \bigcap_{i<\omega} \mathrm{dom}(C_i)$. [Use Theorem 5.3.5 to build up a chain of strong amalgams

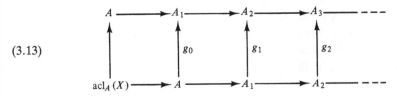

(3.13)

where the g_i are elementary embeddings and the horizontal maps are elementary inclusions. Put $B = \bigcup_{n<\omega} A_n$, $g = \bigcup_{n<\omega} g_n$ and $C_i = g^i B$.]

8. Let L be a first-order language and T a theory in L. Show that the following are equivalent. (a) If A is a model of T, then the intersection of any two elementary substructures of A is again an elementary substructure of A. (b) If A is a model of T, then the intersection of any family of elementary substructures of A is again an elementary substructure of A. (c) If A is a model of T and $(B_i : i < \gamma)$ is a descending sequence of elementary substructures of A then the intersection of the B_i is again an elementary substructure of A. (d) If A is any model of T and X is a set of elements of A then $\mathrm{acl}_A(X)$ is an elementary substructure of A. (e) For any formula $\phi(x, \bar{y})$ of L there are a formula $\psi(x, \bar{y})$ of L and an integer n such that $T \vdash \forall x \bar{y}(\phi \to \exists x(\phi \wedge \psi) \wedge \exists_{\leqslant n} x \, \psi)$.

5.4 Amalgamation and preservation

We have a fair amount of work to do in this section and the next. But fortunately most of the proofs are routine variations, either of each other or of the arguments of section 5.3.

If A and B are L-structures, we write $A \Rightarrow_1 B$ to mean that for every first-order existential sentence ϕ of L, if $A \vDash \phi$ then $B \vDash \phi$. Likewise we write $A \Rightarrow_1^+ B$ to mean the same for every first-order $\exists_1^+$ sentence of L. Note that $\Rightarrow_1$ implies $\Rightarrow_1^+$. Note also that if $f: \langle \bar{a} \rangle_B \to C$ is a homomorphism, then the assertion $(C, f\bar{a}) \Rightarrow_1^+ (B, \bar{a})$ implies that f is an embedding.

Theorem 5.4.1 (*Existential amalgamation theorem*). *Let B and C be L-structures, $\bar{a}$ a sequence of elements of B and $f: \langle \bar{a} \rangle \to C$ a homomorphism such that $(C, f\bar{a}) \Rrightarrow_1 (B, \bar{a})$. Then there exist an elementary extension D of B and an embedding $g: C \to D$ such that $gf\bar{a} = \bar{a}$. In a picture,*

(4.1)

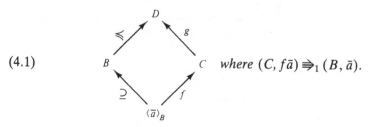

where $(C, f\bar{a}) \Rrightarrow_1 (B, \bar{a})$.

Proof. The assumptions imply that f is an embedding, so that we can replace C by an isomorphic copy and assume that f is the identity on $\langle \bar{a} \rangle_B$, and that $\langle \bar{a} \rangle_B$ is the overlap of dom(B) and dom(C). By the same argument as for Theorem 5.3.1, it suffices to show that the theory $T = \text{eldiag}(B) \cup \text{diag}(C)$ has a model. Again as in the proof of Theorem 5.3.1, if T has no model, then by the compactness theorem there is a conjunction $\phi(\bar{a}, \bar{d})$ of finitely many sentences in diag(C), such that $(B, \bar{a}) \vDash \neg \exists \bar{y}\, \phi(\bar{a}, \bar{y})$. Since $\phi(\bar{a}, \bar{y})$ is quantifier-free and $(C, \bar{a}) \Rrightarrow_1 (B, \bar{a})$, we infer that $(C, \bar{a}) \vDash \neg \exists \bar{y}\, \phi(\bar{a}, \bar{y})$. This contradicts that $\phi(\bar{a}, \bar{d})$ is true in C, and so the proof is complete. $\qquad \square$

Since we allow structures to be empty, the tuple $\bar{a}$ in Theorem 5.4.1 can be the empty tuple. This gives the following.

Corollary 5.4.2. *Let B and C be L-structures such that $C \Rrightarrow_1 B$. Then C is embeddable in some elementary extension of B.* $\qquad \square$

Amalgamation theorems like Theorem 5.4.1 tend to spawn offspring of the following kinds: (i) criteria for a structure to be expandable or extendable in certain ways, (ii) syntactic criteria for a formula or set of formulas to be preserved under certain model-theoretic operations (results of this kind are called **preservation theorems**), (iii) interpolation theorems. Let me illustrate all these.

(i) We give a criterion for a structure to be extendable to a model of a given theory. If T is a theory in a first-order language L, then we write $T_\forall$ for the set of all $\forall_1$ sentences of L which are consequences of T.

Corollary 5.4.3. *If T is a theory in a first-order language L, then the models of $T_\forall$ are precisely the substructures of models of T.*

Proof. Any substructure of a model of T is certainly a model of $T_\forall$, by Corollary 2.4.2. Conversely let C be a model of $T_\forall$. To show that C is a substructure of a model of T, it suffices to find a model B of T such that $C \Rightarrow_1 B$, and then quote Corollary 5.4.2.

We find B as follows. Let U be the set of all $\exists_1$ sentences ϕ of L such that $C \models \phi$. We claim that $T \cup U$ has a model. For if not, then by the compactness theorem there is some finite set $\{\phi_0, \ldots, \phi_{k-1}\}$ of sentences in U such that $T \vdash \neg \phi_0 \vee \ldots \vee \neg \phi_{k-1}$. Now $\neg \phi_0 \vee \ldots \vee \neg \phi_{k-1}$ is logically equivalent to an $\forall_1$ sentence θ, and $T \vdash \theta$, so that $\theta \in T_\forall$ and hence $C \models \theta$. But this is absurd, since $C \models \phi_i$ for each $i < k$. So $T \cup U$ has a model as claimed. $\qquad \square$

(ii) We characterise those formulas or sets of formulas which are preserved in substructures.

Theorem 5.4.4 (Łoś–Tarski theorem). *Let T be a theory in a first-order language L and $\Phi(\bar{x})$ a set of formulas of L. (The sequence of variables $\bar{x}$ need not be finite.) Then the following are equivalent.*
(a) *If A and B are models of T, $A \subseteq B$, $\bar{a}$ is a sequence of elements of A and $B \models \bigwedge \Phi(\bar{a})$, then $A \models \bigwedge \Phi(\bar{a})$. ('$\Phi$ is preserved in substructures for models of T.')*
(b) *Φ is equivalent modulo T to a set $\Psi(\bar{x})$ of $\forall_1$ formulas of L.*

Proof. (b) $\Rightarrow$ (a) is by Corollary 2.4.2(a). For the converse, assume (a). We first prove (b) under the assumption that Φ is a set of sentences. Let Ψ be $(T \cup \Phi)_\forall$. By Corollary 5.4.3, confining ourselves to models of T, the models of Ψ are precisely the substructures of models of Φ. But by (a), every such substructure is itself a model of Φ. So Φ and Ψ are equivalent modulo T.

For the case where $\bar{x}$ is not empty, form the language $L(\bar{c})$ by adding new constants $\bar{c}$ to L. If $\Phi(\bar{x})$ is preserved in substructures for L-structures which are models of T, then it's not hard to see that $\Phi(\bar{c})$ must be preserved in substructures for $L(\bar{c})$-structures which are models of T. But $\Phi(\bar{c})$ is a set of sentences, so the previous argument shows that $\Phi(\bar{c})$ is equivalent modulo T to a set $\Psi(\bar{c})$ of $\forall_1$ sentences of $L(\bar{c})$. Then by the lemma on constants (Lemma 2.3.2), $T \vdash \forall \bar{x}(\bigwedge \Phi(\bar{x}) \leftrightarrow \bigwedge \Psi(\bar{x}))$, so that $\Phi(\bar{x})$ is equivalent to $\Psi(\bar{x})$ modulo T, in the language $L(\bar{c})$ and hence also in the language L. (If there are empty L-structures, they trivially satisfy $\forall \bar{x}(\bigwedge \Phi(\bar{x}) \leftrightarrow \bigwedge \Psi(\bar{x}))$.) $\qquad \square$

If Φ in the Łoś–Tarski theorem is a single formula, then one more application of the compactness theorem boils Ψ down to a single $\forall_1$ formula. In short, modulo any first-order theory T, the formulas preserved in substructures are precisely the $\forall_1$ formulas. The Łoś–Tarski theorem is often quoted in this form.

Since $\exists_1$ formulas are up to logical equivalence just the negations of $\forall_1$ formulas, this version of the theorem immediately implies the following.

Corollary 5.4.5. *If T is a theory in a first-order language L and ϕ is a formula of L, then the following are equivalent:*
(a) ϕ is preserved by embeddings between models of T;
(b) ϕ is equivalent modulo T to an $\exists_1$ formula of L.   □

Actually the full dual of Theorem 5.4.4 is true too, with sets of $\exists_1$ formulas rather than single $\exists_1$ formulas. This can be an exercise (Exercise 1).

(iii) The interpolation theorem associated with Theorem 5.4.1 is a fancy elaboration of the Łoś–Tarski theorem (which it obviously implies, in the case of single formulas).

Theorem 5.4.6. *Let T be a theory in a first-order language L and let $\phi(\bar{x})$, $\chi(\bar{x})$ be formulas of L. Then the following are equivalent.*
(a) Whenever $A \subseteq B$, A and B are models of T, $\bar{a}$ is a tuple in A and $B \vDash \phi(\bar{a})$, then $A \vDash \chi(\bar{a})$.
(b) There is an $\forall_1$ formula $\psi(\bar{x})$ of L such that $T \vdash \forall \bar{x}(\phi \rightarrow \psi) \wedge \forall \bar{x}(\psi \rightarrow \chi)$.
(ψ is an 'interpolant' between ϕ and χ.)

Proof. The obvious adaptation of the proof of Theorem 5.4.4 works.   □

Variants of existential amalgamation

Theorem 5.4.1 has infinitely many variants for different classes of formulas, with only trivial changes in the proof. Each of these variants has its own preservation and interpolation theorems; some are described in the exercises. I quote two variants of Theorem 5.4.1 without proof.

Theorem 5.4.7. *Let L be a first-order language, and let B and C be L-structures, $\bar{a}$ a sequence of elements of C and $f: \langle \bar{a} \rangle_C \rightarrow B$ a homomorphism such that $(C, \bar{a}) \Rightarrow_1^+ (B, f\bar{a})$. Then there exist an elementary extension D of B and a homomorphism $g: C \rightarrow D$ which extends f.*   □

Let L be a first-order language and let A, B be L-structures. We write $A \Rightarrow_2 B$ to mean that for every $\exists_2$ sentence ϕ of L, if $A \vDash \phi$ then $B \vDash \phi$; equivalently, for every $\forall_2$ sentence ϕ of L, if $B \vDash \phi$ then $A \vDash \phi$.

Theorem 5.4.8. *Let L be a first-order language, B and C L-structures, $\bar{a}$ a sequence of elements of B and f: $\langle \bar{a} \rangle_B \rightarrow C$ an embedding such that $(C, f\bar{a}) \Rrightarrow_2 (B, \bar{a})$. Then there exist an elementary extension D of B and an embedding g: $C \rightarrow D$ such that g preserves all $\forall_1$ formulas of L.* $\square$

Theorem 5.4.8 is used to characterise the formulas which are preserved in unions of chains.

Theorem 5.4.9 (Chang–Łoś–Suszko theorem). *Let T be a theory in a first-order language L, and $\Phi(\bar{x})$ a set of formulas of L. Then the following are equivalent.*
(a) $\bigwedge \Phi$ *is preserved in unions of chains $(A_i: i < \gamma)$ whenever $\bigcup_{i<\gamma} A_i$ and all the A_i $(i < \gamma)$ are models of T.*
(b) Φ *is equivalent modulo T to a set of $\forall_2$ formulas of L.*

Proof. (b) $\Rightarrow$ (a) is by Theorem 2.4.4. For the other direction, assume (a). Just as in the proof of Theorem 5.4.4, we can assume that Φ is a set of sentences. Let Ψ be the set of all $\forall_2$ sentences of L which are consequences of $T \cup \Phi$. We have to show that $T \cup \Psi \vdash \Phi$, and for this it will be enough to prove that every model of $T \cup \Psi$ is elementarily equivalent to a union of some chain of models of $T \cup \Phi$ which is itself a model of T.

Let A_0 be any model of $T \cup \Psi$. We shall construct an elementary chain $(A_i: i < \omega)$, extensions $B_i \supseteq A_i$ and embeddings g_i: $B_i \rightarrow A_{i+1}$, so that the following diagram commutes:

(4.2)

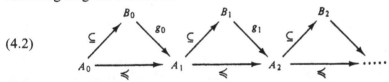

We shall require that for each $i < \omega$,

(4.3) $B_i \models T \cup \Phi$, and $(B_i, \bar{a}_i) \Rrightarrow_1 (A_i, \bar{a}_i)$
         when $\bar{a}_i$ lists all the elements of A_i.

The diagram is constructed as follows. Suppose A_i has been chosen. Then since $A_0 \preccurlyeq A_i$, A_i is a model of all the $\forall_2$ consequences of $T \cup \Phi$. By exactly the argument of Corollary 5.4.3 (with $\forall_2$ in place of $\forall_1$ and Theorem 5.4.8 in place of Theorem 5.4.1), it follows that A_i can be extended to a structure B_i satisfying (4.3). Then by Theorem 5.4.1 and the second part of (4.3), there are an elementary extension A_{i+1} of A_i and an embedding g_i: $B_i \rightarrow A_{i+1}$ such that g is the identity on A_i.

Now in (4.2) we can replace each B_i by its image under g_i, and so assume that all the maps are inclusions. Then $\bigcup_{i<\omega} A_i$ and $\bigcup_{i<\omega} B_i$ are the same structure C. By the Tarski–Vaught elementary chain theorem (Theorem 2.5.2), $A_0 \preccurlyeq C$. So C is a model of T and the union of a chain of models B_i of $T \cup \Phi$, and A_0 is elementarily equivalent to C, as required. $\square$

Just as with the Łoś–Tarski theorem, compactness gives us a finite version: a formula ϕ of L is preserved in unions of chains (where all the structures are models of T) if and only if ϕ is equivalent modulo T to an $\forall_2$ formula of L.

Exercises for section 5.4

1. Let T be a theory in a first-order language L and $\Phi(\bar{x})$ a set of formulas of L. Show that the following are equivalent. (a) If A and B are models of T, $A \subseteq B$, $\bar{a}$ is a sequence of elements of A and $A \models \bigwedge \Phi(\bar{a})$, then $B \models \bigwedge \Phi(\bar{a})$. (b) Φ is equivalent modulo T to a set of $\exists_1$ formulas of L. [Assuming Φ is a set of sentences, let Ψ be the set of $\exists_1$ consequences of $T \cup \Phi$. Take B in Theorem 5.4.1 to be a model of $T \cup \Psi$, and C a model of $T \cup \Phi$ which we can embed into some elementary extension of B. Cf. Corollary 8.3.2 for a different proof using λ-saturated models.]

2. Let L be a first-order language and T a theory in L. Suppose A and B are models of T. Show that the following are equivalent. (a) There is a model C of T such that both A and B can be embedded in C. (b) ϕ and ψ are $\forall_1$ sentences of L such that $T \vdash \phi \lor \psi$, then either (i) A and B are both models of ϕ, or (ii) A and B are both models of ψ.

Let L be a first-order language, and let A, B be L-structures. For any $n < \omega$, we write $A \preccurlyeq_n B$ to mean that $A \subseteq B$ and for every $\exists_n$ formula $\phi(\bar{x})$ of L and every tuple $\bar{a}$ of elements of A, $B \models \phi(\bar{a}) \Rightarrow A \models \phi(\bar{a})$. A $\preccurlyeq_n$-chain is a chain $(A_i: i < \gamma)$ in which $i < j < \gamma$ implies $A_i \preccurlyeq_n A_j$.

3. Let L be a first-order language and let A_0, A_1 be L-structures with $A_0 \subseteq A_1$. Let n be a positive integer. Show that $A_0 \preccurlyeq_{2n-1} A_1$ if and only if there is a chain $A_0 \subseteq \ldots \subseteq A_{2n}$ in which $A_i \preccurlyeq A_{i+2}$ for each i:

(4.4)

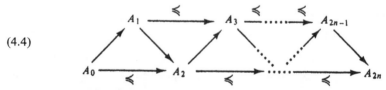

where all the arrows are inclusions.

4. Let L be a first-order language, T a theory in L, n an integer $\geqslant 2$ and $\phi(\bar{x})$ a formula of L. Show that the following are equivalent. (a) ϕ is equivalent modulo T to an $\forall_n$ formula $\psi(\bar{x})$ of L. (b) If A and B are models of T such that $A \preccurlyeq_{n-1} B$, and $\bar{a}$ is a tuple of elements of A such that $B \models \phi(\bar{a})$, then $A \models \phi(\bar{a})$. (c) ϕ is preserved in unions of $\preccurlyeq_{n-2}$-chains of models of T. [Atomise all $\exists_{n-1}$ formulas.]

5. Let L be a first-order language and T a theory in L. Show that the following are equivalent. (a) T is equivalent to a set of sentences of L of the form $\forall x \exists \bar{y}\, \phi(x, \bar{y})$ with ϕ quantifier-free. (b) If A is an L-structure and for every element a of A there is a substructure of A which contains a and is a model of T, then A is a model of T.

6. Let L be a first-order language and T a theory in L. Show that the following are equivalent. (a) Whenever A and B are models of T with $A \preccurlyeq B$, and $A \subseteq C \subseteq B$, then C is also a model of T. (b) Whenever A and B are models of T with $A \preccurlyeq_2 B$, and $A \subseteq C \subseteq B$, then C is also a model of T. (c) T is equivalent to a set of $\exists_2$ sentences.

7. Let T be a first-order theory, and suppose that whenever a model A of T has substructures B and C which are also models of T with non-empty intersection, then $B \cap C$ is also a model of T. Show that T is equivalent to an $\forall_2$ first-order theory. [Let $(D_i : i < \gamma)$ be a chain of models of T with union D. Show that D is a substructure of a model B of T. Let U be the theory consisting of T, $\mathrm{diag}(B)$ and sentences which say 'The elements satisfying Rx are a model of T', 'Rd' for all d in D and '$\neg Rb$' for all b in B but not in D. Show that U has a model A, and take the intersection of B in A with the structure defined by R.]

5.5 Expanding the language

We continue where we left off at the end of section 5.4, except that the next amalgamation result is about expansions rather than extensions.

Theorem 5.5.1. *Let L_1 and L_2 be first-order languages, $L = L_1 \cap L_2$, B an L_1-structure, C an L_2-structure, and $\bar{a}$ a sequence of elements of B and of C such that $(B|L, \bar{a}) \equiv (C|L, \bar{a})$. Then there are an $(L_1 \cup L_2)$-structure D such that $B \preccurlyeq D|L_1$, and an elementary embedding $g : C \to D|L_2$, such that $g\bar{a} = \bar{a}$.*

Proof. We start by noting that an almost invisible alteration of the proof of Theorem 5.3.1 (elementary amalgamation) gives a weak version of the theorem we want, viz.: Under the hypotheses of Theorem 5.5.1, there are an elementary extension D of B and an elementary embedding $g : C|L \to D|L$, such that $g\bar{a} = \bar{a}$. (It suffices to show that $\mathrm{eldiag}(B) \cup \mathrm{eldiag}(C|L)$ has a model.)

Put $B_0 = B$, $C_0 = C$ and use the weak version of the theorem, alternately from this side and from that, to build up a commutative diagram

(5.1)

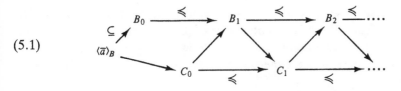

where the maps from B_i to C_i and from C_i to B_{i+1} are elementary embeddings of the L-reducts. The diagram induces an isomorphism $e: \bigcup_{i<\omega}B_i|L \rightarrow \bigcup_{i<\omega}C_i|L$. Now $\bigcup_{i<\omega}B_i|L$ and $\bigcup_{i<\omega}C_i|L$ are respectively an L_1-structure and an L_2-structure. Use e with $\bigcup_{i<\omega}C_i|L$ as a template to expand $\bigcup_{i<\omega}B_i|L$ to an $(L_1 \cup L_2)$-structure D. By the elementary chain theorem (Theorem 2.5.2), D is as required. $\square$

Interests change. When Theorem 5.5.1 first appeared in the 1950s, it wasn't even stated explicitly; but thoughtful readers could extract it from the proof of a purely syntactic result about first-order logic (Robinson's joint consistency lemma, Exercise 1 below).

True to form, Theorem 5.5.1 generates a brood of characterisation and interpolation theorems. If L and L^+ are first-order languages with $L \subseteq L^+$, and T is a theory in L^+, we write T_L for the set of all consequences of T in the language L.

Corollary 5.5.2. *Let L and L^+ be first-order languages with $L \subseteq L^+$ and T a theory in L^+. Let A be an L-structure. Then $A \vDash T_L$ if and only if for some model B of T, $A \preccurlyeq B|L$.*

Proof. As Corollary 5.4.3 followed from Theorem 5.4.1. $\square$

Next an interpolation theorem.

Theorem 5.5.3. *Let L_1, L_2 be first-order languages, $L = L_1 \cap L_2$ and T_1, T_2 theories in L_1, L_2 respectively, such that $T_1 \cup T_2$ has no model. Then there is some sentence ψ of L such that $T_1 \vdash \psi$ and $T_2 \vdash \neg \psi$.*

Proof. Take $\Psi = (T_1)_L$. By the compactness theorem it suffices to show that $\Psi \cup T_2$ has no model. For contradiction let C be a model of $\Psi \cup T_2$. By Corollary 5.5.2 there is an L_1-structure B such that $C|L \preccurlyeq B|L$ and $B \vDash T_1$. Then $B|L \equiv C|L$, and so by Theorem 5.5.1 there are an $(L_1 \cup L_2)$-structure D such that $B \preccurlyeq D|L_1$ and an elementary embedding $g: C \rightarrow D|L_2$. Now on the one hand $B \preccurlyeq D|L_1$, so that $D \vDash T_1$; but on the other hand g is elementary, so that $D \vDash T_2$. Thus $T_1 \cup T_2$ does have a model; contradiction. $\square$

In particular suppose ϕ and χ are sentences of L_1 and L_2 respectively, such that $\phi \vdash \chi$. Then there is a sentence ψ of $L_1 \cap L_2$ such that $\phi \vdash \psi$ and $\psi \vdash \chi$. This case of the theorem is known as **Craig's interpolation theorem**.

And of course there is the expected preservation theorem. It talks about formulas which are preserved under taking off symbols and putting them back on again.

Theorem 5.5.4. *Let L and L^+ be first-order languages with $L \subseteq L^+$, let T be a theory in L^+ and $\phi(\bar{x})$ a formula of L^+. Then the following are equivalent.*
(a) *If A and B are models of T and $A|L = B|L$, then for all tuples $\bar{a}$ in A, $A \vDash \phi(\bar{a})$ if and only if $B \vDash \phi(\bar{a})$.*
(b) *$\phi(\bar{x})$ is equivalent modulo T to a formula $\psi(\bar{x})$ of L.*

Proof. From Theorem 5.5.1 as Theorem 5.4.4 followed from Theorem 5.4.1.
$\square$

Pause for a moment at the implication (a) $\Rightarrow$ (b) in the case where ϕ is an unnested atomic formula $R(x_0, \ldots, x_{n-1})$ or $F(x_0, \ldots, x_{n-1}) = x_n$. This particular case of Theorem 5.5.4 is known as **Beth's definability theorem**.

In more detail: we say that a relation symbol R is **implicitly defined** by T in terms of L if whenever A and B are models of T with $A|L = B|L$, then $R^A = R^B$; and likewise with a function symbol F. As in section 2.6 above, we say that R is **explicitly defined** by T in terms of L if T has some consequence of the form $\forall \bar{x}(R\bar{x} \leftrightarrow \psi)$, where $\psi(\bar{x})$ is a formula of L; and similarly with F. It's immediate that if R (or F) is explicitly defined by T in terms of L, then it is implicitly defined by T in terms of L. Beth's theorem states the converse. In a slogan: *relative to a first-order theory, implicit definability equals explicit definability.*

The notion of implicit definability makes sense in a broader context. Let L and L^+ be languages (not necessarily first-order), T a theory in L^+ and R a relation symbol of L^+. Just as above, we say that R is **implicitly defined** by T in terms of L if whenever A and B are models of T with $A|L = B|L$, then $R^A = R^B$. A person who produces models A and B of T with $A|L = B|L$ but $R^A \neq R^B$ is said to be using **Padoa's method** for proving the undefinability of R by T in terms of L. There were some examples in section 2.6 above; see Exercises 2.6.6 and 2.6.7. If L and L^+ are not first-order, Beth's theorem may fail.

Local theorems

Our last theorem tackles the following question: If 'enough' finitely generated substructures of a structure A belong to a certain class **K**, must A belong to **K** too? Theorems which say that the answer is Yes are known as **local theorems**.

In section 4.2 we defined PC_Δ and PC'_Δ classes. We are now in a position to apply the compactness theorem to them. Many of the local theorems of group theory can be proved in this way.

Theorem 5.5.5. *Let L be a first-order language and $\mathbf{K}$ a PC'_Δ class of L-structures. Suppose that $\mathbf{K}$ is closed under taking substructures. Then $\mathbf{K}$ is axiomatised by a set of $\forall_1$ sentences of L.*

Proof. The theorem refines the Łoś–Tarski theorem (Theorem 5.4.4), and its proof is a refinement of the earlier proof.

Let $\mathbf{K}$ be the PC'_Δ class $\{B_P: B \vDash U\}$. Write T^* for the set of all $\forall_1$ sentences ϕ of L such that $B_P \vDash \phi$ whenever $B \vDash U$. Every structure in $\mathbf{K}$ is a model of T^*. We prove that every model A of T^* is in $\mathbf{K}$. For this, consider the theory

(5.2) $\mathrm{diag}(A) \cup \{Pa: a \in \mathrm{dom}\ A\} \cup U$.

We claim that (5.2) has a model. For if not, then by the compactness theorem there are a conjunction $\psi(\bar{x})$ of literals of L, and a tuple $\bar{a}$ of distinct elements $a_0, \ldots, a_{m-1}$ of A, such that $A \vDash \psi(\bar{a})$ and $U \vdash Pa_0 \wedge \ldots \wedge Pa_{m-1} \rightarrow \neg\, \psi(\bar{a})$. By the lemma on constants,

$$U \vdash \forall\bar{x}(Px_0 \wedge \ldots Px_{m-1} \rightarrow \neg\, \psi(\bar{x})).$$

Hence the sentence $\forall\bar{x}\, \neg\, \psi(\bar{x})$ is in T^*, so it must be true in A. This contradicts the fact that $A \vDash \psi(\bar{a})$. The claim is proved.

By the claim there is a model D of (5.2). By the diagram lemma, A is embeddable in D_P. But D_P is in $\mathbf{K}$ and $\mathbf{K}$ is closed under substructures. Since $\mathbf{K}$ is clearly closed under isomorphic copies, it follows that A is in $\mathbf{K}$.   □

We apply Theorem 5.5.5 at once. The next result was perhaps the first purely algebraic theorem whose proof (by Mal'tsev in 1940) made essential use of model theory. Let n be a positive integer and G a group. We say that G has a **faithful n-dimensional linear representation** if G is embeddable in $GL_n(F)$, the group of invertible n-by-n matrices over some field F.

Corollary 5.5.6. *Let n be a positive integer and G a group. Suppose that every finitely generated subgroup of G has a faithful n-dimensional linear representation. Then G has a faithful n-dimensional linear representation.*

Proof. Let $\mathbf{K}$ be the class of groups with faithful n-dimensional linear representations. We note that $\mathbf{K}$ is closed under substructures. There is a theory U in a suitable first-order language, such that $\mathbf{K}$ is precisely the class $\{B_P: B \vDash U\}$. (If this is not obvious, consult Example 1 of section 4.2.) By the theorem, $\mathbf{K}$ is axiomatised by an $\forall_1$ theory T. If G is not in $\mathbf{K}$, then there is some sentence $\forall \bar{x}\, \psi(\bar{x})$ in T, with ψ quantifier-free, such that $G \vDash \exists \bar{x} \,\neg\, \psi(\bar{x})$. Find a tuple $\bar{a}$ in G so that $G \vDash \neg\, \psi(\bar{a})$; then the subgroup $\langle \bar{a} \rangle_G$ is not in $\mathbf{K}$.

$\qquad\qquad\qquad\qquad\qquad\qquad\qquad\qquad\qquad\qquad\qquad\qquad\qquad$ $\square$

Exercises for section 5.5

1. (Robinson's joint consistency lemma.) Let L_1 and L_2 be first-order languages and $L = L_1 \cap L_2$. Let T_1 and T_2 be consistent theories in L_1 and L_2 respectively, such that $T_1 \cap T_2$ is a complete theory in L. Show that $T_1 \cup T_2$ is consistent.

2. Let L and L^+ be first-order languages with $L \subseteq L^+$, and let $\phi(\bar{x})$ be a formula of L^+ and T a theory in L^+. Suppose that whenever A and B are models of T and $f\colon A|L \to B|L$ is a homomorphism, f preserves ϕ. Show that ϕ is equivalent modulo T to an $\exists_1^+$ formula of L.

3. Let L_1 and L_2 be first-order languages with $L = L_1 \cap L_2$. Suppose ϕ and ψ are sentences of L_1 and L_2 respectively, such that $\phi \vdash \psi$. Show that if every function or constant symbol of L_1 is in L_2, and ϕ is an $\forall_1$ sentence and ψ is an $\exists_1$ sentence, then there is a quantifier-free sentence θ of L such that $\phi \vdash \theta$ and $\theta \vdash \psi$. [If Θ is the set of quantifier-free sentences θ of L such that $\phi \vdash \theta$, consider any L_2-structure A which is a model of Θ; writing A_0 for the substructure of A consisting of elements named by closed terms, embed A_0 in $B|L$ for some model B of ϕ.]

4. Let L_1 and L_2 be first-order languages with $L = L_1 \cap L_2$. Suppose ϕ and ψ are sentences of L_1 and L_2 respectively, such that $\phi \vdash \psi$. Show that if ϕ and ψ are both $\forall_1$ sentences then there is an $\forall_1$ sentence θ of L such that $\phi \vdash \theta$ and $\theta \vdash \psi$.

5. Let L and L^+ be first-order languages with $L \subseteq L^+$, and suppose P is a 1-ary relation symbol of L^+. Let ϕ be a sentence of L^+, and T a theory in L^+ such that for every model A of T, P^A is the domain of a substructure A^* of $A|L$. Suppose that whenever A and B are models of T with $A \vDash \phi$ and $f\colon A^* \to B^*$ is a homomorphism, then $B \vDash \phi$. Show that ϕ is equivalent modulo T to a sentence of the form $\exists y_0 \ldots y_{k-1}(\bigwedge_{i<k} Py_i \wedge \psi(\bar{y}))$ where ψ is a positive quantifier-free formula of L.

6. Let L be the first-order language with relation symbols for 'x is a son of y', 'x is a daughter of y', 'x is a father of y', 'x is a mother of y', 'x is a grandparent of y'. Let T be a first-order theory which reports the basic biological facts about these relations (e.g. that everybody has a unique mother, nobody is both a son and a daughter, etc.). Show (a) in T, 'son of' is definable from 'father of', 'mother of' and 'daughter of', (b) in T, 'son of' is not definable from 'father of' and 'mother of', (c) in T, 'father of' is definable from 'son of' and 'daughter of', (d) in T, 'mother of' is not definable from 'father of' and 'grandparent of', (e) etc. ad lib.

7. Let L be a first-order language, L^+ the language got from L by adding one new relation symbol R, and ϕ a sentence of L^+. Suppose that every L-structure can be

expanded in at most one way to a model of ϕ. Show that there is a sentence θ of L such that an L-structure A is a model of θ if and only if A can be expanded to a model of ϕ.

8. Let T be a first-order theory. Show that the class of groups $\{G: G$ acts faithfully on some model of $T\}$ is axiomatised by an $\forall_1$ first-order theory in the language of groups.

9. Assume it has been proved that every map on the plane, with finitely many countries, can be coloured with just four colours so that two adjacent countries never have the same colour. Use the compactness theorem to show that the same holds even when the map has infinitely many countries (but each country has finitely many neighbours, of course).

10. An **ordered group** is a group whose set of elements is linearly ordered in such a way that $a < b$ implies $c \cdot a < c \cdot b$ and $a \cdot c < b \cdot c$ for all elements a, b, c. A group is **orderable** if it can be made into an ordered group by adding a suitable ordering. (a) Show that an orderable group can't have elements of finite order, except the identity. (b) Show from the structure theorem for finitely generated abelian groups that every finitely generated torsion-free abelian group is orderable. (c) Using the compactness theorem, show that if G is a group and every finitely generated subgroup of G is orderable then G is orderable. (c) Deduce that an abelian group is orderable if and only if it is torsion-free.

5.6 Indiscernibles

This short section introduces a combinatorial result known as Ramsey's theorem. Model theorists often use Ramsey's theorem. It is less usual for them to prove it – they usually leave that to the combinatorialists. But we shall see that it is an easy consequence of the compactness theorem.

In the definitions which follow, remember that an ordinal number is the set of its predecessors, and this set carries a natural linear ordering.

Let X be a set, $<$ a linear ordering of X and k a positive integer. We write $[X]^k$ for the set of all $<$-increasing k-tuples of elements of K. Here '$<$-increasing' will always mean 'strictly increasing in the ordering $(X, <)$'. The notation $[X]^k$ leaves out $<$; we can write $[(X, <)]^k$ when we need to specify it.

Let f be a map whose domain is $[X]^k$. We say that a subset Y of X (or more explicitly a subordering $(Y, <)$ of $(X, <)$) is f-**indiscernible** if for any two $<$-increasing k-tuples $\bar{a}$, $\bar{b}$ from Y, $f(\bar{a}) = f(\bar{b})$; in other words, if f is constant on $[Y]^k$.

Here is an important special case. Let A be an L-structure. Let k be a positive integer and Φ a set of formulas of L, all of the form $\phi(x_0, \ldots, x_{k-1})$. Suppose X is a set of elements of A and $<$ is a linear ordering of X.

Then we say that $(X, <)$ is a Φ-**indiscernible sequence** in A if for every formula $\phi(x_0, \ldots, x_{k-1})$ in Φ and every pair $\bar{a}, \bar{b} \in [X]^k$,

$$(6.1) \qquad\qquad A \vDash \phi(\bar{a}) \leftrightarrow \phi(\bar{b}).$$

We can choose a function $f : [X]^k \to 2^{|L|}$ so that

$$(6.2) \qquad \text{for all } \bar{a} \text{ and } \bar{b} \text{ in } [X]^k,$$
$$f(\bar{a}) = f(\bar{b}) \Leftrightarrow (6.1) \text{ holds for all } \phi \text{ in } \Phi.$$

Then $(X, <)$ is f-indiscernible if and only if it is Φ-indiscernible.

Example 1: *Bases of vector spaces.* Let A be a vector space and X a basis of A. Let $<$ be any linear ordering of X. Then X is $\phi(x_0, \ldots, x_{k-1})$-indiscernible for every formula ϕ in the first-order language L of A. For let $\bar{a}, \bar{b}$ be any two strictly increasing k-tuples from $(X, <)$. Since X is a basis of A, there is an automorphism of A which takes $\bar{a}$ to $\bar{b}$. Hence $A \vDash \phi(\bar{a})$ implies $A \vDash \phi(\bar{b})$.

The set X in this example is indiscernible in a very strong sense. Suppose A is any structure, X a set of elements of A linearly ordered by $<$, and $(X, <)$ is a $\{\phi\}$-indiscernible sequence simultaneously for every first-order formula $\phi(\bar{x})$ of L (with any number of variables); then we say that $(X, <)$ is an **indiscernible sequence** in A. We say that X is an **indiscernible set** in A if $(X, <)$ is an indiscernible sequence for every linear ordering $<$ of X. (***Warning***: set theorists call an indiscernible sequence a **set of indiscernibles**.)

Example 1 shows that a basis of a free algebra in a variety is always an indiscernible set.

Let λ, μ and ν be cardinals and k a positive integer. We write

$$(6.3) \qquad\qquad \lambda \to (\mu)^k_\nu$$

to mean that if X is any linearly ordered set of cardinality λ, and $f : [X]^k \to \nu$, then there is a subset Y of X of cardinality μ such that f is constant on $[Y]^k$. Facts of the form (6.3) are known as **Erdős–Rado partition relations**. The notation was chosen by Paul Erdős and Richard Rado so that if (6.3) holds, then it still holds when λ on the left is raised and μ, ν and k on the right are lowered.

The most important partition relation, and the first to be discovered, runs as follows.

Theorem 5.6.1 (***Ramsey's theorem, infinite form***). *For all positive integers k, n we have $\omega \to (\omega)^k_n$.*

Proof. By induction on k. When $k = 1$, the theorem says that if ω is partitioned into at most m parts (with m finite), then at least one of the parts is infinite. This is true and it has a name: the pigeonhole principle.

Suppose then that $k > 1$, and let a map $f : [\omega]^k \to n$ be given. Let A be the structure built as follows.

(6.4) $\mathrm{dom}(A) = \omega$. There are names 0, 1, ... for the elements of ω. There are a relation symbol $<$ and a function symbol F, to represent the usual ordering of ω and the function f respectively. (Put $F^A(a_0, \ldots, a_{k-1}) = 0$ when $(a_0, \ldots, a_{k-1})$ is not an increasing k-tuple.)

By the compactness theorem there is a proper elementary extension B of A. Now $A \models \forall x (F(x) < n)$, and so the same holds in B. Also $A \models$ '$<$ linearly orders all the elements', and for each natural number m, $A \models$ 'the element m has exactly m $<$-predecessors'. So any element of B which is not in A must come $<$-after all the elements of A. Take such an element and call it ∞.

We shall choose natural numbers $m(0)$, $m(1)$, ... inductively, so that for each $i < \omega$,

(6.5) $m(j) < m(i)$ for each $j < i$,

(6.6)   for all $j_0 < \ldots < j_{k-2} < i$, $B \models$
$F(m(j_0), \ldots, m(j_{k-2}), m(i)) = F(m(j_0), \ldots, m(j_{k-2}), \infty)$.

Suppose $m(0), \ldots, m(i-1)$ have been chosen. Then we can write a first-order formula $\phi(x)$, using the constants and the symbols $<$ and F, which expresses that $x > m(i-1)$, and for all $j_0 < \ldots < j_{k-2} < i$, the value of $F^B(m(j_0), \ldots, m(j_{k-2}), x)$ is $F^B(m(j_0), \ldots, m(j_{k-2}), \infty)$. (There are only finitely many conditions here, and for each choice of $j_0, \ldots, j_{k-2}$ the number $F^B(m(j_0), \ldots, m(j_{k-2}), \infty)$ is $< n$, so that it is named by a constant.) Now clearly $B \models \phi(\infty)$, and so $B \models \exists x \phi(x)$. Since $A \preccurlyeq B$, it follows that $A \models \phi(m)$ for some natural number m. Put $m(i) = m$. This completes the choice of the numbers $m(i)$. Put $W = \{m(i) : i < \omega\}$.

Define a map $g : [W]^{k-1} \to n$ by

(6.7) $g(\bar{b}) = f(\bar{b}^\wedge c)$ for every tuple $\bar{b}^\wedge c \in [W]^k$.

By (6.6), g is well-defined. By induction hypothesis there is an infinite subset Y of W such that g is constant on $[Y]^{k-1}$. Now let $\bar{a} = (y_0, \ldots, y_{k-1})$ and $\bar{b} = (z_0, \ldots, z_{k-1})$ be sets in $[Y]^k$ with $y_0 < \ldots < y_{k-1}$ and $z_0 < \ldots < z_{k-1}$. Taking any c which is $> \max(y_{k-1}, z_{k-1})$,

$$f(\bar{a}) = f(y_0, \ldots, y_{k-2}, c) = g(y_0, \ldots, y_{k-2}) \tag{6.8}$$
$$= g(z_0, \ldots, z_{k-2}) = f(\bar{b}).$$

Hence f is constant on $[Y]^k$ as required. $\square$

Corollary 5.6.2 (Ramsey's theorem, finite form). *For all positive integers k, m, n there is a positive integer l such that $l \to (m)^k_n$.*

Proof. Fix k and n. Suppose for contradiction that there is no positive integer l such that $l \to (m)_n^k$. Let $\mathbb{N}$ be the structure of the natural numbers, $(\omega, 0, S, +, \cdot)$ (where S is the successor function). Let T be the following first-order theory, in the language of $\mathbb{N}$ with extra symbols c and F:

(6.9) $\text{Th}(\mathbb{N}) \cup \{F$ is a function from $[\{0, \ldots, c-1\}]^k$ to $n\}$
        $\cup \{$There is no increasing sequence $\bar{a}$ of m distinct elements,
        all in $\{0, \ldots, c-1\}$, such that F is constant on
        $[\bar{a}]^k\} \cup \{`j < c': j$ a natural number$\}$.

Every finite subset U of T has a model: choose c to be a natural number h greater than every number mentioned in U, and let F be a function which shows that $h \nrightarrow (m)_n^k$. So by the compactness theorem, (6.9) has a model A. The infinite Ramsey theorem applied to $F^A : [\{a : A \vDash a < c^A\}]^k \to \{0^A, \ldots, n^A - 1\}$ gives us an infinite set Y such that F^A is constant on $[Y]^k$, which contradicts (6.9). ☐

A typical application of Ramsey's theorem is the following.

Corollary 5.6.3. *Let $(P, \leqslant)$ be an infinite partially ordered set. Then either P contains an infinite linearly ordered set, or P contains an infinite set of pairwise incomparable elements.*

Proof. Linearly order the elements of P in any way at all, say by an ordering $<$. Define the map $F : [P]^2 \to 3$ as follows, wherever $a < b$ in P: $F(a, b) = 0$ if $a < b$, $F(a, b) = 1$ if $b < a$, $F(a, b) = 2$ if a and b are incomparable. Then apply the infinite Ramsey theorem. ☐

The next corollary is less well known. It has several applications in model theory. If G is a group and H a subgroup, we write $(G : H)$ for the index of H in G.

Corollary 5.6.4. *Let G be a group, and let $H_0, \ldots, H_{n-1}$ be subgroups of G and $a_0, \ldots, a_{n-1}$ elements of G. Suppose that G is the union of the set of cosets $X = \{H_0 a_0, \ldots, H_{n-1} a_{n-1}\}$, but not the union of any proper subset of X. Then for each $i < n$, $(G : H_i) < l$ for some finite l depending only on n; in particular H_i has finite index in G.*

Proof. Quoting the finite Ramsey theorem (Corollary 5.6.2), take l such that $l \to (3)_n^2$. Suppose for contradiction that H_0 has index $\geqslant l$ in G. By the minimality of X there is $g \in G$ which lies in $H_i a_i$ if and only if $i = 0$. Choose distinct coset representatives $y_0, \ldots, y_{l-1}$ of H_0 in G, and define a map $f : [l]^2 \to n$ by:

$$f(i, j) = \text{least } k \text{ such that } y_i y_j^{-1} g \in H_k a_k.$$

By choice of l there are distinct y_0', y_1', y_2' and k such that

$$y_0' \in H_k a_k g^{-1} y_1' \quad \text{and} \quad y_0', y_1' \in H_k a_k g^{-1} y_2'.$$

These two cosets of H_k must be the same since they both contain y_0'. Hence $y_1' \in H_k a_k g^{-1} y_1'$, whence $g \in H_k a_k$, so that $k = 0$ and $H_k a_k g^{-1} = H_0$. But this is impossible, since y_0' and y_1' were in distinct cosets of H_0. $\square$

Exercises for section 5.6

1. Show that $6 \to (3)_2^2$. [Suppose each pair in $[6]^2$ is coloured either red or green. Renumbering if necessary, we can suppose that $(0, 1)$, $(0, 2)$ and $(0, 3)$ all have the same colour. Consider the colours of $(1, 2)$, $(1, 3)$ and $(2, 3)$.] *This result suggests a game: the players RED and GREEN take turns to colour pairs in $[6]^2$, and the first player to complete a triangle in his or her colour loses.*

2. Let A be an infinite structure with finite relational signature. Suppose that for every $n < \omega$, all n-element substructures of A are isomorphic. Show that there is a linear ordering $<$ such that $(\text{dom}(A), <)$ is an indiscernible sequence. [Use the compactness theorem.]

A **tree** *is a partially ordered set* $(P, <)$ *where (i) there is a unique bottom element and (ii) for each element p the set $p^< = \{q \in P : q < p\}$ of predecessors of p is well-ordered by $<$. The order-type of $p^<$ is called the* **height** *of p. The* **height** *of the tree is the least ordinal greater than all the heights of elements of P. The* **immediate successors** *of p are the elements $r > p$ with $\text{height}(r) = \text{height}(p) + 1$. A* **branch** *of the tree is a maximal linearly ordered subset.*

3. We say that a tree has **finite branching** if every element of the tree has at most finitely many immediate successors. Prove **König's tree lemma**: Every tree of height ω with finite branching has an infinite branch. [The usual proof chooses the elements of the branch by induction on height. Model theorists should look for a proof like that of Ramsey's theorem, taking a proper elementary extension of the tree.]

The next result is intermediate between the finite and the infinite forms of Ramsey's theorem. It can be stated in the language of first-order arithmetic but not proved from the first-order Peano axioms.

4. Prove that for all positive integers k, m, n there is a positive integer l such that if $[l]^k = P_0 \cup \ldots \cup P_{n-1}$ then there are $i < n$ and a set $X \subseteq l$ of cardinality at least m, such that $[X]^k \subseteq P_i$ and $|X| \geqslant \min(X)$. [Assuming otherwise for a fixed k, m and n, use König's tree lemma (Exercise 3) to find a partition of $[\omega]^k$ into at most n pieces, such that every l is a counterexample to the theorem. Now apply Ramsey's theorem to this partition.]

Further reading

This chapter has proved several fundamental results in the model theory of first-order languages. One can ask how well these results lift to other languages, for example infinitary or higher-order, or with unusual quantifiers. In a nutshell the answer is: Not well. An authoritative account (if only it had an index!) is the 893-page volume:

Barwise, J. and Feferman, S. (eds) *Model-theoretic logics*. New York: Springer, 1985.

Anatolii Mal'tsev was the first person to state the compactness theorem in full generality. He wrote several papers applying it to group theory. These are not the only reasons why his collected logical works are worth consulting:

Mal'cev, A. I. *The metamathematics of algebraic systems. Collected papers: 1936–1967*. Amsterdam: North-Holland, 1971.

A book that studies Ramsey's theorem in a combinatorial setting:

Graham, R. L., Rothschild, B. L. and Spencer, J. H. *Ramsey theory*. New York: John Wiley and Sons, 1980.

6

The countable case

For eighthly he rubs himself against a post.
For ninthly he looks up for his instructions.
For tenthly he goes in quest of food.
Christopher Smart (1722–71), Jubilate Agno.

The cardinal number ω is the only infinite cardinal which is a limit of finite cardinals. This gives us two reasons why countable structures are good to build. First, a countable structure can be built as the union of a chain of finite pieces. And second, we have infinitely many chances to make sure that the right pieces go in. No other cardinal allows us this amount of control.

So it's not surprising that model theory has a rich array of methods for constructing countable structures. Pride of place goes to Roland Fraïssé's majestic construction, which bestrides this chapter. Next comes the omitting types construction (section 6.2); it was discovered in several forms by several people.

Besides constructions, this chapter contains one section on classification. In section 6.3 we shall see that the countable ω-categorical structures can be recognized from their automorphism groups. These automorphism groups tend to be very rich, and they give plenty of scope for collaboration between model theorists and permutation group theorists.

6.1 Fraïssé's construction

In 1954 Roland Fraïssé published a paper which has become a classic of model theory. He pointed out that we can think of the class of finite linear orderings as a set of approximations to the ordering of the rationals, and he described a way of building the rationals out of these finite approximations. Fraïssé's construction is important because it works in many other cases too. Starting from a suitable set of finite structures, we can build their 'limit', and some of the structures built in this way have turned out to be remarkably interesting. Several will appear later in this book.

Ages

Let L be a signature and D an L-structure. The **age** of D is the class **K** of all finitely generated structures that can be embedded in D. Actually what interests us is not the structures in **K** but their isomorphism types. So we shall also call a class **J** the **age** of D if the structures in **J** are, *up to isomorphism*, exactly the finitely generated substructures of D. Thus it will make sense to say, for example, that D has 'countable age' – it will mean that D has just countably many isomorphism types of finitely generated substructure.

We call a class an **age** if it is the age of some structure. If **K** is an age, then clearly **K** is non-empty and has the following two properties.

(1.1)   **(Hereditary property**, HP for short) If $A \in$ **K** and B is a finitely generated substructure of A then B is isomorphic to some structure in **K**.

(1.2)   **(Joint embedding property**, JEP for short) If A, B are in **K** then there is C in **K** such that both A and B are embeddable in C:

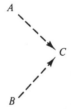

One of Fraïssé's theorems was a converse to this.

Theorem 6.1.1. *Suppose L is a signature and* **K** *is a non-empty finite or countable set of finitely generated L-structures which has the HP and the JEP. Then* **K** *is the age of some finite or countable structure.*

Proof. List the structures in **K**, possibly with repetitions, as $(A_i : i < \omega)$. Define a chain $(B_i : i < \omega)$ of structures isomorphic to structures in **K**, as follows. First put $B_0 = A_0$. When B_i has been chosen, use the joint embedding property to find a structure B' in **K** such that both B_i and A_{i+1} are embeddable in B'. Take B_{i+1} to be an isomorphic copy of B' which extends B_i (recalling Exercise 1.2.4(b)). Finally let C be the union $\bigcup_{i<\omega} B_i$. Since C is the union of countably many structures which are at most countable, C is at most countable. By construction every structure in **K** is embeddable in C. If A is any finitely generated substructure of C, then the finitely many generators of A lie in some B_i, so that A is isomorphic to a structure in **K** (by the hereditary property). So **K** is the age of C. $\qquad\square$

The theorem holds even if L has function symbols. But one way to guarantee that **K** is at most countable is to assume that L is a finite signature with no function symbols – see Exercise 1.2.6 above. When there are no function symbols and only finitely many constant symbols, a finitely generated structure is the same thing as a finite structure.

All infinite linear orderings have exactly the same age, namely the finite linear orderings. (In Fraïssé's own charming terminology, any two infinite linear orderings are 'younger than each other'.) In what sense do the finite linear orderings 'tend to' the rationals rather than, say, the ordering of the integers?

In order to answer this, Fraïssé singled out one further property of the finite linear orderings, to set beside HP and JEP. This property, the amalgamation property, has been of crucial importance in model theory ever since; we met several variants in Chapter 5.

(1.3)   **(Amalgamation property**, AP for short) If A, B, C are in **K** and $e: A \to B, f: A \to C$ are embeddings, then there are D in **K** and embeddings $g: B \to D$ and $h: C \to D$ such that $ge = hf$:

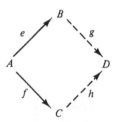

(**Warning**: In general JEP is not a special case of AP. Think of fields).

The class of all finite linear orderings has the amalgamation property. The simplest way to see this is to start with the case where A is a substructure of B and C, the maps e, f are inclusions and A is exactly the overlap of B and C. In this case we can form D as an extension of B, by adding the elements of C one by one in the appropriate places. (Formally, use an induction on the cardinality of C.) The general case then follows by diagram-chasing.

We call a structure D **ultrahomogeneous** if every isomorphism between finitely generated substructures of D extends to an automorphism of D. (Usually one says **homogeneous**; but we shall need this word for a different notion in section 8.1.)

Fraïssé's main conclusion was as follows.

Theorem 6.1.2 (*Fraïssé's theorem*). *Let L be a countable signature and let* **K** *be a non-empty finite or countable set of finitely generated L-structures which* *has HP, JEP and AP. Then there is an L-structure D, unique up to isomorphism, such that*

(1.4)   *D has cardinality* $\leq \omega$,

(1.5)   *K is the age of D, and*

(1.6)   *D is ultrahomogeneous.*

There seems to be no accepted name for the structure D of the theorem. Some people, with Lemma 6.1.3 below in mind, refer to it as the **universal homogeneous structure of age K**. I shall call it the **Fraïssé limit** of the class **K**. Of course it is only determined up to isomorphism.

The rest of this section is devoted to the proof of Theorem 6.1.2.

Uniqueness proof

We begin by saying that a structure D is **weakly homogeneous** if it has the property

(1.7)   if A, B are finitely generated substructures of D, $A \subseteq B$ and $f: A \to D$ is an embedding, then there is an embedding $g: B \to D$ which extends f:

If D is ultrahomogeneous then clearly D is weakly homogeneous.

Lemma 6.1.3. *Let C and D be L-structures which are both at most countable. Suppose the age of C is included in the age of D, and D is weakly homogeneous. Then C is embeddable in D; in fact any embedding from a finitely generated substructure of C into D can be extended to an embedding of C into D.*

Proof. Let $f_0: A_0 \to D$ be an embedding of a finitely generated substructure A_0 of C into D. We shall extend f_0 to an embedding $f_\omega: C \to D$ as follows. Since C is at most countable, it can be written as a union $\bigcup_{n < \omega} A_n$ of a chain of finitely generated substructures, starting with A_0. By induction on n we define an increasing chain of embeddings $f_n: A_n \to D$. The first embedding f_0 is given. Suppose f_n has just been defined. Since the age of D includes that of C, there is an isomorphism $g: A_{n+1} \to B$ where B is a substructure of D.

Then $f_n \cdot g^{-1}$ embeds $g(A_n)$ into D, and by weak homogeneity this embedding extends to an embedding $h: B \to D$. Let $f_{n+1}: A_{n+1} \to D$ be hg. Then $f_n \subseteq f_{n+1}$. This defines the chain of maps f_n. Finally take f_ω to be the union of the f_n $(n < \omega)$.   □

This lemma justifies some terminology. We say that a countable structure D of age $\mathbf{K}$ is **universal** (for $\mathbf{K}$) if every finite or countable structure of age $\subseteq \mathbf{K}$ is embeddable in D. Lemma 6.1.3 tells us that countable weakly homogeneous structures are universal for their age.

When C and D are both weakly homogeneous and have the same age, we can throw the argument of this lemma to and fro between C and D to prove that C is isomorphic to D.

Lemma 6.1.4. (a) *Let C and D be L-structures with the same age. Suppose that C and D are both at most countable and are both weakly homogeneous. Then C is isomorphic to D. In fact if A is a finitely generated substructure of C and $f: A \to D$ is an embedding, then f extends to an isomorphism from C to D.*

(b) *A finite or countable structure is ultrahomogeneous (and hence is the Fraïssé limit of its age) if and only if it is weakly homogeneous.*

Proof. (a) Express C and D as the unions of chains $(C_n: n < \omega)$ and $(D_n: n < \omega)$ of finitely generated substructures. Define a chain $(f_n: n < \omega)$ of isomorphisms between finitely generated substructures of C and D, so that for each n, the domain of f_{2n} includes C_n and the image of f_{2n+1} includes D_n. The machinery for doing this is just as in the proof of the previous lemma. Then the union of the f_n is an isomorphism from C to D.

To get the last sentence of (a), take C_0 to be A and D_0 to be $f(A)$.

(b) We have already noted that ultrahomogeneous structures are weakly homogeneous. The converse follows at once from (a), taking $C = D$.   □

What if C and D are not countable? Then part (a) of the lemma fails. For example let η be the order-type of the rationals, and consider the order-type $\eta \cdot \omega_1$ $(= \omega_1$ copies of η laid in a row) and its mirror image ξ. Both $\eta \cdot \omega_1$ and ξ are weakly homogeneous, and they have the same age, namely the set of all finite linear orderings. But clearly they are not isomorphic, since in $\eta \cdot \omega_1$ but not in ξ every element has uncountably many successors. The best we can say is the following.

Lemma 6.1.5. *Suppose C and D are weakly homogeneous L-structures with the same age. Then C is back-and-forth equivalent to D, so that $C \equiv_{\infty\omega} D$. If moreover $C \subseteq D$ then for every tuple $\bar{c}$ in C, $(C, \bar{c}) \equiv_{\infty\omega} (D, \bar{c})$, so that $C \preccurlyeq D$.*

The *proof* is by the proof of Lemma 6.1.4. Use Theorem 3.2.4 for the connection with $L_{\infty\omega}$.   □

Existence proof

Lemma 6.1.4 takes care of the uniqueness of Fraïssé limits.

If the signature L is finite and has no function symbols, the statement 'The age of D is a subset of **K**' can be written as an $\forall_1$ first-order theory T. To see this, take the set of all those finite L-structures which *don't* occur in **K**, and for each such structure A write an $\forall_1$ sentence χ_A which says 'No substructure is isomorphic to A'. Then T is the set of all these sentences χ_A.

If L is not finite, or has function symbols, there is no guarantee that the statement 'The age is a subset of **K**' can be written as a first-order theory, even when the structures in **K** are all finite. (Consider for instance the class **K** of all finite groups; see Example 1 below.) This will cause fewer difficulties than one might have feared. But it does remind us that we are working with a class of structures where the upward Löwenheim–Skolem theorem need not apply; Theorem 6.1.2 will give us a structure of cardinality $\leqslant \omega$, but it may be much harder a find a similar structure of uncountable cardinality.

We first note an easy fact about ages.

Lemma 6.1.6. *Let* **J** *be a set of finitely generated L-structures, and* $(D_i : i < \alpha)$ *a chain of L-structures. If for each* $i < \alpha$ *the age of* D_i *is included in* **J***, then the age of the union* $\bigcup_{i<\alpha} D_i$ *is also included in* **J***. If each* D_i *has age* **J***, then* $\bigcup_{i<\alpha} D_i$ *has age* **J***.*   □

Henceforth we assume that **K** is non-empty, has HP, JEP and AP, and contains at most countably many isomorphism types of structure. We can suppose without loss that **K** is closed under taking isomorphic copies.

We shall construct a chain $(D_i : i < \omega)$ of structures in **K**, such that the following holds:

(1.8)   if A and B are structures in **K** with $A \subseteq B$, and there is an embedding $f : A \to D_i$ for some $i < \omega$, then there are $j > i$ and an embedding $g : B \to D_j$ which extends f.

We take D to be the union $\bigcup_{i<\omega} D_i$. Then the age of D is included in **K** by Lemma 6.1.6. In fact the age of D is exactly **K**. For suppose A is in **K**; then by JEP there is B in **K** such that $A \subseteq B$ and D_0 is embeddable in B. By (1.8) the identity map on D_0 extends to an embedding of B in some D_j, so that B and A lie in the age of D. Thus (1.8) tells us that D is weakly homogeneous, and so by Lemma 6.1.4(b) it is ultrahomogeneous of age **K** as required.

It remains to construct the chain. Let **P** be a countable set of pairs of structures (A, B) such that $A, B \in \mathbf{K}$ and $A \subseteq B$; we can choose $\mathbf{P}$ so that it includes a representative of each isomorphism type of such pairs. Take a bijection $\pi: \omega \times \omega \to \omega$ such that $\pi(i, j) \geq i$ for all i and j. Let D_0 be any structure in $\mathbf{K}$. The rest is by induction, as follows. When D_k has been chosen, list as $((f_{kj}, A_{kj}, B_{kj}): j < \omega)$ the triples (f, A, B) where $(A, B) \in \mathbf{P}$ and $f: A \to D_k$. Construct D_{k+1} by the amalgamation property, so that if $k = \pi(i, j)$ then f_{ij} extends to an embedding of B_{ij} into D_{k+1}.

☐ Theorem 6.1.2

All the conditions on $\mathbf{K}$ in Theorem 6.1.2 were necessary, according to the next theorem.

Theorem 6.1.7. *Let L be a countable signature and D a finite or countable structure which is ultrahomogeneous. Let $\mathbf{K}$ be the age of D. Then $\mathbf{K}$ is non-empty, $\mathbf{K}$ has at most countably many isomorphism types of structure, and $\mathbf{K}$ satisfies HP, JEP and AP.*

Proof. We already know everything except that $\mathbf{K}$ satisfies the amalgamation property. For this we can assume without loss that $\mathbf{K}$ contains all finitely generated substructures of D. Suppose that A, B, C are in $\mathbf{K}$ and $e: A \to B$, $f: A \to C$ are embeddings. Then there are isomorphisms $i_A: A \to A'$, $i_B: B \to B'$ and $i_C: C \to C'$ where A', B', C' are substructures of D. So $i_A \cdot e^{-1}$ embeds $e(A)$ into D, and by weak homogeneity there is an embedding $j_B: B \to D$ which extends $i_A \cdot e^{-1}$, so that the bottom left quadrilateral in (1.9) commutes; likewise with the bottom right quadrilateral:

(1.9)

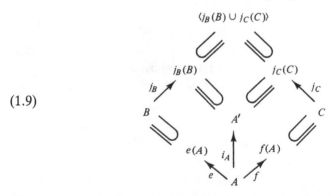

The top square commutes, and hence the outer maps in (1.9) give the needed amalgam.   ☐

We shall return to Fraïssé's construction with some more examples in section 6.4 below, after we have discussed ω-categoricity.

Exercises for section 6.1

1. Suppose **K** is a class of L-structures, and **K** contains a structure which is embeddable in every structure in **K**. Show that if **K** has the AP then **K** has the JEP too.

The next three exercises give algebraic objects which are Fraïssé limits.

2. Let p be a prime and let **K** be the class of all finite fields of characteristic p. Show that **K** has HP, JEP and AP, and that the Fraïssé limit of **K** is the algebraic closure of the prime field of characteristic p.

3. Let **K** be the class of finitely generated torsion-free abelian groups. Show that **K** has HP, JEP and AP, and that the Fraïssé limit of **K** is the direct sum of countably many copies of the additive group of rationals.

4. Let **K** be the class of finite boolean algebras. Show that **K** has HP, JEP and SAP, and that the Fraïssé limit of **K** is the countable atomless boolean algebra.

5. Show that the abelian group $Z(4) \oplus \bigoplus_{i<\omega} Z(2)$ is not elementarily equivalent to any ultrahomogeneous structure.

6. Let L be a finite signature with no function symbols, and **K** a class of finite L-structures which has HP, JEP and AP. Show that there is a first-order theory T in L such that (a) the countable models of T are exactly the countable ultrahomogeneous structures of age **K**, and (b) every sentence in T is either $\forall_1$ or of the form $\forall \bar{x} \exists y\, \phi(\bar{x}, y)$ where ϕ is quantifier-free.

6.2 Omitting types

Let L be a first-order language and T a theory in L. Let $\Phi(\bar{x})$ be a set of formulas of L, with $\bar{x} = (x_0, \ldots, x_{n-1})$. As in section 2.3, we say that Φ is **realised in** an L-structure A if there is a tuple $\bar{a}$ of elements of A such that $A \vDash \bigwedge \Phi(\bar{a})$; we say A **omits** Φ if Φ isn't realised in A.

When does T have a model that omits Φ?

Suppose for example that

(2.1)   there is a formula $\theta(\bar{x})$ of L such that $T \cup \{\exists \bar{x}\, \theta\}$ has a model, and for every formula $\phi(\bar{x})$ in Φ, $T \vdash \forall \bar{x}(\theta \to \phi)$.

If T is a complete theory, then (2.1) implies that $T \vdash \exists \bar{x}\, \theta$, and so T certainly has no model that omits Φ. Our next theorem will imply that when the language L is countable, the converse holds too, even if T is not a complete theory: if every model of T realises Φ then (2.1) is true.

Example 1: *A type omitted.* Let L be a first-order language whose signature consists of 1-ary relation symbols P_i ($i < \omega$). Let T be the theory in L which consists of all the sentences $\exists x\, P_0(x)$, $\exists x\, \neg P_0(x)$, $\exists x(P_0(x) \wedge P_1(x))$, $\exists x(P_0(x) \wedge \neg P_1(x))$, $\exists x(\neg P_0(x) \wedge P_1(x))$ etc. (through all the possible combinations). Then T is complete. (One can show this by quantifier elimination; Exercise 9 below suggests another proof.) If s is any subset of ω, let $\Phi_s(x)$ be the set $\{P_i(x): i \in s\} \cup \{\neg P_i(x): i \notin s\}$. Now T has a countable model A, which must omit at least one of the continuum many sets Φ_s ($s \subseteq \omega$). So by symmetry, if $s \subseteq \omega$ there must be a countable model of T which omits Φ_s. However, a model of T can't omit all the sets Φ_s, or it would be empty.

Note that if Φ is Φ_s for some $s \subseteq \omega$, then there is obviously no formula θ as in (2.1) – it takes infinitely many first-order formulas to specify Φ_s.

When (2.1) holds, we say that θ is a **support** of Φ over T. When (2.1) holds and θ is a formula in Φ, we say that θ **generates** Φ over T. We say that a set of first-order formulas $\Phi(\bar{x})$ is a **supported type over** T if Φ has a support over T; we say that Φ is a **principal type over** T if Φ has a generator over T. The set Φ is said to be **unsupported** (resp. **non-principal**) over T if it is not a supported (resp. principal) type over T.

Note that if $p(\bar{x})$ is a complete type over the empty set, then a formula $\phi(\bar{x})$ of L is a support of p if and only if it generates p; so a complete type p is principal if and only if it is supported. We say that a formula $\phi(\bar{x})$ is **complete** (for T) if it generates a complete type of T.

Theorem 6.2.1 *(Countable omitting types theorem).* *Let L be a countable first-order language, T a theory in L which has a model, and for each $m < \omega$ let Φ_m be an unsupported set over T in L. Then T has a model which omits all the sets Φ_m.*

Proof. The theorem is trivial when T has an empty model, so that we can assume that T has a non-empty model. Let L^+ be the first-order language which comes from L by adding countably many new constants c_i ($i < \omega$), to be known as **witnesses**. We shall define an increasing chain $(T_i: i < \omega)$ of finite sets of sentences of L^+, such that for every i, $T \cup T_i$ has a model. To prime the pump we take T_{-1} to be the empty theory. Then (since T has a non-empty model) $T \cup T_{-1}$ has a model which is an L^+-structure.

The intention is that the union of the chain, call it T^+, will be a Hintikka set for L^+ (see section 2.3), and the canonical model of the atomic sentences in T^+ will be a model of T which omits all the types Φ_m. To ensure that T^+ will have these properties, we carry out various tasks as we build the chain. These tasks are as follows.

(2.2)   Ensure that for every sentence ϕ of L^+, either ϕ or $\neg \phi$ is
          in T^+.

$(2.3)_{\psi(x)}$ (For each formula $\psi(x)$ of L^+:) Ensure that if $\exists x\ \psi(x)$ is in T^+ then there are infinitely many witnesses c such that $\psi(c)$ is in T^+.

$(2.4)_m$ (For each $m < \omega$:) Ensure that for every tuple $\bar{c}$ of distinct witnesses (of appropriate length) there is a formula $\phi(\bar{x})$ in Φ_m such that the formula $\neg\,\phi(\bar{c})$ is in T^+.

If these tasks are all carried out, then Theorem 2.3.4 tells us that T^+ will be a Hintikka set. Write A^+ for the canonical model of the atomic sentences in T^+ (see Theorem 1.5.2). Then A^+ is a model of T^+ in which every element is named by a closed term. By the tasks (2.3) where $\psi(x)$ are the formulas $x = t$ (t a closed term), every element of A^+ is named by infinitely many witnesses, and so every tuple of elements is named by a tuple of distinct witnesses. Given this, the tasks (2.4) make sure that A^+ omits all the types Φ_m. The required model of T will be the reduct $A^+|L$.

There are countably many tasks in the list: (2.2) is one, and we have a task $(2.3)_{\psi(x)}$ for each formula $\psi(x)$ and a task $(2.4)_m$ for each $m < \omega$. Metaphorically speaking, we shall hire countably many experts and give them one task each. We partition ω into infinitely many infinite sets, and we assign one of these sets to each expert. When T_{i-1} has been chosen, if i is in the set assigned to some expert E, then E will choose T_i. It remains to tell the experts how they should go about their business. (Experts in this kind of task are female. This comes from a useful convention in model-theoretic games; see the remarks after this proof.)

First consider the expert who handles task (2.2), and let X be her subset of ω. Let her list as $(\phi_i : i \in X)$ all the sentences of L^+. When T_{i-1} has been chosen with i in X, she should consider whether $T \cup T_{i-1} \cup \{\phi_i\}$ has a model. If it has, she should put $T_i = T_{i-1} \cup \{\phi_i\}$. If not, then every model of $T \cup T_i$ is a model of $\neg\phi_i$, and she can take T_i to be $T_{i-1} \cup \{\neg\phi_i\}$. In this way she can be sure of carrying out task (2.2) by the time the chain is complete.

Next consider the expert who deals with task $(2.3)_\psi$. She waits until she is given a set T_{i-1} which contains $\exists x\ \psi(x)$. Every time this happens, she looks for a witness c which is not used anywhere in T_{i-1}; there is such a witness, because T_{i-1} is finite. Then a model of $T \cup T_{i-1}$ can be made into a model of $\psi(c)$ by choosing a suitable interpretation for c. Let her take T_i to be $T_{i-1} \cup \{\psi(c)\}$. Otherwise she should do nothing. This strategy works, because her subset of ω contains arbitrarily large numbers.

Finally consider the expert who handles task $(2.4)_m$, where Φ_m is a type in n variables. Let Y be her assigned subset of ω. She begins by listing as $\{\bar{c}_i : i \in Y\}$ all the n-tuples $\bar{c}$ of distinct witnesses. When T_{i-1} has been given, with i in Y, she writes $\bigwedge T_{i-1}$ as a sentence $\chi(\bar{c}_i, \bar{d})$ where $\chi(\bar{x}, \bar{y})$ is in L and $\bar{d}$ lists the distinct witnesses which occur in T_{i-1} but not in $\bar{c}_i$. By assumption the theory $T \cup \{\exists \bar{x} \exists \bar{y}\, \chi(\bar{x}, \bar{y})\}$ has a model. But Φ_m is unsupported, and it follows that there is some formula $\phi(\bar{x})$ in Φ_m such that $T \nvdash \forall \bar{x} (\exists \bar{y}\, \chi(\bar{x}, \bar{y}) \to \phi(\bar{x}))$. By the lemma on constants (Lemma 2.3.2) it follows that $T \nvdash \chi(\bar{c}_i, \bar{d}) \to \phi(\bar{c}_i)$. Hence she can put $T_i = T_{i-1} \cup \{\neg \phi(\bar{c}_i)\}$. Thus she fulfils her task. $\square$

In the proof of Theorem 6.2.1, each expert has to make sure that the theory T^+ has some particular property π. The proof shows that the expert can make T^+ have π, provided that she is allowed to choose T_i for infinitely many i. We can express this in terms of a game, call it $G(\pi, X)$. There are two players, $\forall$ (male, sometimes called $\forall$belard) and $\exists$ (female, $\exists$loise), and X is an infinite subset of ω with $\omega \backslash X$ infinite and $0 \notin X$. The players have to pick the sets T_i in turn; player $\exists$ makes the choice of T_i if and only if $i \in X$. Player $\exists$ wins if T^+ has property π; otherwise $\forall$ wins. We say that π is **enforceable** if player $\exists$ has a winning strategy for this game. (One can show that the question whether π is enforceable is independent of the choice of X, provided that both X and $\omega \backslash X$ are infinite and $0 \notin X$.) Some properties of T^+ are really properties of the canonical model A^+ – for example, that every element of A^+ is named by infinitely many witnesses. So we can talk of 'enforceable properties' of A^+ too.

The proof of Theorem 6.2.1 amounted to showing that the properties described in (2.2), (2.3)$_\psi$ and (2.4)$_m$ are enforceable. The main advantage of this viewpoint is that it breaks down the overall task into infinitely many smaller tasks, and these can be carried out independently without interfering with each other.

Atomic and prime models

We say that a structure A is **atomic** if for every tuple $\bar{a}$ of elements of A, the complete type of $\bar{a}$ in A is principal. A model A of a theory T is said to be **prime** if A be elementarily embedded in every model of T. (**Warning**: Don't confuse this with Abraham Robinson's notion of 'prime model' in section 7.3 below.)

Recall from section 5.2 that $S_n(T)$ is the set of complete first-order types $p(x_0, \ldots, x_{n-1})$ over the empty set with respect to models of T.

Theorem 6.2.2. *Let L be a countable first-order language and T a complete theory in L which has infinite models.*

(a) *If for every $n < \omega$, $S_n(T)$ is at most countable, then T has a countable atomic model.*

(b) *If B is a countable atomic L-structure which is a model of T, then B is a prime model of T.*

Proof. (a) There are only countably many non-principal complete types, so by Theorem 6.2.1 we can omit the lot of them in some model A of T. Since T is complete and has infinite models, A can be found with cardinality ω.

(b) Let A be any model of T. We shall show

(2.5)   if $\bar{a}, \bar{b}$ are n-tuples realising the same complete type in A, B respectively, and d is any element of B, then there is an element c of A such that $\bar{a}c$, $\bar{b}d$ realise the same complete $(n + 1)$-type in A, B respectively.

Since the complete type of $\bar{b}d$ is principal by assumption, it has a generator $\psi(\bar{x}, y)$. Since $\bar{a}$ and $\bar{b}$ realise the same complete type, and $B \vDash \exists y\, \psi(\bar{b}, y)$, we infer that $A \vDash \exists y\, \psi(\bar{a}, y)$, and hence there is an element c in A such that $A \vDash \psi(\bar{a}, c)$. Then $\bar{a}c$ realises the same complete type as $\bar{b}d$. This proves (2.5).

Now let $b_0, b_1, \ldots$ list all the elements of B. By induction on n, use (2.5) to find elements $a_0, a_1, \ldots$ of A so that for each $n, (A, a_0, \ldots, a_{n-1}) \equiv (B, b_0, \ldots, b_{n-1})$. Then the map $b_i \mapsto a_i$ is an elementary embedding of B into A, by the elementary diagram lemma (Lemma 2.5.3). $\qquad\square$

In short, if T is complete and all the sets $S_n(T)$ are countable then there is a 'smallest' countable model of T. It happens that under the same assumptions there is also a 'biggest' countable model of T, into which all other countable models of T can be elementarily embedded; see Exercise 6.

While the proof of Theorem 6.2.2 is fresh at hand, let me adapt it to prove another useful result which has nothing to do with countable structures.

Theorem 6.2.3. *Let L be a countable first-order language. Let A and B be two elementarily equivalent L-structures, both of which are atomic. Then A and B are back-and-forth equivalent.*

Proof. We show that if $\bar{a}$ and $\bar{b}$ are tuples in A and B respectively, such that $(A, \bar{a}) \equiv (B, \bar{b})$, then for every element c of A there is an element d of B such that $(A, \bar{a}, c) \equiv (B, \bar{b}, d)$, and conversely for every element d of B there is an element c of A such that $(A, \bar{a}, c) \equiv (B, \bar{b}, d)$. The argument is exactly the same as for (2.5) above. $\qquad\square$

Exercises for section 6.2

The first two questions refer to the games $G(\pi, X)$ described after the proof of Theorem 6.2.1.

1. Let X be an infinite subset of $\omega \backslash \{0\}$ such that $\omega \backslash X$ is infinite, and let Y be the set of odd positive integers. Show that player $\exists$ has a winning strategy for the game $G(\pi, X)$ if and only if she has one for $G(\pi, Y)$.

2. Show that if π_i is an enforceable property for each $i < \omega$, then the property $\bigwedge_{i<\omega}\pi_i$ (which T^+ has if it has all the properties π_i) is also enforceable.

3. Let T be a complete theory in a countable first-order language L, and suppose that T has infinite models. Show that T has a countable model A such that for every tuple $\bar{a}$ of elements of A there is a formula $\phi(\bar{x})$ with $A \vDash \phi(\bar{a})$, such that either (a) ϕ supports a complete type over T, or (b) no principal complete type over T contains ϕ.

4. (a) Let A be a finite or countable structure of countable signature. Show that A is atomic if and only if A is a prime model of $\text{Th}(A)$. (b) Deduce that any two prime models of a countable complete theory are isomorphic.

*Anticipating Chapter 8, an L-structure A is said to be ω-**saturated** if for every L-structure B and all tuples $\bar{a}, \bar{b}$ of elements of A, B respectively, if $(A, \bar{a}) \equiv (B, \bar{b})$ and d is an element of B, then there is an element c of A such that $(A, \bar{a}, c) \equiv (B, \bar{b}, d)$.*
5. (a) Show that if A and B are elementarily equivalent ω-saturated structures then A is back-and-forth equivalent to B. (b) Show that if A is ω-saturated and B is a countable structure elementarily equivalent to A, then B is elementarily embeddable in A.

6. Show that the following are equivalent, for any countable complete first-order theory T with infinite models. (a) T has a countable ω-saturated model. (b) T has a countable model A such that every countable model of T is elementarily embeddable in A. (c) For every $n < \omega$, $S_n(T)$ is at most countable.

7. Give an example of a countable first-order theory which has a countable prime model but no countable ω-saturated model. [Take infinitely many equivalence relations E_i so that each E_{i+1} refines E_i.]

8. Let L be a countable first-order language, T a theory in L, and $\Phi(\bar{x})$ and $\Psi(\bar{y})$ sets of formulas of L. Show that (a) implies (b): (a) for every formula $\sigma(\bar{x}, \bar{y})$ of L there is a formula $\psi(\bar{y})$ in Ψ such that for all $\phi_1(\bar{x}), \ldots, \phi_n(\bar{x})$ in Φ, if $T \cup \{\exists\bar{x}\bar{y}(\sigma \wedge \phi_1 \wedge \ldots \wedge \phi_n)\}$ has a model then $T \cup \{\exists\bar{x}\bar{y}(\sigma \wedge \phi_1 \wedge \ldots \wedge \phi_n \wedge \neg\psi)\}$ has a model; (b) T has a model which realises Φ and omits Ψ.

9. Show that the theory T of Example 1 is complete as follows. If A is a model of T and s is a subset of ω, $|\Phi_s(A)|$ is the number of elements of A which realise Φ_s. (a) Show that if A is a model of T, then A is determined up to isomorphism by the cardinals $|\Phi_s(A)|$ ($s \subseteq \omega$). (b) Show that if $s \subseteq \omega$ and A is a model of T of cardinality $\leqslant 2^\omega$, then A has an elementary extension B of cardinality 2^ω with $|\Phi_s(B)| = 2^\omega$. (c) By iterating (b), show that every model of T of cardinality $\leqslant 2^\omega$ has an elementary extension C of cardinality 2^ω with $|\Phi_s(C)| = 2^\omega$ for each $s \subseteq \omega$.

6.3 Countable categoricity

A complete theory which has exactly one countable model up to isomorphism is said to be ω-**categorical**. We also say that a structure A is ω-**categorical** if $\mathrm{Th}(A)$ is ω-categorical. People with Hebrew on their word processors sometimes say $\aleph_0$-**categorical**.

Our first result is a handful of characterisations of ω-categoricity. It rests on the countable omitting types theorem (Theorem 6.2.1 in the previous section).

Theorem 6.3.1 (*Theorem of Engeler, Ryll-Nardzewski and Svenonius*). *Let L be a countable first-order language and T a complete theory in L which has infinite models. Then the following are equivalent.*

(a) *Any two countable models of T are isomorphic.*

(b) *If A is any countable model of T, then $\mathrm{Aut}(A)$ is oligomorphic (i.e. for every $n < \omega$, $\mathrm{Aut}(A)$ has only finitely many orbits in its action on n-tuples of elements of A).*

(c) *T has a countable model A such that $\mathrm{Aut}(A)$ is oligomorphic.*

(d) *Some countable model of T realises only finitely many complete n-types for each $n < \omega$.*

(e) *For each $n < \omega$, $S_n(T)$ is finite.*

(f) *For each $\bar{x} = (x_0, \ldots, x_{n-1})$, there are only finitely many pairwise non-equivalent formulas $\phi(\bar{x})$ of L modulo T.*

(g) *For each $n < \omega$, every type in $S_n(T)$ is principal.*

Proof. (b) $\Rightarrow$ (c) is immediate, since T has a countable model (by the downward Löwenheim–Skolem theorem, Corollary 3.1.4). (c) $\Rightarrow$ (d) is also direct: automorphisms preserve all formulas.

(d) $\Rightarrow$ (e). Let A be a countable model of T realising only finitely many complete n-types for each $n < \omega$. For a fixed n, let $p_0, \ldots, p_{k-1}$ be the distinct types in $S_n(T)$ which are realised in A. For each of these types p_i there is a formula $\phi_i(\bar{x})$ of L which is in p_i but not in p_j when $j \neq i$. Since $A \vDash \forall \bar{x} \bigvee_{i<k} \phi_i(\bar{x})$, this sentence $\forall \bar{x} \bigvee_{i<k} \phi_i(\bar{x})$ must be a consequence of the complete theory T. Likewise if $\psi(\bar{x})$ is any formula of L, then A is a model of the sentence $\forall \bar{x} \bar{y} (\phi_i(\bar{x}) \wedge \phi_i(\bar{y}) \rightarrow (\psi(\bar{x}) \leftrightarrow \psi(\bar{y})))$, for every $i < k$, and these sentences must also be consequences of T. It follows that $p_0, \ldots, p_{k-1}$ are the only types in $S_n(T)$.

(e) $\Rightarrow$ (f). If two formulas $\phi(\bar{x})$ and $\psi(\bar{x})$ of L lie in exactly the same types $\in S_n(T)$, then ϕ and ψ must be equivalent modulo T. So if $S_n(T)$ has finite cardinality k, then there are at most 2^k non-equivalent formulas $\phi(\bar{x})$ of L modulo T.

(f) $\Rightarrow$ (g). For any $n < \omega$ and $\bar{x} = (x_0, \ldots, x_{n-1})$, take a maximal family of pairwise non-equivalent formulas $\phi(\bar{x})$ of L modulo T. Assuming (f), this family is finite. If p is any type $\in S_n(T)$, let θ be the conjunction of all formulas of the family which also lie in p. Then θ is a support of p.

(a) $\Rightarrow$ (g). Suppose (g) fails. Then for some $n < \omega$, there is a non-principal type q in $S_n(T)$. By the omitting types theorem (Theorem 6.2.1), T has a model A which omits q. By the definition of types, T also has a model B which realises q. Since T is complete and has infinite models, both A and B are infinite. By the downward Löwenheim–Skolem theorem we can suppose that both A and B are countable. Hence T has two countable models which are not isomorphic, and thus (a) fails.

(g) $\Rightarrow$ (a). By (g) all models of T are atomic, and hence back-and-forth equivalent by Theorem 6.2.3. It follows by Theorem 3.2.3(b) that all countable models of T are isomorphic.

(g) $\Rightarrow$ (b). Again we deduce from (g) that all models of T are atomic. But rather than quote Theorem 6.2.3, we extract part of its proof:

(3.1)   if A, B are countable models of T and $\bar{a}, \bar{b}$ are n-tuples in A, B respectively, such that $(A, \bar{a}) \equiv (B, \bar{b})$, then there is an isomorphism from A to B which takes $\bar{a}$ to $\bar{b}$.

Now let A be a countable model of T and let $\bar{a}, \bar{b}$ be n-tuples which realise the same complete type in A. Then by (3.1), $\bar{a}$ and $\bar{b}$ lie in the same orbit of Aut(A). So to deduce (b), we need only show that (g) implies (e).

Assume $S_n(T)$ is infinite. Suppose $S_n(T)$ contains λ principal types, and let $\theta_i(\bar{x})$ $(i < \lambda)$ be supports of these types. Take an n-tuple $\bar{c}$ of distinct new constants, and let T' be the theory

(3.2) $T \cup \{\neg \theta_i(\bar{c}): i < \lambda\}.$

If $(A, \bar{a})$ is a model of T', then A is a model of T in which $\bar{a}$ realises a non-principal type. So it suffices to prove that T' has a model. If $\Phi(\bar{x})$ is a finite subset of $\{\theta_i(\bar{x}): i < \lambda\}$, then since $S_n(T)$ is infinite, there is a type $p(\bar{x})$ in $S_n(T)$ distinct from the types generated by the formulas in Φ. Hence every finite subset of T' has a model, and the compactness theorem does the rest.

□

Where do ω-categorical structures occur in nature?

As soon as Theorem 6.3.1 was proved, model theorists went searching through the mathematical archives for examples of ω-categorical structures. It was natural to ask, for example, what countable rings are ω-categorical. There was also a good hope of interesting mathematicians outside model theory, because one can explain what it means for Aut(A) to be oligomorphic without mentioning any notions from logic. The following corollary was a handy starting point.

Corollary 6.3.2. *If A is an ω-categorical structure, then A is locally finite. In fact there is a (unique) function $f: \omega \to \omega$, depending only on $\mathrm{Th}\,(A)$, with the property that for each $n < \omega$, $f(n)$ is the least number m such that every n-generator substructure of A has at most m elements.*

Proof. Let $\bar{a}$ be an n-tuple of elements of A. If c and d are two distinct elements of the substructure $\langle \bar{a} \rangle_A$ generated by $\bar{a}$, then the complete types of $\bar{a}c$ and $\bar{a}d$ over the empty set must be different, because they say how c and d are generated. So by (e) of the theorem for $n + 1$, $\langle \bar{a} \rangle_A$ is finite. This proves the first sentence.

Now let B be the unique countable structure elementarily equivalent to A. By (b) of the theorem, for each $n < \omega$ there are finitely many orbits of n-tuples in B; let $\bar{b}_0, \ldots, \bar{b}_{k-1}$ be representatives of these orbits, and write m_i for the number of elements of the substructure $\langle \bar{b}_i \rangle_B$ generated by $\bar{b}_i$. Then we can put $f(n) = \max(m_i: i < k)$. This choice of f works for A as well as B, since A and B realise exactly the same types in $S_n(T)$, namely all of them. $\qquad\square$

Example 1: ω-categorical groups. By the corollary, every countable ω-categorical group is locally finite and has finite exponent. For abelian groups, this is the end of the story: any abelian group A of finite exponent is a direct sum of finite cyclic groups, and we can write down a first-order theory which says how often each cyclic group occurs in the sum (where the number of times is either $0, 1, 2, \ldots$ or infinity). So an infinite abelian group is ω-categorical if and only if it has finite exponent. For groups in general the situation is much more complicated.

Interpretations and ω-categorical structures

In the course of proving Theorem 6.3.1, we found the property which I numbered (3.1). This property is worth extracting.

Corollary 6.3.3. *Let L be a countable first-order language. Let A be an L-structure which is either finite, or countable and ω-categorical. Then for any positive integer n, a pair $\bar{a}, \bar{b}$ of n-tuples from A are in the same orbit under $\mathrm{Aut}\,(A)$ if and only if they satisfy the same formulas of L.*

Proof. As in the proof of Theorem 6.3.1, it suffices to show that A is atomic. We know this by (g) of the theorem when A is countable and ω-categorical. When A is finite, we deduce it by the argument of (d) $\Rightarrow$ (e) in the proof of the theorem. $\qquad\square$

Since there are finitely many types of $S_n(T)$, and all of them are principal, Corollary 6.3.3 can be rephrased as follows (recalling that a formula is **principal** if it generates a complete type).

Corollary 6.3.4. *Let L be a countable first-order language. Let A be an L-structure which is either finite, or countable and ω-categorical. Then for each n there are finitely many complete formulas $\phi_i(x_0, \ldots, x_{n-1})$ $(i < k_n)$ of L for* Th(A), *and the orbits of* Aut(A) *on* $(\text{dom } A)^n$ *are exactly the sets $\phi_i(A^n)$* $(i < k_n)$. $\qquad\qquad\square$

This tells us in particular that we can almost recover A from the permutation group Aut(A).

Theorem 6.3.5. *Let A be a countable ω-categorical L-structure with domain Ω, and let B be the canonical structure for* Aut(A) *on Ω (see section 4.1 above). Then the relations on Ω which are first-order definable in A without parameters are exactly the same as those definable in B without parameters; in other words, A and B are definitionally equivalent (see section 2.6).*

Proof. By definition of the canonical structure B, it has the same automorphism group as A; write G for this group. Write L' for the language of B; we assume it is disjoint from L. If R is an n-ary relation symbol of L, then R^A is a union of finitely many orbits of G on Ω, and so R can be defined by a disjunction of formulas of L' which define these orbits. The same argument works in the other direction too. $\qquad\qquad\square$

We note that interpretations always preserve ω-categoricity:

Theorem 6.3.6. *Let K and L be countable first-order languages, Γ an interpretation of L in K, and A an ω-categorical K-structure. Then ΓA is ω-categorical.*

Proof. Let A' be a countable structure which is elementarily equivalent to A. Then $\Gamma A' \equiv \Gamma A$ by the reduction theorem (Theorem 4.3.1), and so it suffices to show that $\Gamma A'$ is ω-categorical. By the construction in Theorem 4.3.2, every element of $\Gamma A'$ is an equivalence class of the relation $=_\Gamma$ on dom (A'); write $\bar{a}^=$ for the equivalence class containing the tuple $\bar{a}$. Each automorphism α of A' induces an automorphism $\Gamma(\alpha)$ of $\Gamma A'$, by the rule $\Gamma(\alpha)(\bar{a}^=) = (\alpha \bar{a})^=$. Since Aut$(A')$ is oligomorphic, it follows at once that Aut$(\Gamma A')$ is oligomorphic too. $\qquad\qquad\square$

In particular, relativised reducts of ω-categorical structures are ω-categorical.

Exercises for section 6.3

1. Show that if A is an infinite structure and $\bar{a}$ is a tuple of elements of A, then $\text{Th}(A, \bar{a})$ is ω-categorical if and only if $\text{Th}(A)$ is ω-categorical.

2. Show that if A is a countable structure, then A is ω-categorical if and only if for every tuple $\bar{a}$ of elements of A, the number of orbits of $\text{Aut}(A, \bar{a})$ on single elements of A is finite.

3. Show that if T is a first-order theory with countable models, then T is ω-categorical if and only if all the models of T are pairwise back-and-forth equivalent.

4. Let T be a complete and countable first-order theory. Show that T is ω-categorical if and only if every countable model of T is atomic.

5. Give an example of an ω-categorical first-order theory T such that no skolemisation of T is ω-categorical. [Rationals!]

6. Let B be a countable boolean algebra. Show that B is ω-categorical if and only if B has finitely many atoms.

6.4 ω-categorical structures by Fraïssé's method

Fraïssé's construction from section 6.1 above has proved to be a very versatile way of building ω-categorical structures.

The trick is to make sure that if **K** is the class whose Fraïssé limit we are taking, the sizes of the structures in **K** are kept under control by the number of generators. For this, we say that a structure A is **uniformly locally finite** if there is a function $f: \omega \to \omega$ such that

(4.1)   for every substructure B of A, if B has a generator set of cardinality $\leq n$ then B itself has cardinality $\leq f(n)$.

We say that a class **K** of structures is **uniformly locally finite** if there is a function $f: \omega \to \omega$ such that (4.1) holds for every structure A in **K**.

Note that if the signature of **K** is finite and has no function symbols, then **K** is uniformly locally finite. (See Exercise 1.2.6.)

Theorem 6.4.1. *Suppose that the signature L is finite and K is a countable uniformly locally finite set of finitely generated L-structures with HP, JEP and AP. Let M be the Fraïssé limit of K, and let T be the first-order theory $\text{Th}(M)$ of M.*
   (a) T is ω-categorical,
   (b) T has quantifier elimination.

Proof. First we show that there is an $\forall_2$ theory U in L whose models are precisely the weakly homogeneous structures (see (1.7) in section 6.1) of age **K**. There are two crucial points here. The first is that by our assumption on L, if A is any finite L-structure with n generators $\bar{a}$, then there is a quantifier-free formula $\psi = \psi_{A,\bar{a}}(x_0, \ldots, x_{n-1})$ such that for any L-structure B and n-tuple $\bar{b}$ of elements of B,

(4.2) $B \vDash \psi(\bar{b})$ if and only if

there is an isomorphism from A to $\langle \bar{b} \rangle_B$ which takes $\bar{a}$ to $\bar{b}$.

In fact $\psi_{A,\bar{a}}$ is a conjunction of literals satisfied by $\bar{a}$ in A. The second is that by the uniform local finiteness, for each $n < \omega$ there are only finitely many isomorphism types of structures in **K** with n generators.

These two facts can both be checked by inspection. Given them, we take U_0 to be the set of all sentences of the form

(4.3) $\forall \bar{x}(\psi_{A,\bar{a}}(\bar{x}) \to \exists y \, \psi_{B,\bar{a}b}(\bar{x}, y))$

where B is a structure in **K** generated by a tuple $\bar{a}b$ of distinct elements, and A is the substructure generated by $\bar{a}$. This includes the case where $\bar{a}$ is empty, so that the sentence (4.3) reduces to $\exists y \, \psi_{B,b}(y)$. We take U_1 to be the set of all sentences of the form

(4.4) $\forall \bar{x} \bigvee_{A,\bar{a}} \psi_{A,\bar{a}}(\bar{x})$

where the disjunction ranges over all pairs $A, \bar{a}$ such that A is in **K** and $\bar{a}$ is a tuple of the same length as $\bar{x}$ which generates A. Uniform local finiteness implies that this is a finite disjunction (up to logical equivalence). We write U for the union $U_0 \cup U_1$. Clearly M is a model of U.

Suppose D is any countable model of U. When $\bar{a}$ is empty, the sentences (4.3) say that every one-generator structure in **K** is embeddable in D. In general the sentences (4.3) say that

(4.5)   if A, B are finitely generated substructures of D, $A \subseteq B$, B
          comes from A by adding one more generator, and $f: A \to D$
          is an embedding, then there is an embedding $g: B \to D$ which
          extends f.

It's not hard to see, using induction on the number of generators, that these two facts imply that every structure in **K** is embeddable in D; so together with the sentences (4.4), they tell us that the age of D is exactly **K**. Using the sentences (4.3) again, an induction on the size of $\text{dom}(B)\backslash\text{dom}(A)$ tells us that D is weakly homogeneous. So by Lemma 6.1.4, D is isomorphic to M. Hence U is ω-categorical, and U is a set of axioms for T.

Suppose now that $\phi(\bar{x})$ is a formula of L with $\bar{x}$ non-empty, and let X be the set of all tuples $\bar{a}$ in M such that $M \vDash \phi(\bar{a})$. If $\bar{a}$ is in X, and $\bar{b}$ is a tuple of elements such that there is an isomorphism $e: \langle \bar{a} \rangle_M \to \langle \bar{b} \rangle_M$ taking $\bar{a}$ to $\bar{b}$, then e extends to an automorphism of M, so that $\bar{b}$ is in X too. It follows that ϕ is equivalent modulo T to the disjunction of all the formulas $\psi_{\langle \bar{a} \rangle, \bar{a}}(\bar{x})$ with $\bar{a}$ in X. This is a finite disjunction of quantifier-free formulas. Finally if ϕ is a sentence of L, then since T is complete, ϕ is equivalent modulo T to either $\neg \bot$ or $\bot$. So T has quantifier elimination. $\square$

Corollary 6.4.2. *Let L be a finite signature and M a countable L-structure. Then the following are equivalent.*
(a) *M is ultrahomogeneous and uniformly locally finite.*
(b) *$\mathrm{Th}(M)$ is ω-categorical and has quantifier elimination.*

Proof. (a) $\Rightarrow$ (b) is by Theorem 6.4.1. (b) $\Rightarrow$ (a) is by Corollaries 6.3.2 and 6.3.3 above. $\square$

There are so many attractive applications of Theorem 6.4.1 that it's hard to know which to mention first. I describe two examples in detail and leave others to the exercises.

First application: the random graph

A **graph** is a structure consisting of a set X with an irreflexive symmetric binary relation R defined on X (see Example 1 in section 1.1). The elements of X are called the **vertices**; an **edge** is a pair of vertices $\{a, b\}$ such that aRb. We say that two vertices a, b are **adjacent** if $\{a, b\}$ is an edge. A **path of length** n is a sequence of edges $\{a_0, a_1\}, \{a_1, a_2\}, \ldots, \{a_{n-2}, a_{n-1}\}$, $\{a_{n-1}, a_n\}$; the path is a **cycle** if $a_n = a_0$. A **subgraph** of a graph G is simply a substructure of G. We write L for the first-order language appropriate for graphs; its signature consists of just one binary relation symbol R.

Let **K** be the class of all finite graphs. The following facts hardly need proof.

Lemma 6.4.3. *The class **K** has HP, JEP and AP. Also **K** contains arbitrarily large finite structures and is uniformly locally finite. The signature of **K** contains only finitely many symbols.* $\square$

So by Theorems 6.1.2 and 6.4.1, **K** has a Fraïssé limit A; $\mathrm{Th}(A)$ is ω-categorical and has quantifier elimination. The structure A is a countable graph known as the **random graph**. We shall denote it by Γ.

Some Fraïssé limits are hard to describe in detail. Not so this one.

Theorem 6.4.4. *Let A be a countable graph. The following are equivalent.*
(a) *A is the random graph Γ.*

(b) *Let X and Y be disjoint finite sets of vertices of A; then there is an element $\notin X \cup Y$ which is adjacent to all vertices in X and no vertices in Y.*

Proof. (a) $\Rightarrow$ (b). Let A be the random graph Γ and let X, Y be disjoint finite sets of vertices of Γ. There is a finite graph G as follows: the vertices of G are the vertices in $X \cup Y$ together with one new vertex v, and vertices in $X \cup Y$ are adjacent in G if they are adjacent in Γ, while v is adjacent to all the vertices in X and none of the vertices in Y. Since Γ is the Fraïssé limit of **K**, there is an embedding $f: G \to \Gamma$. The restriction of f to $X \cup Y$ is an isomorphism between finite substructures of Γ, and so it extends to an automorphism g of Γ. Then $g^{-1}f(v)$ is the element described in (b).

(b) $\Rightarrow$ (a). Assume (b). We make the following claim.

(4.6)   Suppose $G \subseteq H$ are finite graphs and $f: G \to A$ is an embedding. Then f extends to an embedding $g: H \to A$.

This is proved by induction on the number n of vertices which are in H but not in G. Clearly we only need worry about the case $n = 1$. Let w be the vertex which is in H but not in G. Let X be the set of vertices $f(x)$ such that x is in G and adjacent to w in H, and let Y be the set of vertices $f(y)$ such that y is in G but not adjacent to w. By (b) there is a vertex v of A which is adjacent to all of X and none of Y. We extend f to g by putting $g(w) = v$. This proves the claim.

Taking G to be the empty structure, it follows that every finite graph is embeddable in A, so that the age of A is **K**. Taking G to be a substructure of A, it follows that A is weakly homogeneous. So by Lemma 6.1.4, A is the Fraïssé limit of **K**, proving (a). $\square$

Second application: the random structure

The main interest of our next example is that it leads directly to a theorem about finite models. Until recently the model theory of finite structures was fallow ground. A team of Russians from Gorki planted the first seed by proving Theorem 6.4.6 in 1969.

Let L be a non-empty finite signature with no function symbols, and let **K** be the class of all finite L-structures. Then clearly **K** has HP, JEP and AP, and (by Exercise 1.2.6) there are just countably many isomorphism types of structures in **K**. So **K** has a countable Fraïssé limit, which is known as the **random structure** of signature L. Let us call it $\mathrm{Ran}(L)$.

Let T be the set of all sentences of L of the form

(4.7) $\forall \bar{x}(\psi(\bar{x}) \to \exists y \, \chi(\bar{x}, y))$

such that for some finite L-structure B and some listing of the elements of B without repetition as $\bar{b}, c$, the formula $\psi(\bar{x})$ (resp. $\chi(\bar{x}, y)$) lists the literals satisfied by $\bar{b}$ (resp. $\bar{b}, c$) in B. Inspection shows that T consists of exactly the sentences (4.3) of the proof of Theorem 6.4.1. The sentences (4.4) of that proof are trivially true in this case, so they add nothing. It follows that T is a set of axioms for the theory of Ran(L). In particular T is complete.

Consider any $n < \omega$, any formula $\phi(x_0, \ldots, x_{n-1})$ of L and any tuple $\bar{a}$ of objects a_i with $i < n$. We write $\kappa_n(\phi)$ for the number of non-isomorphic L-structures B whose distinct elements are $a_0, \ldots, a_{n-1}$, such that $B \vDash \phi(\bar{a})$. We write $\mu_n(\phi(\bar{a}))$ for the ratio $\kappa_n(\phi(\bar{a}))/\kappa_n(\forall x \, x = x)$, i.e. the proportion of those L-structures with elements $a_0, \ldots, a_{n-1}$ for which $\phi(\bar{a})$ is true.

Lemma 6.4.5. *Let ϕ be any sentence in T. Then $\lim_{n\to\infty} \mu_n(\phi) = 1$.*

Proof. Let ϕ be the sentence $\forall \bar{x}(\psi(\bar{x}) \to \exists y \, \chi(\bar{x}, y))$. We shall show that $\lim_{n\to\infty} \mu_n(\neg \phi) = 0$. Since $\mu_n(\neg \phi) = 1 - \mu_n(\phi)$, this will prove the lemma.

Suppose $\bar{x}$ is $(x_0, \ldots, x_{m-1})$, and $n > m$. Consider those structures B whose distinct elements are $a_0, \ldots, a_{n-1}$, such that $B \vDash \psi(a_0, \ldots, a_{m-1})$. What is the probability p that $B \vDash \forall y \, \neg\chi(a_0, \ldots, a_{m-1}, y)$? It must be the $(n - m)$-th power of the probability that $B \vDash \neg\chi(a_0, \ldots, a_{m-1}, a_m)$, since the $n - m$ elements $a_m, \ldots, a_{n-1}$ have equal and independent chances of serving for y. Since the signature of L is not empty, there is some positive real $k < 1$ such that $B \vDash \neg\chi(a_0, \ldots, a_{m-1}, a_m)$ with probability k, and so $p = k^{n-m}$.

Next, consider those L-structures C whose distinct elements are $a_0, \ldots, a_{n-1}$. We estimate the probability $\mu_n(\neg \phi)$ that $C \vDash \neg \phi$. This probability q is at most the probability that $C \vDash \psi(\bar{c}) \land \forall y \, \neg\chi(\bar{c}, y)$ for a tuple $\bar{c}$ of distinct elements of C, times the number of ways of choosing $\bar{c}$ in C. So $\mu_n(\neg \phi) \leqslant n^m \cdot k^{n-m} = \gamma \cdot n^m \cdot k^n$ where $\gamma = k^{-m}$.

Now since $0 < k < 1$, we have that $n^m \cdot k^n \to 0$ as $n \to \infty$ (see Exercise 9). It follows that $\lim_{n\to\infty} \mu_n(\neg \phi) = 0$ as claimed. $\qquad\square$

Theorem 6.4.6 *(Zero–one law).* *Let ϕ be any first-order sentence of a finite relational signature. Then $\lim_{n\to\infty} \mu_n(\phi)$ is either 0 or 1.*

Proof. We have already seen that T is a complete theory. If ϕ is a consequence of T, then it follows from the lemma that $\lim_{n\to\infty} \mu_n(\phi)$ is 1. If ϕ is not a consequence of T, then T implies $\neg \phi$, and so $\lim_{n\to\infty} \mu_n(\neg \phi)$ is 1, whence $\lim_{n\to\infty} \mu_n(\phi)$ is 0. $\qquad\square$

Exercises for section 6.4

1. Show that if M is a countable ω-categorical structure, then there is a definitional expansion of M which is ultrahomogeneous. [Atomise.]

2. Let L be the first-order language whose signature consists of one 2-ary relation symbol R. For each integer $n \geqslant 2$, let A_{n-2} be the L-structure with domain n, such that $A_n \vDash \neg R(i, j)$ iff $i + 1 \equiv j \pmod{n}$. If S is any infinite subset of ω, we write $\mathbf{J}_S$ for the class $\{A_n : n \in S\}$ and $\mathbf{K}_S$ for the class of all finite L-structures C such that no structure in $\mathbf{J}_S$ is embeddable in C. Show (a) for each infinite $S \subseteq \omega$, the class $\mathbf{K}_S$ has HP, JEP and AP, and is uniformly locally finite, (b) if S and S' are distinct infinite subsets of ω then the Fraïssé limits of $\mathbf{K}_S$ and $\mathbf{K}_{S'}$ are non-isomorphic countable ω-categorical L-structures, (c) if L is the signature with one binary relation symbol, there are continuum many non-isomorphic ultrahomogeneous ω-categorical structures of signature L.

3. We define a graph A on the set of vertices $\{c_i : i < \omega\}$ as follows. When $i < j$, first write j as a sum of distinct powers of 2, and then make c_i adjacent to c_j iff 2^i occurs in this sum. Show that A is the random graph.

4. Show that if Γ is the random graph, then the vertices of Γ can be listed as $\{v_i : i < \omega\}$ in such a way that for each i, v_i is joined to v_{i+1}.

5. Show that if the set of vertices of the random graph Γ is partitioned into finitely many sets X_i $(i < n)$, then there is some $i < n$ such that the restriction of Γ to X_i is isomorphic to Γ.

The **complete graph on** n **vertices**, K_n, *is the graph with n vertices, such that vertex v is joined to vertex w iff $v \neq w$.*
6. Let n be an integer $\geqslant 3$ and let $\mathbf{K}_n$ be the class of finite graphs which do not have K_n as a subgraph. (a) Show that $\mathbf{K}_n$ has HP, JEP and AP. (b) If Γ_n is the Fraïssé limit of $\mathbf{K}_n$, show that Γ_n is the unique countable graph with the following two properties: (i) every finite subgraph of Γ_n is in $\mathbf{K}_n$, and (ii) if X and Y are disjoint finite sets of vertices of Γ_n, and K_{n-1} is not embeddable in the restriction of Γ_n to X, then there is a vertex in Γ_n which is joined to every vertex in X and to no vertex in Y.

7. Show that the zero–one law (Theorem 6.4.6) still holds if the signature L is empty.

Theorem 6.4.6 fails if the language has function symbols.
8. Let L be the first-order language whose signature consists of one 1-ary function symbol F. Show that $\lim_{n \to \infty} \mu_n(\forall x \, F(x) \neq x) = 1/e$. $[\mu_n(\forall x \, F(x) \neq x) = (n-1)^n/n^n.]$

9. Prove that if $0 < k < 1$ and m is a positive integer, then $n^m . k^n \to 0$ as $n \to \infty$. [Put $k = 1/(1 + p)$. If $n > 2(m + 1)$ then by considering the $(m + 1)$-th term in the binomial expansion of $(1 + p)^n$, $(1 + p)^n > ((n/2)^{(m+1)} . p^{(m+1)})/(m + 1)!$. So $n^m . k^n < (n^m . (m + 1)!)/((n/2)^{(m+1)} . p^{(m+1)}) = \beta/n$ for some $\beta > 0$ independent of n.]

Further reading

We now recognize Roland Fraïssé as one of the major founders of model theory. But for some decades he and his group worked in semi-isolation from the mainstream. This helps to account for his highly original vision on the one hand, and his unusual terminology on the other. One can sample both in his book:

Fraïssé, R. *Theory of relations*. Amsterdam: North-Holland, 1986.

Robert Vaught's influential paper on omitting types is still a joy to read:

Vaught, R. L. Denumerable models of complete theories. In *Infinitistic methods, Proceedings of a Symposium in Foundations of Mathematics, Warsaw 1959*, pp. 303–321. Warsaw: Państwowe Wydawnictwo Naukowe, 1961.

The paper announced 'Vaught's conjecture', that the number of non-isomorphic countable models of a complete countable first-order theory is always either 2^ω or $\leq \omega$. Two papers report the state of this conjecture, which is still open:

Steel, J. On Vaught's conjecture. In *Cabal Seminar 76–77*, Kechris, A. S. and Moschovakis, Y. N. (eds), Lecture Notes in Mathematics 689, pp. 193–208. Berlin: Springer, 1978.

Buechler, S. Vaught's conjecture for superstable theories of finite rank. *Annals of Pure and Applied Logic*, to appear.

One can study ω-categorical theories through their automorphism groups, which are the same as oligomorphic permutation groups on countable sets:

Cameron, P. J. *Oligomorphic permutation groups*. Cambridge: Cambridge University Press, 1990.

Zero-one laws on finite structures matter in theoretical computer science:

Gurevich, Y. Zero-one laws. *Bulletin of the European Association for Theoretical Computer Science* **46** (1992), 90–106.

7

The existential case

J'ay trouvé quelques élémens d'une nouvelle caracteristique, tout à fait différent de l'Algebre, et qui aura des grands avantages pour representer à l'esprit exactement et au naturel, quoyque sans figures, tout ce qui depend de l'imagination. ... L'utilité principale consiste dans les consequences et raisonnements, qui se peuvent fair par les operations des caracteres, qui ne sçauroient exprimer par des figures (et encor mois par des modelles). ... Cette caracteristique servira beaucoup à trouver de belles constructions, parceque le calcul et la construction s'y trouvent tout à la fois.

Leibniz, Letters to Huygens (1679).

Abraham Robinson referred to these letters of Leibniz in his address to the International Congress of Mathematicians in 1950. It seemed to Robinson that Leibniz hinted at Robinson's own ambition: to make logic useful for 'actual mathematics, more particularly for the development of algebra and ... algebraic geometry'.

Robinson introduced many of the fundamental notions of model theory. He gave us diagrams (section 1.4), preservation of formulas (section 2.5), model-completeness and model companions (both in this chapter), model-theoretic forcing and nonstandard methods (not covered in this book). We shall explore model-completeness and some related notions. They allow us to develop a tidy piece of pure model theory for its own sake, and at the same time they fit snugly onto various pieces of 'actual mathematics'. Often they lead to neater and more conceptual proofs of known mathematical theorems.

Of course not everything in this chapter is the work of Abraham Robinson. The property of quantifier elimination harks back to the work of Skolem and Tarski in the years after the first world war. Existentially closed structures are one of those happy ideas which occur to several different mathematicians at different times and places, beginning with those Renaissance scholars who invented imaginary numbers.

The ideas in this chapter are perhaps the main methods of applied model theory. I have given plenty of examples along the way; but to understand these things properly one should apply the methods vigorously to some concrete branch of mathematics. Some opportunities for doing this are listed at the end of the chapter.

7.1 Existentially closed structures

Take a first-order language L without relation symbols and a class **K** of L-structures. For example L might be the language of rings and **K** the class of fields; or L might be the language of groups and **K** the class of groups. We say that a structure A in **K** is **existentially closed in K** (or more briefly, **e.c. in K**) if the following holds:

(1.1) if E is a finite set of equations and inequations with para-
        meters from A, and E has a simultaneous solution in some
        extension B of A with B in **K**, then E has a solution already
        in A.

We can rewrite this definition in a model-theoretic form. The model-theoretic version has the advantage that it covers languages which have relation symbols too.

A formula is **primitive** if it has the form $\exists \bar{y} \bigwedge_{i<n} \psi_i(\bar{x}, \bar{y})$, where n is a positive integer and each formula ψ_i is a literal. (In a language without relation symbols, each literal is either an equation or an inequation; so a primitive formula expresses that a certain finite set of equations and inequations has a solution.) We say that a structure A in the class **K** of L-structures is **e.c. in K** if

(1.2) for every primitive formula $\phi(\bar{x})$ of L and every tuple $\bar{a}$ in
        A, if there is a structure B in **K** such that $A \subseteq B$ and
        $B \vDash \phi(\bar{a})$, then already $A \vDash \phi(\bar{a})$.

This definition agrees with (1.1).

When **K** is the class of fields, an e.c. structure in **K** is known as an **e.c. field**; likewise **e.c. lattice** when **K** is the class of lattices, and so on. When **K** is the class of all models of a theory T, we refer to e.c. structures in **K** as **e.c. models of** T.

The next lemma explains the name 'existentially closed'. If L is a first-order language and A, B are L-structures, we write $A \leqslant_1 B$ to mean that for every existential formula $\phi(\bar{x})$ of L and every tuple $\bar{a}$ in A, $B \vDash \phi(\bar{a})$ implies $A \vDash \phi(\bar{a})$.

Lemma 7.1.1. *Let* **K** *be a class of L-structures and A a structure in* **K**. *Then A is e.c. in* **K** *if and only if* (1.2) *holds with 'primitive' replaced by 'existential'. In particular if A and B are structures in* **K**, $A \subseteq B$ *and A is e.c., then $A \leqslant_1 B$.*

Proof. By disjunctive normal form, every $\exists_1$ formula is logically equivalent to a disjunction of primitive formulas. $\qquad\qquad\square$

Example 1: *Algebraically closed fields.* What are the e.c. fields? Certainly an e.c. field A must be algebraically closed. For suppose $F(y)$ is a polynomial of positive degree with coefficients from A. Using the language of rings, we can rewrite $F(y)$ as $p(\bar{a}, y)$ where $p(\bar{x}, y)$ is a term and $\bar{a}$ is a tuple of elements of A. Replacing F by an irreducible factor if necessary, we have a field $B = A[y]/(F)$ which extends A and contains a root of F. So $B \vDash \exists y\, p(\bar{a}, y) = 0$. Since A is an e.c. field, $A \vDash \exists y\, p(\bar{a}, y) = 0$ too, so that F has a root already in A. Thus every e.c. field is algebraically closed.

The converse holds as well, using results proved below:

(1.3)   if A is an algebraically closed field, then every finite system
         of equations and inequations over A which is solvable in
         some field extending A already has a solution in A.

This is one form of Hilbert's Nullstellensatz; cf. Exercise 9 for another form. If $E(\bar{x})$ is a finite system of equations and inequations over A, then we can write the statement 'E has a solution' as a primitive formula $\phi(\bar{a})$ where $\bar{a}$ are the coefficients of E in A. To prove (1.3), suppose E has a solution in some field B which extends A. Extend B to an e.c. field C by Theorem 7.2.1 below. Then $B \vDash \phi(\bar{a})$ by assumption, so $C \vDash \phi(\bar{a})$ since $\phi(\bar{a})$ is an $\exists_1$ formula (see Theorem 2.4.1). But C is algebraically closed by the argument of the previous paragraph, and Example 2 in section 7.4 below tells us that the theory of algebraically closed fields is model-complete. Since $A \subseteq C$, it follows that $A \vDash \phi(\bar{a})$, and so E already has a solution in A. In short, *the e.c. fields are precisely the algebraically closed fields*.

A remark of Rabinowitsch is worth recalling here. Note first that every equation with parameters $\bar{a}$ from a field A can be written as $p(\bar{a}, \bar{y}) = 0$, where $p(\bar{x}, \bar{y})$ is a polynomial whose indeterminates are variables from $\bar{x}$, $\bar{y}$, with integer coefficients. So we can assume that any primitive formula has the form

(1.4) $\exists \bar{y}(p_0(\bar{a}, \bar{y}) = 0 \wedge \ldots \wedge p_{k-1}(\bar{a}, \bar{y}) = 0$
$\wedge\, q_0(\bar{a}, \bar{y}) \neq 0 \wedge \ldots \wedge q_{m-1}(\bar{a}, \bar{y}) \neq 0).$

In a field, $x \neq 0$ says the same as $\exists z\, x \cdot z = 1$. Hence (and this is Rabinowitsch's observation) we can eliminate the inequations in (1.4), bringing the new existential quantifiers $\exists z$ forward to join $\exists \bar{y}$. This reduces (1.4) to the form

(1.5) $\exists \bar{y}(p_0(\bar{a}, \bar{y}) = 0 \wedge \ldots \wedge p_{k-1}(\bar{a}, \bar{y}) = 0).$

In short: to show that a field A is existentially closed, we only need to consider *equations* in the definition (1.1), not both equations and inequations. But this is just a happy fact about fields; in general one has to live with the inequations in (1.1).

Are there interesting e.c. models of other theories besides that of fields? Yes, certainly there are. In the next section we shall see two very general methods for proving the existence of e.c. models. But before that, let me note that we have already found a bunch of e.c. models in Chapter 6.

Example 2: *Fraïssé limits*. Let L be a countable signature and **J** a countable set of finitely generated L-structures which has HP, JEP and AP (see section 6.1). Let **K** be the class of all L-structures with age $\subseteq$ **J**, and let A be the Fraïssé limit of **K**. Then A is existentially closed in **K**. For suppose B is in **K**, $A \subseteq B$, $\bar{a}$ is a tuple of elements of A and $\psi(\bar{x}, \bar{y})$ is a conjunction of literals of L such that $B \vDash \exists \bar{y}\, \psi(\bar{a}, \bar{y})$. Take a tuple $\bar{b}$ in B such that $B \vDash \psi(\bar{a}, \bar{b})$; let C be the substructure of A generated by $\bar{a}$, and D the substructure of B generated by $\bar{a}\bar{b}$. Then $C \subseteq D$, and both C and D are in the age **J** of A. Since A is weakly homogeneous (see (1.7) in section 6.1), the inclusion map $f: C \to A$ extends to an embedding $g: D \to A$. So $A \vDash \psi(\bar{a}, g\bar{b})$, and hence $A \vDash \exists \bar{y}\, \psi(\bar{a}, \bar{y})$ as required. It's not hard to show that if **J** above is a set of finite structures, then the Fraïssé limit of **J** is the unique countable e.c. structure in **K** (see Exercise 1).

The influence of JEP and AP

Example 2 above is interesting because it's untypical. JEP and AP are quite strong properties, and we shall see that there are plenty of examples of e.c. structures in classes where the finitely generated structures fail to have one or other of these properties. Both JEP and AP exert a strong influence on the behaviour of e.c. structures. This is a convenient moment to prove two theorems which illustrate the point.

If T is a first-order theory, we say that T has the **joint embedding property** (JEP) if, given any two models A, B of T, there is a model of T in which both A and B are embeddable.

Theorem 7.1.2. *Let L be a first-order language and T a theory in L. Suppose that T has JEP, and let A, B be e.c. models of T. Then every $\forall_2$ sentence of L which is true in A is true in B too.*

Proof. By assumption there is a model C of T in which both A and B are embeddable; we lose nothing by assuming that A and B are substructures of C. Since A is e.c. in the class of models of T, we have $A \preccurlyeq_1 C$. It follows by the existential amalgamation theorem (Theorem 5.4.1) that there is an elementary extension D of A with $C \subseteq D$. Suppose $A \vDash \forall \bar{x} \exists \bar{y}\, \phi(\bar{x}, \bar{y})$ where ϕ is a quantifier-free formula of L, and let $\bar{b}$ be a tuple of elements of B. We must show that $B \models \exists \bar{y} \, \phi(\bar{b}, \bar{y})$. But $D \models \exists \bar{y} \, \phi(\bar{b}, \bar{y})$ since $A \preccurlyeq D$. Now B is an e.c. model, so that $B \preccurlyeq_1 D$, whence $B \models \exists \bar{y} \, \phi(\bar{b}, \bar{y})$ as required. $\qquad\square$

Theorem 7.1.3. *Let L be a first-order language. Let $\mathbf{K}$ be a class of L-structures which is closed under isomorphic copies, and suppose that the class of all substructures of structures in $\mathbf{K}$ has AP. Then for every $\exists_1$ formula $\phi(\bar{x})$ of L there is a quantifier-free formula $\chi(\bar{x})$ (possibly infinitary) which is equivalent to ϕ in all e.c. structures in $\mathbf{K}$. In particular if there is a first-order theory T such that the e.c. structures in $\mathbf{K}$ are exactly the models of T, then ϕ is equivalent modulo T to a quantifier-free formula of L.*

Proof. Let us say that a pair $(A, \bar{a})$ is **good** if A is an e.c. structure in $\mathbf{K}$, $\bar{a}$ is a tuple of elements of A, and $A \models \phi(\bar{a})$. For each good pair $(A, \bar{a})$, let $\theta_{(A, \bar{a})}(\bar{x})$ be the conjunction $\bigwedge \{ \psi(\bar{x}) : \psi$ is a literal of L and $A \models \psi(\bar{a}) \}$. Let $\chi(\bar{x})$ be the disjunction of all the formulas $\theta_{(A, \bar{a})}$ where $(A, \bar{a})$ ranges over good pairs. Clearly if B is any e.c. structure in $\mathbf{K}$ and $\bar{b}$ a tuple in B such that $B \models \phi(\bar{b})$, then $B \models \chi(\bar{b})$ since $(B, \bar{b})$ is a good pair. Conversely if B is an e.c. structure in $\mathbf{K}$ and $B \models \chi(\bar{b})$, then there is a good pair $(A, \bar{a})$ with an isomorphism $e \colon \langle \bar{a} \rangle_A \to B$ taking $\bar{a}$ to $\bar{b}$. By the amalgamation property for the class of substructures of structures in $\mathbf{K}$, there is a structure C in $\mathbf{K}$ with embeddings $g \colon A \to C$ and $h \colon B \to C$ such that $g\bar{a} = h\bar{b}$. Since $\mathbf{K}$ is closed under isomorphic copies, we can suppose that h is an inclusion map. Now $A \models \phi(\bar{a})$ by assumption, so $C \models \phi(g\bar{a})$ since ϕ is an $\exists_1$ formula. It follows that $B \models \phi(\bar{b})$ since B is e.c. in $\mathbf{K}$.

Finally suppose that the models of T are exactly the e.c. structures in $\mathbf{K}$. Then $T \vdash \forall \bar{x}(\phi \leftrightarrow \chi)$. Two applications of the compactness theorem reduce χ to a first-order quantifier-free formula. $\qquad\square$

Warning. Theorem 7.1.3 is false if we exclude empty structures. Consider the theory $\forall xy(P(x) \leftrightarrow P(y))$, and let ϕ be the formula $\exists x \, P(x)$. Readers who do exclude empty structures can rescue the theorem by requiring ϕ to have at least one free variable. See Exercise 7 for more details.

Theorem 7.1.3 gives useful information. Let $\mathbf{K}$ be the class of fields. Thanks to Example 1, we already know that the class of e.c. fields is first-order axiomatisable. It's a well-known fact of algebra that $\mathbf{K}$ has the amalgamation property. From this we easily deduce that the class of integral domains has the amalgamation property too (by taking fields of fractions). But in the signature of rings, a substructure of a field is the same thing as an integral domain. Hence we have the following.

Corollary 7.1.4. *Let T be the theory of algebraically closed fields. Then T has quantifier elimination.*

Proof. By Theorem 7.1.3, the argument above shows that every $\exists_1$ formula is equivalent to a quantifier-free formula modulo T. This suffices, by Lemma 2.3.1. ☐

We shall come back to applications of Theorem 7.1.3 in sections 7.2 and 7.4 below.

Exercises for section 7.1

1. Let L be a signature with just finitely many symbols, and **J** an at most countable set of finite (NB: not just finitely generated) L-structures which has HP, JEP and AP. Let **K** be the class of all L-structures whose age is $\subseteq$ **J**. If C is a finite or countable structure in **K**, show that C is existentially closed in **K** if and only if C is the Fraïssé limit of **J**.

2. Show that there is a unique countable e.c. linear ordering, namely the ordering of the rationals.

3. Show (a) an e.c. integral domain is the same thing as an e.c. field, (b) an e.c. field is never an e.c. commutative ring.

4. Show that if A is an e.c. abelian group, then A is divisible and has infinitely many p-ary direct summands for each prime p. [Don't try to prove there are infinitely many torsion-free direct summands – it's false!]

5. Give an example of a first-order theory with e.c. models, where there are e.c. models which satisfy different quantifier-free sentences. [Try the theory of fields.]

6. Let T be the theory $\forall x \, \neg(\exists y \, Rxy \land \exists y \, Ryx)$. Describe the e.c. models of T. Give an example of an $\exists_1$ first-order formula which is not equivalent to a quantifier-free formula in e.c. models of T.

We say that a class **K** *has* **amalgamation over non-empty structures** *if the amalgamation property holds whenever the structures involved are all non-empty.*
7. In Theorem 7.1.3, suppose that we weaken the assumption of AP to amalgamation over non-empty structures. Show that the conclusion still holds, provided that we require ϕ to have at least one free variable.

The following example is a freak, but I include it to prevent careless conjectures.
8. Let T be complete first-order arithmetic, i.e. the first-order theory of the natural numbers with symbols 0, 1, $+$, $\cdot$. Show that the natural numbers form an e.c. model of T.

9. Using purely algebraic arguments, show the equivalence between (1.3) above and the following form of Hilbert's Nullstellensatz. Suppose A is an algebraically closed field, I is an ideal in the polynomial ring $A[x_0, \ldots, x_{n-1}]$ and $p(x_0, \ldots, x_{n-1})$ is a polynomial $\in A[x_0, \ldots, x_{n-1}]$ such that for all $\bar{a}$ in A, if $q(\bar{a}) = 0$ for all $q \in I$ then $p(\bar{a}) = 0$. Then for some positive integer k, $p^k \in I$. [$\Rightarrow$: if not, there is a prime ideal P of $A[x_0, \ldots, x_{n-1}]$ which contains I and not p; consider the field of fractions of the integral domain $A[x_0, \ldots, x_{n-1}]/P$. $\Leftarrow$: use Rabinowitsch to eliminate the inequations, and turn the equations into generators of an ideal I of $A[x_0, \ldots, x_{n-1}]$; if the equations have no solution in A then there is no $\bar{a}$ in A such that $q(\bar{a}) = 0$ for all $q \in I$.]

7.2 Constructing e.c. structures

In this section we describe a way of finding e.c. structures. It is based on Steinitz's proof that every field has an algebraically closed extension.

Let L be a signature and **K** a class of L-structures. We say that **K** is **inductive** if (1) K is closed under taking unions of chains, and (2) (for tidiness) every structure isomorphic to a structure in **K** is also in **K**.

For example if T is an $\forall_2$ first-order theory in L and **K** is the class of all models of T, then Theorem 2.4.4 says that **K** is inductive. Thus the class of all groups is inductive, as is the class of all fields.

Local properties give us some more examples. Let π be a structural property which an L-structure might have. We say that an L-structure A **has π locally** if all the finitely generated substructures of A have property π. Then the class of all L-structures which have π locally is an inductive class. Thus for example a structure is **locally finite** if all its finitely generated substructures are finite. The class of all locally finite groups is inductive. Likewise the class of all groups without elements of infinite order is inductive. Neither of these two classes is first-order axiomatisable.

Theorem 7.2.1. *Let* **K** *be an inductive class of L-structures and A a structure in* **K**. *Then there is an e.c. structure B in* **K** *such that $A \subseteq B$.*

Proof. We begin by showing a weaker result: there is a structure A^* in **K** such that $A \subseteq A^*$, and if $\phi(\bar{x})$ is an $\exists_1$ formula of L, $\bar{a}$ a tuple in A and there is a structure C in **K** such that $A^* \subseteq C$ and $C \vDash \phi(\bar{a})$, then $A^* \vDash \phi(\bar{a})$.

List as $(\phi_i, \bar{a}_i)_{i < \lambda}$ all pairs $(\phi, \bar{a})$ where ϕ is an $\exists_1$ formula of L and $\bar{a}$ is a tuple in A. By induction on i, define a chain of structures $(A_i : i \leqslant \lambda)$ in **K** by

(2.1) $A_0 = A$,

(2.2)   when δ is a limit ordinal $\leqslant \lambda$, $A_\delta = \bigcup_{i < \delta} A_i$,

$$(2.3) \qquad A_{i+1} = \begin{cases} \text{some structure } C \text{ in } \mathbf{K} \text{ such that } A_i \subseteq C \\ \qquad \text{and } C \models \phi_i(\bar{a}_i), \text{ if there is such a structure } C, \\ A_i \text{ otherwise.} \end{cases}$$

Put $A^* = A_\lambda$. To show that A^* is as required, take any $\exists_1$ formula $\phi(\bar{x})$ of L and any tuple $\bar{a}$ of elements of A. Then $(\phi, \bar{a})$ is $(\phi_i, \bar{a}_i)$ for some $i < \lambda$. Suppose C is a structure in $\mathbf{K}$ such that $A^* \subseteq C$ and $C \models \phi(\bar{a})$. Then $A_i \subseteq C$, and so $A_{i+1} \models \phi(\bar{a})$ by (2.3). Since ϕ is an $\exists_1$ formula and $A_{i+1} \subseteq A^*$, we infer by Theorem 2.4.1 that $A^* \models \phi(\bar{a})$.

Now define a chain of structures $A^{(n)}$ ($n < \omega$) in $\mathbf{K}$ as follows, by induction on n:

$$(2.4) \qquad\qquad\qquad A^{(0)} = A,$$

$$(2.5) \qquad\qquad\qquad A^{(n+1)} = A^{(n)*}.$$

Put $B = \bigcup_{n<\omega} A^{(n)}$. Then B is in $\mathbf{K}$ since $\mathbf{K}$ is inductive, and certainly $A \subseteq B$. Suppose $\phi(\bar{x})$ is an $\exists_1$ formula of L and $\bar{a}$ is a tuple from B, such that $C \models \phi(\bar{a})$ for some C which is in $\mathbf{K}$ and extends B. Since $\bar{a}$ is finite, it lies within $A^{(n)}$ for some $n < \omega$. Hence $A^{(n+1)} \models \phi(\bar{a})$ since $A^{(n+1)}$ is $A^{(n)*}$. But ϕ is an $\exists_1$ formula and $A^{(n+1)} \subseteq B$, so that again it follows that $B \models \phi(\bar{a})$. $\qquad \square$

Often we can say more about the size of the e.c. structure.

Corollary 7.2.2. *Let K be an inductive class, A a structure in $\mathbf{K}$ and λ an infinite cardinal $\geq |A|$. Suppose also that for every structure C in $\mathbf{K}$ and every set X of $\leq \lambda$ elements of C, there is a structure B in $\mathbf{K}$ such that $X \subseteq \mathrm{dom}(B)$, $B \leq C$ and $|B| \leq \lambda$. (For example, suppose $\mathbf{K}$ is the class of all models of some $\forall_2$ theory in a first-order language of cardinality $\leq \lambda$.) Then there is an e.c. structure B in $\mathbf{K}$ such that $A \subseteq B$ and $|B| \leq \lambda$.*

Proof. Counting the number of pairs $(\phi, \bar{a})$, we find that we can use λ as the cardinal λ in the proof of the theorem. In (2.3) of that proof, choose each structure A_{i+1} so that it has cardinality $\leq \lambda$. For example if there is a structure C in $\mathbf{K}$ such that $A_i \subseteq C$ and $C \models \phi_i(\bar{a}_i)$, choose A_{i+1} in $\mathbf{K}$ so that $A_{i+1} \leq C$, $\mathrm{dom}(A_i) \subseteq \mathrm{dom}(A_{i+1})$ and $|A_{i+1}| \leq \lambda$. Then $A_i \subseteq A_{i+1}$ and $A_{i+1} \models \phi_i(\bar{a}_i)$ as required. With these choices, A^* also has cardinality $\leq \lambda$, and hence so does B in the theorem. $\qquad \square$

E.c. models of first-order theories

Theorem 7.2.1 and its corollary more or less exhaust what one can say about e.c. structures in arbitrary inductive classes. So for the rest of this section we turn to the most interesting case, namely the e.c. models of an $\forall_2$ first-order theory T.

In section 5.4, $T_\forall$ was defined to be the set of all $\forall_1$ sentences ϕ of L such that $T \vdash \phi$. By Corollary 5.4.3, every model of $T_\forall$ can be extended to a model of T.

Corollary 7.2.3. *Let L be a first-order language, T an $\forall_2$ theory in L and A an infinite model of $T_\forall$. Then there is an e.c. model B of T such that $A \subseteq B$ and $|B| = \max(|A|, |L|)$.*

Proof. By Corollary 5.4.3 there is a model C of T such that $A \subseteq C$. By the downward Löwenheim–Skolem theorem we can take C to be of cardinality $\max(|A|, |L|)$. The rest follows from Corollary 7.2.2.   □

There is a wealth of information about e.c. models of $\forall_2$ first-order theories. Our next result takes us straight to the inner personality of these models. It says that in such a model, no $\exists_1$ formula is ever false unless some true $\exists_1$ formula compels it to be.

Theorem 7.2.4. *Let L be a first-order language, T an $\forall_2$ theory in L and A a model of T. Then the following are equivalent.*
(a) *A is an e.c. model of T.*
(b) *A is a model of $T_\forall$, and for every $\exists_1$ formula $\phi(\bar{x})$ of L and every tuple $\bar{a}$ in A, if $A \vDash \neg \phi(\bar{a})$ then there is an $\exists_1$ formula $\chi(\bar{x})$ of L such that $A \vDash \chi(\bar{a})$ and $T \vdash \forall \bar{x}(\chi \rightarrow \neg \phi)$.*
(c) *A is an e.c. model of $T_\forall$.*

Proof. (a) $\Rightarrow$ (b). Assume (a). Then certainly A is a model of $T_\forall$. Suppose $\phi(\bar{x})$ is an $\exists_1$ formula of L and $\bar{a}$ is a tuple in A such that $A \vDash \neg \phi(\bar{a})$. Then since A is an e.c. model of T, it follows that there is no model C of T such that $A \subseteq C$ and $C \vDash \phi(\bar{a})$. Add distinct new constants $\bar{c}$ to the language as names for the elements of $\bar{a}$. Since there is no model C as described, the diagram lemma (Lemma 1.4.2) tells us that the theory

$$(2.6) \qquad\qquad \mathrm{diag}(A) \cup T \cup \{\phi(\bar{c})\}$$

has no model. So by the compactness theorem (Theorem 5.1.1) there are a tuple $\bar{d}$ of distinct elements of A and a quantifier-free formula $\theta(\bar{x}, \bar{y})$ of L such that $A \vDash \theta(\bar{c}, \bar{d})$ and

$$(2.7) \qquad\qquad T \vdash \theta(\bar{c}, \bar{d}) \rightarrow \neg \phi(\bar{c}).$$

Now we use the lemma on constants (Lemma 2.3.2), noting that even if the tuple $\bar{a}$ contains repetitions, we had the foresight to introduce a tuple $\bar{c}$ of distinct constants. The result is that $T \vdash \forall \bar{x}(\exists \bar{y}\, \theta(\bar{x}, \bar{y}) \rightarrow \neg \phi(\bar{x}))$. To infer (b), let χ be $\exists \bar{y}\, \theta$.

(b) $\Rightarrow$ (c). Assume (b). Then A is a model of $T_\forall$. Suppose $\phi(\bar{x})$ is an $\exists_1$

formula of L and $\bar{a}$ is a tuple in A such that for some model C of $T_\forall$, $A \subseteq C$ and $C \vDash \phi(\bar{a})$. We must show that $A \vDash \phi(\bar{a})$. If not, then by (b) there is an $\exists_1$ formula $\chi(\bar{x})$ of L such that $A \vDash \chi(\bar{a})$ and $T \vdash \forall\bar{x}(\chi \rightarrow \neg\phi)$. Since χ is $\exists_1$ and $A \subseteq C$, we have $C \vDash \chi(\bar{a})$. The sentence $\forall\bar{x}(\chi \rightarrow \neg\phi)$ is $\forall_1$ (after some trivial rearrangement), and so it lies in $T_\forall$. Since C is a model of $T_\forall$, we infer that $C \vDash \neg\phi(\bar{a})$; contradiction.

(c) $\Rightarrow$ (a). Assume (c). First we must show that A is a model of T. Since T is an $\forall_2$ theory, a typical sentence in T can be written $\forall\bar{x}\exists\bar{y}\,\psi(\bar{x}, \bar{y})$ with ψ quantifier-free. Let $\bar{a}$ be any tuple in A; we must show that $A \vDash \exists\bar{y}\,\psi(\bar{a}, \bar{y})$. By Corollary 5.4.3, since $A \vDash T_\forall$ there is a model C of T such that $A \subseteq C$. Then $C \vDash \exists\bar{y}\,\psi(\bar{a}, \bar{y})$, and so $A \vDash \exists\bar{y}\,\psi(\bar{a}, \bar{y})$ since A is an e.c. model of $T_\forall$. Thus A is a model of T. It follows easily that A is an e.c. model of T, since every model of T extending A is in fact a model of $T_\forall$ too. $\square$

Resultants

What systems of equations and inequations over A can be solved in some extension of A? Our next lemma answers this question, at least in general terms. Of course, for particular kinds of structure (fields, groups etc.) one has to work harder to get a specific answer.

Let T be an $\forall_2$ theory in a first-order language L, and let $\phi(\bar{x})$ be an $\exists_1$ formula of L. We write $\mathrm{Res}_\phi(\bar{x})$ for the set of all $\forall_1$ formulas $\psi(\bar{x})$ of L such that $T \vdash \forall\bar{x}(\phi \rightarrow \psi)$. The set Res_ϕ is called the **resultant** of ϕ.

Lemma 7.2.5. *Let L be a first-order language, T an $\forall_2$ theory in L and A an L-structure. Suppose $\phi(\bar{x})$ is an $\exists_1$ formula of L and $\bar{a}$ is a tuple from A. Then the following are equivalent.*
(a) *There is a model B of T such that $A \subseteq B$ and $B \vDash \phi(\bar{a})$.*
(b) *$A \vDash \bigwedge\mathrm{Res}_\phi(\bar{a})$.*

Proof. (a) $\Rightarrow$ (b). Suppose (a) holds, and let $\psi(\bar{x})$ be a formula in Res_ϕ. Then $B \vDash \psi(\bar{a})$ since B is a model of T. But ψ is an $\forall_1$ formula and $A \subseteq B$, so that $A \vDash \psi(\bar{a})$.

(b) $\Rightarrow$ (a). Assuming that (a) fails, we shall contradict (b). Introduce a tuple $\bar{c}$ of distinct new constants to name the elements $\bar{a}$. The diagram lemma (Lemma 1.4.2) tells us that if (a) fails, then the following theory has no model:

(2.8) $T \cup \mathrm{diag}(A) \cup \{\phi(\bar{c})\}$.

So by the compactness theorem there are a quantifier-free formula $\theta(\bar{x}, \bar{y})$ of L and distinct elements $\bar{d}$ of A such that $A \vDash \theta(\bar{a}, \bar{d})$ and

(2.9) $T \vdash \phi(\bar{c}) \rightarrow \neg\theta(\bar{c}, \bar{d})$.

Applying the lemma on constants (Lemma 2.3.2), we find

(2.10) $T \vdash \forall \bar{x}(\phi \rightarrow \forall \bar{y} \,\neg\, \theta)$

so that $\forall \bar{y} \,\neg\, \theta$ is a formula in Res_ϕ. But this contradicts (b), since $A \vDash \exists \bar{y} \, \theta(\bar{a}, \bar{y})$.   □

The formula $\bigwedge \text{Res}_\phi(\bar{x})$ in (b) of the lemma is generally an infinitary formula. The following example is typical.

Example 1: *Nilpotent elements in commutative rings.* Let A be a commutative ring and a an element of A. When is there a commutative ring $B \supseteq A$ containing a non-zero idempotent b (i.e. $b^2 = b$) which is divisible by a? In other words, when can A be extended to a commutative ring B in which the formula $\exists z \,(az \neq 0 \wedge (az)^2 = az)$ is true? The answer is that there is such a ring B if and only if a is not nilpotent (i.e. if and only if there is no $n < \omega$ such that $a^n = 0$). In one direction, if $ab \neq 0$ and $(ab)^2 = ab$, then for every positive integer n, $0 \neq (ab)^n = a^n b^n$ and so $a^n \neq 0$. In the other direction, if a is not nilpotent, consider the ring $A[x]/I$ where I is the ideal generated by $a^2 x^2 - ax$. To show that $A[x]/I$ will serve for B with x/I as b, we need to check that I doesn't contain either ax or any non-zero element of A. Suppose for example that

(2.11) $ax = \left(\sum_{i<n} c_i x^i\right)(a^2 x^2 - ax).$

Multiplying out, $(-c_0 a - 1)x + \sum_{2 \leq i \leq n+1} (c_{i-2}a^2 - c_{i-1}a)x^i + c_n a^2 x^{n+2} = 0$. From this we infer

(2.12) $c_0 a = -1, \; c_{i-2}a^2 = c_{i-1}a \; (2 \leq i \leq n + 1), \; c_n a^2 = 0.$

Then $0 = c_n a^2 = c_{n-1}a^3 = \ldots = c_0 a^{n+2} = -a^{n+1}$, contradiction. The argument to show $A \cap I = \{0\}$ is similar but easier.

Thus in commutative rings, the resultant of the formula $\exists z(xz \neq 0 \wedge (xz)^2 = xz)$ is a set of $\forall_1$ formulas which is equivalent (modulo the theory of commutative rings) to the set $\{x^n \neq 0: n > \omega\}$. There will be no harm in identifying the resultant with this set, or with the infinitary formula $\bigwedge_{n<\omega} x^n \neq 0$.

Theorem 7.2.6. *Let L be a first-order language, T an $\forall_2$ theory in L and A a model of T. The following are equivalent.*
(a) *A is an e.c. model of T.*
(b) *For every $\exists_1$ formula $\phi(\bar{x})$ of L, $A \vDash \forall \bar{x}(\phi(\bar{x}) \leftrightarrow \bigwedge \text{Res}_\phi(\bar{x}))$.*

Proof. By definition of Res_ϕ, every model of T satisfies the implication $\forall \bar{x}(\phi(\bar{x}) \rightarrow \bigwedge \text{Res}_\phi(\bar{x}))$. The implication in the other direction is just a rewrite of clause (b) in Theorem 7.2.4.   □

Example 1: *continued.* By the theorem, if A is an e.c. commutative ring, then an element of A is nilpotent if and only if it doesn't divide any nonzero idempotent. It follows that the condition 'x is nilpotent' can be expressed in A by a *first-order* formula, and moreover the same first-order formula works for any other e.c. commutative ring. This condition certainly isn't first-order for commutative rings in general (see Exercise 10).

The forms of resultants are closely related to other properties of the theory T, as the next theorem bears witness. We say that a theory T has the **amalgamation property** (AP) if the class of models of T has AP (cf. section 6.1).

Theorem 7.2.7. *Let T be an $\forall_2$ theory in a first-order language L. Then the following are equivalent.*
(a) *$T_\forall$ has the amalgamation property.*
(b) *For every $\exists_1$ formula $\phi(\bar{x})$ of L, Res_ϕ is equivalent modulo T to a set $\Phi(\bar{x})$ of quantifier-free formulas of L.* $\qquad\square$

Proof. The direction (a) $\Rightarrow$ (b) follows from Exercise 13 and Theorem 7.1.3. In the other direction, assume (b) and suppose A, B and C are models of $T_\forall$ and A is a substructure of both B and C. We can extend B and C to e.c. models of T, call them B' and C' respectively. Suppose $\phi(\bar{x})$ is an $\exists_1$ formula of L and $\bar{a}$ is a tuple in A such that $C' \vDash \phi(\bar{a})$. Then, $A \vDash \bigwedge\mathrm{Res}_\phi(\bar{a})$, and so $B' \vDash \bigwedge\mathrm{Res}_\phi(\bar{a})$ since Res_ϕ is quantifier-free. Then $B' \vDash \phi(\bar{a})$, and the existential amalgamation theorem (Theorem 5.4.1) does the rest. $\qquad\square$

Exercises for section 7.2

1. If **K** is an inductive class of L-structures, and **J** is the class of e.c. structures in **K**, show that **J** is closed under unions of chains.

2. A **near-linear space** is a structure with two kinds of elements, 'points' and 'lines', and a symmetric binary relation of 'incidence' relating points and lines in such a way that (i) any line is incident with at least two points, and (ii) any two distinct points are incident with at most one line. The near-linear space is a **projective plane** if (ii′) any two distinct points are incident with exactly one line, and moreover (iii) any two distinct lines are incident with at least one point, and (iv) there is a set of four points, no three of which are all incident with a line. (a) Show that the class of near-linear spaces is inductive. (b) Show that every e.c. near-linear space is a projective plane. (c) Deduce that every near-linear space can be embedded in a projective plane.

3. Show that in Theorem 7.2.4(b) we can replace '$\exists_1$' by 'primitive' (both times).

4. Let L be a first-order language and T an $\forall_2$ theory in L. (a) Show that if B is a model of T and $A \preccurlyeq_1 B$, then A is a model of T. (b) Show that if B is an e.c. model of T and $A \preccurlyeq_1 B$, then A is an e.c. model of T.

5. Let L be a first-order language, T an $\forall_2$ theory in L and A an L-structure. Show that the following are equivalent. (a) A is an e.c. model of T. (b) A is a model of $T_\forall$, and for every model C of T such that $A \subseteq C$, we have $A \preccurlyeq_1 C$. (c) For some e.c. model B of T, $A \preccurlyeq_1 B$.

6. Let L be a countable first-order language and T an $\forall_2$ theory in L. Show that there is a sentence ϕ of $L_{\omega_1\omega}$ such that the models of ϕ are exactly the e.c. models of T.

7. Let L be a first-order language and T an $\forall_2$ theory in L. (a) Show that if A is an e.c. model of T and B is a model of T with $A \subseteq B$, then for every $\forall_2$ formula $\phi(\bar{x})$ of L and every tuple $\bar{a}$ in A, if $B \vDash \phi(\bar{a})$ then $A \vDash \phi(\bar{a})$. (b) Show that if B in (a) is also an e.c. model of T then the same holds with $\forall_3$ in place of $\forall_2$.

8. Let L be a first-order language and U an $\forall_2$ theory in L. Show that among the $\forall_2$ theories T in L such that $T_\forall = U_\forall$, there is a unique maximal one under the ordering $\subseteq$. Writing U_0 for this maximal T, show that U_0 is the set of those $\forall_2$ sentences of L which are true in every e.c. model of U. (U_0 is known as the **Kaiser hull** of U.)

A model A of a theory T is said to be **simple** *if for every homomorphism $h: A \to B$ with B a model of T, either h is an embedding or h is a constant map. It is known that every group can be embedded into a simple group.*
9. Let L be a first-order language (not necessarily countable) and T an $\forall_2$ theory in L. (a) Show that if every model of T is embeddable in a simple model, then every e.c. model of T is simple. [Suppose A is an e.c. model and the homomorphism $h: A \to B$ is a counterexample to the conclusion. By assumption A extends to a simple model C. Now quote Theorem 5.4.7 to find a homomorphism contradicting the simplicity of C.] (b) Deduce that every e.c. group is simple.

10. In commutative rings, write (*) for the condition 'For each positive integer n, there is an element a such that $a^n = 0$ but $a^i \neq 0$ for all $i < n$'. (a) Show that if A is a commutative ring in which (*) holds, and $\phi(x)$ is any formula (maybe with parameters from A) such that $\phi(A)$ is the set of nilpotent elements in A, then A has an elementary extension in which some non-nilpotent element satisfies ϕ. (b) By iterating (a) to form an elementary chain, show that there is a commutative ring in which the set of nilpotent elements is not first-order definable with parameters. (c) Show that every e.c. commutative ring satisfies (*).

11. Let A be a commutative ring, a an element of A and n a positive integer. Show that the following are equivalent. (a) For every element b of A, if $ba^{n+1} = 0$ then $ba^n = 0$. (b) There is a commutative ring B which extends A and contains an element b such that $(a^2 b - a)^n = 0$.

12. Show that the amalgamation property fails for the class of commutative rings without nonzero nilpotents. [Let A be $\mathbb{Q}[x]$; let B be a field containing A, and let C be $A[y]/(xy)$. A zero-divisor can never be amalgamated with an invertible element.]

The next exercise is a partial converse to Theorem 7.2.6.

13. Let T be an $\forall_2$ theory in a first-order language L. Suppose $\phi(\bar{x})$ is an $\exists_1$ formula of L, and $\chi(\bar{x})$ is a quantifier-free formula of $L_{\infty\omega}$ such that for every e.c. model A of T, $A \vDash \forall\bar{x}(\phi \leftrightarrow \chi)$. Show that χ is equivalent to Res_ϕ modulo T, and hence that Res_ϕ is equivalent modulo T to a set of quantifier-free formulas of L.

7.3 Model-completeness

In the early 1950s Abraham Robinson noticed that certain maps studied by algebraists are in fact elementary embeddings. If you choose a map at random, the chances of its being an elementary embedding are negligible. So Robinson reckoned that there must be a systematic reason for the appearance of these elementary embeddings, and he set out to find what the reason was. In the course of his quest he introduced the notions of model-complete theories, companionable theories and model companions. These notions have become essential tools for the model theory of algebra. In this section we shall examine them.

Model-completeness

In section 2.6 we defined a theory T in a first-order language L to be **model-complete** if every embedding between L-structures which are models of T is elementary.

Theorem 7.3.1. *Let T be a theory in a first-order language L. The following are equivalent.*

(a) *T is model-complete.*

(b) *Every model of T is an e.c. model of T.*

(c) *If L-structures A, B are models of T and $e: A \to B$ is an embedding then there are an elementary extension D of A and an embedding $g: B \to D$ such that ge is the identity on A.*

(d) *If $\phi(\bar{x}, \bar{y})$ is a formula of L which is a conjunction of literals, then $\exists\bar{y}\,\phi$ is equivalent modulo T to an $\forall_1$ formula $\psi(\bar{x})$ of L.*

(e) *Every formula $\phi(\bar{x})$ of L is equivalent modulo T to an $\forall_1$ formula $\psi(\bar{x})$ of L.*

Proof. (a) $\Rightarrow$ (b) is immediate from the definition of e.c. model.

(b) $\Rightarrow$ (c). Assume (b). Let $e: A \to B$ be an embedding between models of T, and let $\bar{a}$ be a sequence listing all the elements of A. Then

$(B, e\bar{a}) \Rrightarrow_1 (A, \bar{a})$ since A is an e.c. model of T. (See section 5.4 for the notion $\Rrightarrow_1$.) The conclusion of (c) follows by the existential amalgamation theorem, Theorem 5.4.1.

(c) $\Rightarrow$ (d). We first claim that if (c) holds, then every embedding between models of T preserves $\forall_1$ formulas of L. For if $e\colon A \to B$ is such an embedding, $\bar{a}$ is in A and $\phi(\bar{x})$ is an $\forall_1$ formula of L such that $A \vDash \phi(\bar{a})$, then taking D and g as in (c) we have $D \vDash \phi(ge\bar{a})$ and so $B \vDash \phi(e\bar{a})$ since ϕ is an $\forall_1$ formula. This proves the claim. It follows by Corollary 5.4.5, taking negations, that every $\exists_1$ formula $\phi(\bar{x})$ of L is equivalent modulo T to an $\forall_1$ formula $\psi(\bar{x})$ of L; hence (d) holds.

(d) $\Rightarrow$ (e). Assume (d), and let $\phi(\bar{x})$ be any formula of L; we can assume that ϕ is in prenex form, say as $\exists \bar{x}_0 \forall \bar{x}_1 \ldots \ \forall \bar{x}_{n-2} \exists \bar{x}_{n-1} \theta_n(\bar{x}_0, \bar{x}_1, \ldots, \bar{x}_{n-1}, \bar{x})$ where θ_n is quantifier-free. By (d), $\exists \bar{x}_{n-1} \theta_n(\bar{x}_0, \bar{x}_1, \ldots, \bar{x}_{n-1}, \bar{x})$ is equivalent modulo T to a formula $\forall \bar{z}_{n-1} \theta_{n-1}(\bar{x}_0, \bar{x}_1, \ldots, \bar{x}_{n-2}, \bar{z}_{n-1}, \bar{x})$ with θ_{n-1} quantifier-free, and so ϕ is equivalent to $\exists \bar{x}_0 \forall \bar{x}_1 \ldots \ \forall \bar{x}_{n-2} \bar{z}_{n-1} \theta_{n-1}(\bar{x}_0, \bar{x}_1, \ldots, \bar{x}_{n-2}, \bar{z}_{n-1}, \bar{x})$. By (d) again, taking negations, the formula $\forall \bar{x}_{n-2} \bar{z}_{n-1} \theta_{n-1}(\bar{x}_0, \bar{x}_1, \ldots, \bar{x}_{n-2}, \bar{z}_{n-1}, \bar{x})$ is equivalent modulo T to a formula $\exists \bar{z}_{n-2} \theta_{n-2}(\bar{x}_0, \bar{x}_1, \ldots, \bar{z}_{n-2}, \bar{x})$ with θ_{n-2} quantifier-free, so that ϕ is equivalent modulo T to $\exists \bar{x}_0 \forall \bar{x}_1 \ldots \ \exists \bar{x}_{n-3} \bar{z}_{n-2} \theta_{n-2}(\bar{x}_0, \bar{x}_1, \ldots, \bar{z}_{n-2}, \bar{x})$. After n steps in this style, all the quantifiers will have been gathered up into a universal quantifier $\forall \bar{z}_0$.

(e) $\Rightarrow$ (a) follows from the fact (Corollary 2.4.2(a)) that $\forall_1$ formulas are preserved in substructures. $\qquad\square$

Corollary 7.3.2 *(Robinson's test). For the first-order theory T to be model-complete, it is necessary and sufficient that if A and B are any two models of T with $A \subseteq B$ then $A \preccurlyeq_1 B$.*

Proof. This is a restatement of (b) in the theorem. $\qquad\square$

The next fact is elementary, but it should be mentioned.

Theorem 7.3.3. *Let T be a model-complete theory in a first-order language L. Then T is equivalent to an $\forall_2$ theory in L.*

Proof. Every chain of models of T is elementary, and so its union is a model of T by Theorem 2.5.2. Hence T is equivalent to an $\forall_2$ theory by the Chang–Łoś–Suszko theorem, Theorem 5.4.9. $\qquad\square$

Ways of proving model-completeness

How does one show in practice that a theory is model-complete? It's not a property that can be checked by inspection; serious mathematical work may be involved.

One method is already to hand. Let L be a first-order language and T a theory in L. We say that T **has quantifier elimination** if for every formula $\phi(\bar{x})$ of L there is a quantifier-free formula $\phi^*(\bar{x})$ of L which is equivalent to ϕ modulo T. Every quantifier-free formula is an $\forall_1$ formula. So by condition (e) in Theorem 7.3.1, *every theory with quantifier elimination is model-complete*.

This observation gives us plenty of examples, thanks to the quantifier elimination technique of section 2.7 above. Thus the theory of infinite vector spaces over a fixed field is model-complete (Exercise 2.7.9); so is the theory of real-closed fields in the language of ordered fields (Theorem 2.7.2); and so is the theory of dense linear orderings without endpoints (Theorem 2.7.1).

This method is to hand, but it is also rather heavy. However, there are other methods. One of the simplest – when it applies – is the approach of Per Lindström. Recall that a theory is λ-**categorical** if it has models of cardinality λ and all its models of cardinality λ are isomorphic.

Theorem 7.3.4 (*Lindström's test*). *Let L be a first-order language and T an $\forall_2$ theory in L which has no finite models. If T is λ-categorical for some cardinal $\lambda \geqslant |L|$ then T is model-complete.*

Proof. We use Robinson's test. Suppose $\lambda \geqslant |L|$ and T is λ-categorical, and for contradiction assume that T has models A and B such that $A \subseteq B$, but there are an $\exists_1$ formula $\phi(\bar{x})$ of L and a tuple $\bar{a}$ in A such that $A \vDash \neg\phi(\bar{a})$ and $B \vDash \phi(\bar{a})$. Extend L to a language L^+ by adding a new 1-ary relation symbol P, and expand B to an L^+-structure B^+ by interpreting P as $\mathrm{dom}(A)$. Let T^+ be $\mathrm{Th}(B^+)$. Note that T^+ contains the sentence

(3.1) $\qquad \exists\bar{x}(P(x_0) \wedge \ldots \wedge P(x_{n-1}) \wedge \phi(\bar{x}) \wedge \neg\phi^P(\bar{x}))$

by the relativisation theorem (Theorem 4.2.1).

Since $|L^+| \leqslant \lambda$ and T has no finite models, an easy argument with the compactness theorem (see e.g. Exercise 5.1.10) shows that T^+ has a model D^+ of cardinality λ such that $|P^{D^+}| = \lambda$. Let D be $D^+|L$. Since $D^+ \equiv B^+$, there is an L-structure C with domain P^{D^+} such that $C \subseteq D$. Using the relativisation theorem, C is a model of T of cardinality λ. But by Corollary 7.2.3 there is an e.c. model of T of cardinality λ, and so C is an e.c. model of T since T is λ-categorical. It follows that for every tuple $\bar{c}$ in C, if $D \vDash \phi(\bar{c})$ then $C \vDash \phi(\bar{c})$. This contradicts the fact that D^+ is a model of (3.1). $\qquad\square$

For example, the theory of dense linear orderings without endpoints ((2.32) in section 2.2) is $\forall_2$ by inspection and ω-categorical by Example 3 in section 3.2. So by Lindström's test it is model-complete.

Completeness

A proof of model-completeness can sometimes be the first step towards a proof of completeness.

Let L be a first-order language and T a theory in L. We say that a model A of T is an **algebraically prime** model if A is embeddable in every model of T. (Abraham Robinson said simply 'prime', but this terminology clashes with the notion of a prime model in stability theory.)

Theorem 7.3.5. *Let L be a first-order language and T a theory in L. If T is model-complete and has an algebraically prime model then T is complete.*

Proof. Let A and B be any two models of T, and C an algebraically prime model of T. Then C is embeddable in A and in B. Since A and C are both models of T and T is model-complete, it follows that $C \preccurlyeq A$. Similarly $C \preccurlyeq B$. It follows that $A \equiv C \equiv B$. $\qquad\square$

The results above say nothing about whether the theory T is decidable. But there is an easy general argument to show that if T is complete and has a recursively enumerable set of axioms, then T is decidable – we simply list the consequences of T and wait for the relevant sentence or its negation to turn up (see Exercise 5.1.14). One should avoid trying to impress number theorists with this kind of algorithm: it is hopelessly inefficient.

Model companions

Let T be a theory in a first-order language L. We say that a theory U in L is a **model companion** of T if

(3.2) U is model-complete,

(3.3)   every model of T has an extension which is a model of U, and

(3.4)   every model of U has an extension which is a model of T.

(By Corollary 5.4.3, (3.3) and (3.4) together are equivalent to the equation $T_\forall = U_\forall$.) Maybe T has no model companion; if it does have one, we say that T is **companionable**.

Our next theorem shows that model companions are intimately connected with e.c. models.

Theorem 7.3.6. *Let T be an $\forall_2$ theory in a first-order language L.*

(a) *T is companionable if and only if the class of e.c. models of T is axiomatisable by a theory in L.*

(b) *If T is companionable, then up to equivalence of theories, its model companion is unique and is the theory of the class of e.c. models of T.*

Proof. Suppose first that T is companionable, with a model companion T'. We show that the e.c. models of T are precisely the models of T'. First assume A is a model of T'. Then by Theorem 7.2.1 some extension B of A is an e.c. model of T. Since T' is a model companion of T, some extension C of B is a model of T', and $A \preccurlyeq C$. If $\phi(\bar{x})$ is an $\exists_1$ formula of L and $\bar{a}$ is a tuple of elements of A such that $B \vDash \phi(\bar{a})$, then $C \vDash \phi(\bar{a})$ since $B \subseteq C$, and hence $A \vDash \phi(\bar{a})$ since $A \preccurlyeq C$. So $A \preccurlyeq_1 B$, and it follows (see Exercise 7.2.4(b)) that A is an e.c. model of T. Conversely if A is an e.c. model of T, then some extension B of A is a model of T'. Since T' is equivalent to an $\forall_2$ theory (by Theorem 7.3.3), it easily follows that A is a model of T' (see Exercise 7.2.7(a)).

This proves (b) and left to right in (a). For the other direction in (a), suppose that the class of e.c. models of T is the class of all models of a theory U in L. Then (3.3) and (3.4) hold, so that $T_\forall = U_\forall$. Since the class of e.c. models of T is closed under unions of chains, we can assume by the Chang–Łoś–Suszko theorem (Theorem 5.4.9) that U is an $\forall_2$ theory. Every model A of U is an e.c. model of T, and hence of $T_\forall$ and of U by two applications of Theorem 7.2.4. By Theorem 7.3.1(a $\Leftrightarrow$ b) it follows that U is model-complete. So U is a model companion of T. $\qquad\square$

Corollary 7.3.7. *Let L be a first-order language and T an $\forall_2$ theory in L with a model companion T^*. Then for every $\exists_1$ formula $\phi(\bar{x})$ of L, $\mathrm{Res}_\phi(\bar{x})$ is equivalent modulo T^* to a single $\forall_1$ formula $\psi(\bar{x})$ of L.*

Proof. This follows from the theorem and Theorem 7.2.6. Note that in general $\mathrm{Res}_\phi(\bar{x})$ won't be equivalent to a single $\forall_1$ formula of L modulo T; Exercise 12 gives a counterexample. $\qquad\square$

In practice, one often finds that the easiest way to show that a theory T is not companionable is to show that some e.c. model has an elementary extension which is not e.c. For example, in Example 1 of section 7.2 we saw that an element of an e.c. commutative ring is nilpotent if and only if it doesn't divide any nonzero idempotent. But Exercise 7.2.10 showed that every e.c. commutative ring has an elementary extension where this equivalence fails. So the property of being an e.c. commutative ring is not preserved by elementary extension, and hence by Theorem 7.3.6 the theory of commutative rings is not companionable.

The classic example of a model companion is the theory of algebraically closed fields, which is the model companion of the theory of fields. (In fact it is the model completion, a stronger notion which we shall meet in the next section.) Abraham Robinson hoped that the notion would be useful for identifying classes of structure which play the role of algebraically closed fields in other branches of algebra. This hope has borne its best fruit in the model theory of fields.

Exercises for section 7.3

1. Show that if L is a first-order language, T and U are theories in L, $T \subseteq U$ and T is model-complete, then U is also model-complete.

2. Show that if a first-order theory T is model-complete and has the joint embedding property, then T is complete.

3. In the first-order language whose signature consists of one 1-ary function symbol F, let T be the theory which consists of the sentences $\forall x \, F^n(x) \neq x$ (for all positive integers n) and $\forall x \exists_{=1} y \, F(y) = x$. Apply Lindström's test to show that T is model-complete.

4. Let T be the theory of (non-empty) linear orderings in which each element has an immediate predecessor and an immediate successor, in a language with relation symbols $<$ for the ordering and $S(x, y)$ for the relation 'y is the immediate successor of x'. Show that T can be written as an $\forall_2$ theory. Show by Lindström's test that T is model-complete. Deduce that T is complete.

5. Give an example of a theory T in a first-order language L, such that T is not model-complete but every complete theory in L containing T is model-complete.

6. Suppose T is a theory in a first-order language, and every completion of T is equivalent to a theory of the form $T \cup U$ for some set U of $\exists_1$ sentences. Suppose also that every completion of T is model-complete. Show that T is model-complete.

7. Give an example of a theory T in a countable first-order language, such that T has 2^ω pairwise non-isomorphic algebraically prime models. [Let Ω be a signature consisting of countably many 1-ary function symbols. Write a theory T which says that for each $i < \omega$, the set of elements satisfying $P_i(x)$ is a term algebra of Ω. In a model A of T, consider the number of components of each P_i^A.]

There are many examples of non-companionable theories, but most of them depend on substantial mathematics. Here is an elementary example.

8. Let T be the theory which says the following. All elements satisfy exactly one of $P(x)$ and $Q(x)$; for every element a satisfying $P(x)$ there are unique elements b and c such that $R(b, a)$ and $R(a, c)$; $R(x, y)$ implies $P(x)$ and $P(y)$; there are no finite

R-cycles; $S(x, y)$ implies $P(x)$ and $Q(y)$; if $R(x, y)$ and $Q(z)$ then $S(x, z)$ iff $S(y, z)$. Show that in an e.c. model of T, elements a, b satisfying $P(x)$ are connected by R if and only if there is no element c such that $S(a, c) \leftrightarrow \neg S(b, c)$. Deduce that T has e.c. models with elementary extensions which are not e.c., and hence that T is not companionable.

9. Let G be a group and a, b two elements of G. (a) Show that the following are equivalent. (i) There is a group $H \supseteq G$ with an element h such that $h^{-1}ah = b$. (ii) The elements a and b have the same order. (b) Explain this as an instance of Lemma 7.2.5. (c) Prove that the theory of groups is not companionable.

10. Give an example of an ω_1-categorical theory T in a countable first-order language L, such that no definitional expansion of T by adding finitely many symbols is model-complete. [$T = \mathrm{Th}(A)$, where $\mathrm{dom}(A) = \omega$ and A carries equivalence relations E_i ($i < \omega$) as follows: the classes of E_i are $\{0, \dots, 2^{i+1} - 2\}$ and $\{(2^{i+1} - 1) + 2^{i+1}k, \dots, (2^{i+1} - 1) + 2^{i+1}(k + 1) - 1\}$ ($k < \omega$).]

11. Give an example of an ω_1-categorical countable first-order theory which is not companionable. [First consider the following structure A. The elements are the pairs (m, n) of natural numbers with $m \geqslant -1$ and $n \geqslant 0$. There is a 1-ary relation symbol *Diagonal* picking out the elements (m, m). There are two symmetric 2-ary relation symbols *Horizontal* and *Vertical*. The relation *Horizontal* is an equivalence relation; its classes are the sets $\{(m, b) : m \geqslant -1\}$ with b fixed. The relation *Vertical* holds between (m, n) and (m, m) whenever m, $n \geqslant 0$. Note that A has elementary extensions B and C such that $B \subseteq C$ but C is not an elementary extension of B. The required theory is $\mathrm{Th}(A, \bar{a})$ where $\bar{a}$ is a sequence listing all the elements of A. Suppose T is a model companion. Then T says that infinitely many named elements are not paired with other elements by *Vertical*, so every model of T has an elementary extension with new such elements.]

12. Give an example of a companionable $\forall_1$ theory T in a first-order language L, and an $\exists_1$ formula $\phi(\bar{x})$ in L such that Res_ϕ is not equivalent modulo T to any finite set of $\forall_1$ formulas. [Consider the set of sentences $\forall xy \, \neg (P_0 x \wedge P_i x)$ ($0 < i < \omega$), and let ϕ be the sentence $\exists x \, P_0 x$.]

7.4 Quantifier elimination revisited

A theory T in a first-order language L is said to have **quantifier elimination** if every formula $\phi(\bar{x})$ of L is equivalent modulo T to a quantifier-free formula $\psi(\bar{x})$ of L. In section 2.7 we described a procedure which can be used to show that certain theories T have quantifier elimination. The main idea of the procedure is to find a formula which has a certain form and is equivalent modulo T to some given formula.

Abraham Robinson urged a different approach. For many interesting theories T, we have a mass of good structural information about the models of T: for example decomposition theorems, facts about algebraic or other closures of sets of elements, or results about embedding one model in another. It's hard to use these facts in an argument which concentrates on deducibility from T in the first-order predicate calculus.

So Robinson's message was: to prove quantifier elimination, use model theory rather than syntax, when you can fit the model theory onto the known algebraic structure theory. Of course we can only follow suit when we know some model-theoretic criteria for a theory to have quantifier elimination. The next theorem states some.

Criteria for quantifier elimination

If A and B are L-structures, we write $A \equiv_0 B$ to mean that exactly the same quantifier-free sentences of L are true in A as in B; as before, we write $A \Rightarrow_1 B$ to mean that for every $\exists_1$ sentence ϕ of L, if $A \vDash \phi$ then $B \vDash \phi$. Recall that $T_\forall$ is the set of $\forall_1$ first-order consequences of T.

We say that a first-order theory T has the **amalgamation property** (AP) if the class **K** of all models of T has the AP; in other words, if the following holds:

(4.1)   if A, B, C are models of T and $e: A \to B$, $f: A \to C$ are
          embeddings, then there are D in **K** and embeddings
          $g: B \to D$ and $h: C \to D$ such that $ge = hf$.

(Cf. (1.3) in section 6.1.)

Theorem 7.4.1. *Let L be a first-order language and T a theory in L. The following are equivalent.*
(a) *T has quantifier elimination.*
(b) *If A and B are models of T, and $\bar{a}$, $\bar{b}$ are tuples from A, B respectively such that $(A, \bar{a}) \equiv_0 (B, \bar{b})$, then $(A, \bar{a}) \Rightarrow_1 (B, \bar{b})$.*
(c) *If A and B are models of T, $\bar{a}$ a sequence from A and $e: \langle \bar{a} \rangle_A \to B$ is an embedding, then there are an elementary extension D of B and an embedding $f: A \to D$ which extends e:*

(4.2)

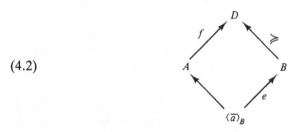

(d) *T is model-complete and $T_\forall$ has the amalgamation property.*

(e) *For every quantifier-free formula $\phi(\bar{x}, y)$ of L, $\exists y\, \phi$ is equivalent modulo T to a quantifier-free formula $\psi(\bar{x})$.*

Proof. (a) $\Rightarrow$ (b) is immediate.

(b) $\Rightarrow$ (c). Assume (b). For every tuple $\bar{a}'$ inside $\bar{a}$, the hypothesis of (c) implies that $(A, \bar{a}') \equiv_0 (B, e\bar{a}')$ and hence $(A, \bar{a}') \Rightarrow_1 (B, e\bar{a}')$. Since every sentence of $L(\bar{a})$ (L with parameters $\bar{a}$ added) mentions just finitely many elements, it follows that $(A, \bar{a}) \Rightarrow_1 (B, e\bar{a})$. Hence the conclusion of (c) holds by the existential amalgamation theorem (Theorem 5.4.1).

(c) $\Rightarrow$ (d). Assume (c). If $e: B \to A$ is any embedding between models of T, put $\langle \bar{a} \rangle_A = B$ in (4.2) and deduce that T is model-complete by Theorem 7.3.1(c) (with A, B transposed). To prove that $T_\forall$ has the amalgamation property, let $g: C \to A'$ and $e: C \to B'$ be embeddings between models of $T_\forall$. By Corollary 5.4.3, A' and B' can be extended to models A, B of T respectively. Now apply (c) with the embedding $C \to A$ in place of the inclusion $\langle \bar{a} \rangle_A \subseteq A$.

(d) $\Rightarrow$ (e) is by Theorem 7.1.3, taking **K** to be the class of all models of T, and noting that all models of T are e.c. by Theorem 7.3.1. (e) $\Rightarrow$ (a) is by Lemma 2.3.1. $\qquad\square$

Warning. The theorem fails if L has no constants and we require that every L-structure is non-empty. (See the remarks after Theorem 7.1.3.)

An L-structure A is said to be **quantifier-eliminable (q.e.** for short) or to **have quantifier elimination** if A is a model of a theory in L which has quantifier elimination; this is equivalent to saying that $\mathrm{Th}(A)$ has quantifier elimination. If the structure A is finite, the corollary below gives a purely algebraic criterion for A to be q.e. As in section 6.1, we say that a structure A is **ultrahomogeneous** if every isomorphism between finitely generated substructures of A extends to an automorphism of A.

Corollary 7.4.2. *Let A be a finite L-structure. Then A is q.e. if and only if A is ultrahomogeneous.*

Proof. Since A is finite, $\mathrm{Th}(A)$ says how many elements A has. So in (4.2) with $A = B$, D must be A and f must be an isomorphism. $\qquad\square$

Let us put these techniques to work by proving that some theories have quantifier elimination. In our first two illustrations, Lindström's test (Theorem 7.3.4) gives a quick proof that the theory is model-complete, and we use (d) of Theorem 7.4.1 to climb from model-completeness to quantifier elimination. The third example has to work harder to prove model-completeness, but it also finishes the argument by quoting (d) of Theorem 7.4.1.

Example 1: *Vector spaces*. Let R be a field and T the theory of infinite (left) vector spaces over R. The axioms of T consist of (2.21) in section 2.2 together with the sentences $\exists_{\geqslant n} x \, x = x$ for each positive integer n; inspection shows that these axioms are $\forall_2$ first-order sentences. Certainly T is λ-categorical for any infinite cardinal $\lambda > |R|$, and by definition T has no finite models. So by Lindström's test, T is model-complete. Elementary algebra shows that $T_\forall$, which is the theory of vector spaces over R, has the amalgamation property. So by (d) of the theorem, T has quantifier elimination.

Now by Corollary 7.4.2, every finite vector space is q.e.; so we have shown that *every vector space is q.e.*

Example 2: *Algebraically closed fields and finite fields*. By a theorem of Steinitz, any two algebraically closed fields of the same characteristic and the same uncountable cardinality are isomorphic. Also algebraically closed fields are infinite. So by Lindström's test, the theory of algebraically closed fields of a fixed characteristic is model-complete, whence (see Exercise 7.3.6) the theory of algebraically closed fields is model-complete. We saw already in Corollary 7.1.4 how this implies that every algebraically closed field is q.e.; it's the same proof as by Theorem 7.4.1 (d $\Rightarrow$ e).

Finite fields are q.e. too – in other words, by Corollary 7.4.2, they are ultrahomogeneous. Let A be a finite field; let B and C be subfields of A with $f: B \to C$ an isomorphism. We must show that f extends to an automorphism of A. Since B and C have the same cardinality, their multiplicative groups are the same subgroup of the multiplicative group of A, which is a finite cyclic group; it follows that $B = C$. Thus we need only show that every automorphism of B extends to one of A. We do this by a counting argument. Write D for the prime field of the same characteristic p as B. Then D is rigid, so the Galois group $\mathrm{Gal}(B/D)$ is $\mathrm{Aut}(B)$; likewise $\mathrm{Gal}(A/D)$ is $\mathrm{Aut}(A)$. Write H for the group of all those automorphisms of B which extend to an automorphism of A. Then using Galois theory,

$$|\mathrm{Aut}(A)_{(B)}| \cdot H = |\mathrm{Aut}(A)| = (A:D) = (A:B) \cdot (B:D)$$
$$= |\mathrm{Aut}(A)_{(B)}| \cdot |\mathrm{Aut}(B)|.$$

So $|\mathrm{Aut}(B)| = |H|$. Since $H \subseteq \mathrm{Aut}(B)$, this proves that $H = \mathrm{Aut}(B)$, i.e. every automorphism of B extends to one of A.

Our third example needs a more substantial argument and a new heading.

Real-closed fields

The necessary and sufficient conditions of Theorem 7.4.1 are not always the best route to a proof of quantifier elimination. Sometimes a sufficient condition ties in better with the known algebra. For example, we have the following.

Corollary 7.4.3. *Let L be a first-order language and T a theory in L. Suppose that T satisfies the following conditions.*

(a) *For any two models A and B of T, if $A \subseteq B$, $\phi(\bar{x}, y)$ is a quantifier-free formula of L and $\bar{a}$ is a tuple of elements of A such that $B \vDash \exists y \, \phi(\bar{a}, y)$, then $A \vDash \exists y \, \phi(\bar{a}, y)$. ('$T$ is 1-model-complete.')*

(b) *For every model A of T and every substructure C of A there is a model A' of T such that (i) $C \subseteq A' \subseteq A$, and (ii) if B is another model of T with $C \subseteq B$ then there is an embedding of A' into B over C.*

Then T has quantifier elimination.

Proof. Assuming (a) and (b), we prove Theorem 7.4.1(b). Suppose A and B are models of T, $\bar{a}$ and $\bar{b}$ are tuples of elements of A and B respectively, and $(A, \bar{a}) \equiv_0 (B, \bar{b})$. Let $\phi(\bar{x}, \bar{y})$ be a quantifier-free formula of L such that $A \vDash \exists \bar{y} \, \phi(\bar{a}, \bar{y})$; we must show that $B \vDash \exists \bar{y} \, \phi(\bar{b}, \bar{y})$. Without loss we can suppose that $\bar{b}$ is $\bar{a}$.

For this, suppose $A \vDash \phi(\bar{a}, \bar{c})$ where $\bar{c}$ is $(c_0, \ldots, c_{k-1})$. We claim that there is an element d_0 in some elementary extension B_0 of B, such that $(A, \bar{a}, c_0) \equiv_0 (B_0, \bar{a}, d_0)$.

Write $\Psi_0(\bar{x}, y)$ for the set of all quantifier-free formulas $\psi(\bar{x}, y)$ such that $A \vDash \psi(\bar{a}, c_0)$. Since $(A, \bar{a}) \equiv_0 (B, \bar{a})$, we can write $C = \langle \bar{a} \rangle_A = \langle \bar{a} \rangle_B$. By (b) there is a model A' of T such that $C \subseteq A' \subseteq A$ and there is an embedding of A' into B over C; without loss we can suppose that A' is a substructure of B. Since $\bar{a}$ is in A' and each formula $\psi(\bar{a}, y)$ in $\Psi_0(\bar{a}, y)$ has just one free variable, we infer by (a) that $A' \vDash \exists y \, \psi(\bar{a}, y)$, since $A \vDash \exists y \, \psi(\bar{a}, y)$. But since $A' \subseteq B$, it follows that $B \vDash \exists y \, \psi(\bar{a}, y)$ too. Hence every finite subset of $\Psi_0(\bar{a}, y)$ is satisfied by an element of B. By compactness it follows that there is an elementary extension B_0 of B with an element d_0 which realises the type $\Psi_0(\bar{a}, y)$. Thus $(A, \bar{a}, c_0) \equiv_0 (B_0, \bar{a}, d_0)$, proving the claim.

Now we repeat to find an elementary extension B_1 of B_0 with an element d_1 such that $(A, \bar{a}, c_0, c_1) \equiv_0 (B_1, \bar{a}, d_0, d_1)$, and so on. Eventually we reach an elementary extension B_{n-1} of B and elements $\bar{d}$ such that $(A, \bar{a}, \bar{c}) \equiv_0 (B_{n-1}, \bar{a}, \bar{d})$. In particular we have $B_{n-1} \vDash \phi(\bar{a}, \bar{d})$, and so $B_{n-1} \vDash \exists \bar{y} \, \phi(\bar{a}, \bar{y})$. Thus $B \vDash \exists \bar{y} \, \phi(\bar{a}, \bar{y})$ as required. $\qquad \square$

To illustrate this, we give a Robinson-style proof of a result of Tarski which was quoted without proof as Theorem 2.7.2.

Theorem 7.4.4. *The theory T of real-closed fields in the first-order language L of ordered fields has quantifier elimination.*

Proof. The proof borrows two facts from the algebraists.

Fact 7.4.5. *The intermediate value theorem holds in real-closed fields for all functions defined by polynomials $p(x)$, possibly with parameters. (I.e. if $p(a) \cdot p(b) < 0$ then $p(c) = 0$ for some c strictly between a and b.)*

Fact 7.4.6. *If A is a real-closed field and C an ordered subfield of A, then there is a smallest real-closed field B such that $C \subseteq B \subseteq A$. Moreover if A' is any real-closed field $\supseteq C$ then B is embeddable in A' over A. (A' is called the* **real closure** *of C in A.)*

We use Corollary 7.4.3. First we prove (a). Let A and B be real-closed fields with $A \subseteq B$, and let $\phi(x)$ be a quantifier-free formula of L with parameters from A, such that $B \vDash \exists x \, \phi$. We must show that $A \vDash \exists x \, \phi$. After bringing ϕ to disjunctive normal form and distributing the quantifier through it, we can assume that ϕ is a conjunction of literals. Now $y \neq z$ is equivalent to $y < z \lor z < y$, and $\neg y < z$ is equivalent to $y = z \lor z < y$. So we can suppose that ϕ has the form

(4.3) $p_0(x) = 0 \land \ldots \land p_{k-1}(x) = 0 \land q_0(x) > 0 \land \ldots \land q_{m-1}(x) > 0$

where $p_0, \ldots, p_{k-1}, q_0, \ldots, q_{m-1}$ are polynomials with coefficients in A.

If ϕ contains a non-trivial equation $p_i(x) = 0$, then any element of B satisfying ϕ is algebraic over A and hence is already in A. Suppose on the other hand that $k = 0$. There are finitely many points $c_0 < \ldots < c_{n-1}$ in A which are zeros of one or more of the polynomials q_j $(j < m)$. By the intermediate value property (Fact 7.4.5), none of the q_j changes sign except at the points c_i $(i < n)$. So it suffices to take a point b of B such that $B \vDash \phi(b)$, and choose a point a in A which lies in the same interval of the c_is as b. This proves (a) of Corollary 7.4.3.

Next we prove (b). Let A and B be real-closed fields and C a common substructure of A and B. Then C is an ordered integral domain. It is not hard to show that the quotient field of C in A is isomorphic over C to the quotient field of C in B; so we can identify these quotient fields and suppose that C is itself an ordered field. By Fact 7.4.6 we can take A' to be the real closure of C in A, and (b) is proved.   □

Exercises for section 7.4

1. In Theorem 7.4.1, show that T has quantifier elimination if and only if condition (c) holds whenever $\bar{a}$ is a tuple of elements of A.

2. Show that in Corollary 7.4.3, condition (a) can be replaced by (a'): if B is a model of T and A is a proper substructure of B, then there are an element b of B which is not in A, and a set $\Phi(x)$ of quantifier-free formulas of L with parameters in A, such that $B \vDash \bigwedge \Phi(b)$, Φ determines the quantifier-free type of b over A, and for every finite subset Φ_0 of Φ, $A \vDash \exists x \bigwedge \Phi_0$.

In section 3.3 above we saw a way to use Ehrenfeucht–Fraïssé games in order to find elimination sets. The following exercise translates that discussion into a criterion for quantifier elimination.

3. Let L be a first-order language with finite signature, and T a theory in L. Show that the following are equivalent. (a) T has quantifier elimination. (b) If A and B are any models of T, then for each $n < \omega$, any pair of tuples $(\bar{a}, \bar{b})$ from A, B respectively, such that $(A, \bar{a}) \equiv_0 (B, \bar{b})$, is a winning position for player $\exists$ in the game $G_n[A, B]$.

This exercise shows the link between the back-and-forth test of Exercise 3 and the amalgamation criterion of Theorem 7.4.1(c). It uses the notion of λ-saturation, which is defined in section 8.1 below.

4. Let L be a first-order language and T a theory in L. Show that the following are equivalent. (a) T has quantifier elimination. (b) If A and B are any ω-saturated models of T and $\bar{a}$ a tuple of elements of T such that $(A, \bar{a}) \equiv_0 (B, \bar{a})$, then $(A, \bar{a})$ and $(B, \bar{a})$ are back-and-forth equivalent. (c) If A is a model of T, B is a λ-saturated model of T for some infinite cardinal $\lambda > |A|$, and $\bar{a}$ is a tuple of elements of T such that $(A, \bar{a}) \equiv_0 (B, \bar{a})$, then there is an elementary embedding $f: A \to B$ such that $f\bar{a} = \bar{b}$.

5. Let L be a first-order language, T a theory in L and $\phi(\bar{x})$ a formula of L. Show that the following are equivalent. (a) ϕ is equivalent modulo T to a quantifier-free formula $\psi(\bar{x})$. (b) If A and B are any two models of T and $\bar{a}$, $\bar{b}$ are tuples of elements of A, B respectively such that $(A, \bar{a}) \equiv_0 (B, \bar{b})$, then $A \vDash \phi(\bar{a})$ implies $B \vDash \phi(\bar{b})$. (c) If A and B are any two models of T, $\bar{a}$ is a tuple of elements of A such that $A \vDash \phi(\bar{a})$, and $f: \langle \bar{a} \rangle_A \to B$ is an embedding, then $B \vDash \phi(f\bar{a})$. [(b) and (c) are different ways of saying exactly the same thing.]

The next exercise assumes the existence of strongly ω-homogeneous elementary extensions; see section 8.2 below.

6. Let L be a first-order language and T a complete theory in L. Show that T has quantifier elimination if and only if every model of T has an ultrahomogeneous elementary extension.

7. Let A be an integral domain. Show that if A is q.e. then A is a field. [If A is infinite, use Exercise 6.] *In fact every q.e. field is either finite or algebraically closed.*

8. Let T be a first-order theory with a model companion U. Show that U has quantifier elimination if and only if $T_\forall$ has the amalgamation property.

9. Let T be a theory in a first-order language L, and U a model companion of T. Show that the following are equivalent. (a) T has the amalgamation property. (b) For every model A of T, $U \cup \mathrm{diag}(A)$ is a complete theory in $L(A)$. (A theory U satisfying (a) or (b) is said to be a **model-completion** of T.)

10. Let L be the first-order language whose signature consists of one 1-ary function symbol. Show that the empty theory in L has a model completion.

11. An **ordered abelian group** is an abelian group with a 2-ary relation $\leqslant$ satisfying the laws '$\leqslant$ is a linear ordering' and $\forall xyz(x \leqslant y \to x + z \leqslant y + z)$. Let T_{oa} be the first-order theory of ordered abelian groups, and T_{doa} the first-order theory of ordered abelian groups which are divisible as abelian groups. (a) Show that T_{doa} is the model companion of T_{oa}. (b) Show that T_{doa} has quantifier elimination and is complete. (c) Show that $T_{\mathrm{doa}} = \mathrm{Th}(\mathbb{Q}, \leqslant)$ where $\mathbb{Q}$ is the additive group of rationals and $\leqslant$ is the usual ordering.

12. In the notation of the previous exercise, show that all ordered abelian groups satisfy the same $\exists_1$ first-order sentences. [Show this for $(\mathbb{Q}, \leqslant)$ and $(\mathbb{Z}, \leqslant)$, and trap all other groups between these.]

13. Let L be a first-order language and T an $\forall_1$ theory in L. Show that if $\phi(\bar{x})$ is a formula of L such that both ϕ and $\neg\phi$ are preserved by all embeddings between models of T, then ϕ is equivalent modulo T to a quantifier-free formula $\psi(\bar{x})$.

The next result is known to logic programmers (and others) as Herbrand's theorem.
14. Let L be a first-order language, T an $\forall_1$ theory in L and $\phi(\bar{x})$ a quantifier-free formula of L. Show that if $T \vdash \exists \bar{x}\, \phi$, then for some $m < \omega$ there are tuples of terms $\bar{t}_0(\bar{y}), \ldots, \bar{t}_{m-1}(\bar{y})$ such that $T \vdash \forall \bar{y} \bigvee_{i<m} \phi(\bar{t}_i(\bar{y}))$. *In particular the terms can be chosen to be ground terms when L has at least one constant; in this case the result is a kind of interpolation theorem, finding a quantifier-free interpolant between an $\forall_1$ premise and an $\exists_1$ conclusion.*

Further reading

For model-completeness in general the survey by Macintyre is still well worth reading:

Macintyre, A. Model completeness. In *Handbook of Mathematical Logic*, ed. Barwise, pp. 139–180. Amsterdam: North-Holland, 1977.

After this one must go to writings on particular kinds of structure. A good place to study quantifier elimination for real-closed fields is:

Prestel, A. *Lectures on Formally Real Fields*. Lecture Notes in Mathematics 1093. Berlin: Springer-Verlag, 1984.

One can read about the model theory of some other kinds of field in the following papers, and also in the paper of van den Dries referenced at the end of Chapter 2 above.

Chatzidakis, Z., van den Dries, L. and Macintyre, A. Definable sets over finite fields. *J. reine und angewandte Math.* **427** (1992) 107–135.

Delon, F. Idéaux et types sur les corps séparablement clos. *Supplément au Bulletin de la Société Mathématique de France*, Mémoire 33, 1988.

Wilkie, A. Model completeness results for expansions of the ordered field of real numbers by restricted Pfaffian functions and the exponential function. *Journal of the American Mathematical Society* **9** (1996) 1051–1094.

The main omission in this chapter is model-theoretic forcing, which Robinson introduced in 1970. It turned out to be a powerful tool for constructing and analysing e.c. groups. The definitive account is:

Ziegler, M. Algebraisch abgeschlossene Gruppen. In *Word problems II, The Oxford book*, ed. Adian *et al.*, pp. 449–576. Amsterdam: North-Holland, 1980.

There is also an exposition of forcing and its applications in

Hodges, W. *Building models by games*. Cambridge: Cambridge University Press, 1985.

8

Saturation

I have made numerous composites of various groups of convicts ... The first set of portraits are those of criminals convicted of murder, manslaughter, or robbery accompanied with violence. It will be observed that the features of the composites are much better looking than those of the components. The special villainous irregularities have disappeared, and the common humanity that underlies them has prevailed.

Francis Galton, Inquiries into human faculty and its development (1907)

A saturated model of a complete theory T is a 'most typical' model of T. It has no avoidable asymmetries; unlike the man of Devizes, it's not short at one side and long at the other. (Or like the man of the Nore, it's the same shape behind as before.) We create saturated models by amalgamating together all possible models, rather in the spirit of Fraïssé's construction from section 6.1.

Although in a sense these are typical models, in another sense every saturated model A has some quite remarkable properties. Every small enough model of T is elementarily embeddable in A – this is called *universality*. Every type-preserving map between small subsets of A extends to an automorphism of A – this is *strong homogeneity*. We can expand A to a model of any theory consistent with Th(A) – this is a form of *resplendence*.

These properties, particularly the resplendence and the strong homogeneity, make saturated models a valuable tool for proving facts about the theory T. For example, saturated models can be used for proving preservation and interpolation theorems. The universality makes saturated models suitable as work-spaces: we choose a large saturated model M, and all the other models that we consider are regarded as elementary substructures of M. A model M used as a work-space is said to be a *big* or *monster* model of T.

Sadly there is no guarantee that T has a saturated model, unless T happens to be stable. One response to this is to declare yourself a formalist and announce that infinitary mathematics is meaningless anyway. For those of us unwilling to take this easy escape route, it's more sensible to weaken the notion of saturation, and the first natural way to do this is to think of λ-*saturated* models. As section 8.1 explains, these models are saturated with respect to any set of fewer than λ parameters. The good news is that every structure has λ-saturated elementary extensions, for any cardinal λ. The bad news is that λ-saturation is too weak for many applications; it doesn't guarantee resplendence or any degree of strong homogeneity.

So we have to try again, aiming somewhere between saturation and λ-saturation. In this chapter I describe λ-*bigness*, which lies in the right area. The professional model theorist needs to know about *special models*, which also live in this middle ground; sadly there is no space to include them here.

8.1 The great and the good

In arguments which involve several structures and maps between them, things usually go smoother when the maps are inclusions. There are at least two good mathematical reasons for this. First, *if the maps are inclusions, then diagrams automatically commute.* And second, *if A is a substructure of B, then we can specify A by giving B and* $\mathrm{dom}(A)$; *there is no need to describe the relations of A as well as those of B.*

Thoughts of this kind have led to the use of *big models*, sometimes known as *monster models*. Informally, a big model is a structure M such that every commutative diagram of structures and maps that we want to consider is isomorphic to a diagram of inclusions between substructures of M. Of course a structure M with this property can't exist. It would have to contain isomorphic copies of all structures, and so its domain would be a proper class and not a set.

So we draw in our horns and demand something less. For the moment let us say that M is **splendid** if the following holds.

(1.1)   Suppose L^+ is a first-order language got by adding a new relation symbol R to L. If N is an L^+-structure such that $M \equiv N|L$, then we can interpret R by a relation S on the domain of M so that $(M, S) \equiv N$.

Informally this says that M is compatible with any extra structural features which are consistent with $\mathrm{Th}(M)$.

Example 1: *Equivalence relations.* Let M be a structure consisting of an equivalence relation with two equivalence classes, whose cardinalities are ω and ω_1. Then M is not splendid, because we can take an elementary extension N where the two equivalence classes have the same size, and add a bijection between these classes.

For any cardinal λ, we shall say that M is λ-**big** if $(M, \bar{a})$ is splendid whenever $\bar{a}$ is a sequence of fewer than λ elements of M. Thus splendid $= 1$-big.

One can define a **big model** (or **monster model**) to be a model which is λ-big for some cardinal λ which is 'large enough to cover everything interesting'. This is vague, but in practice there is no need to make it more precise. In stability theory one is interested in the models of some complete first-order theory T; the usual habit is to choose a big model of T without specifying how large λ is.

It will emerge that every structure has λ-big elementary extensions for any λ. The proof is technical and doesn't illuminate the use of these structures. So I postpone it to the next section, and turn to some important properties which are closely related to bigness.

λ-saturation and λ-homogeneity

I recall some definitions from section 5.2. Let A be an L-structure and X a set of elements of A; write $L(X)$ for the first-order language formed from L by adding constants for the elements of X. If $n < \omega$, then a **complete n-type** over X with respect to A is a set of the form $\{\phi(x_0, \ldots, x_{n-1}): \phi \text{ is in } L(X)$ and $B \models \phi(\bar{b})\}$ where B is an elementary extension of A and $\bar{b}$ is an n-tuple of elements of B. We write this n-type as $\mathrm{tp}_B(\bar{b}/X)$, and we say that $\bar{b}$ **realises** this n-type **in** B. We write $S_n(X; A)$ for the set of all complete n-types over X with respect to A. A **type** is an n-type for some $n < \omega$.

Now let λ be a cardinal. We say that A is λ-**saturated** if

(1.2)   for every set X of elements of A, if $|X| < \lambda$ then all complete 1-types over X with respect to A are realised by elements in A.

We say that A is **saturated** if A is $|A|$-saturated. We say that A is λ-**homogeneous** if

(1.3)   for every pair of sequences $\bar{a}$, $\bar{b}$ of length less than λ, if $(A, \bar{a}) \equiv (A, \bar{b})$ and d is any element of A, then there is an element c such that $(A, \bar{a}, c) \equiv (A, \bar{b}, d)$.

We say that A is **homogeneous** if A is $|A|$-homogeneous.

The literature contains far too many different concepts called homogeneity, and muddles have occurred. One should probably refer to λ-homogeneous structures as **elementarily λ-homogeneous**. But this is a mouthful, and throughout this chapter there is no danger of confusion.

Finally we say that A is λ-**universal** if

(1.4)   if B is any L-structure of cardinality $< \lambda$ and $B \equiv A$, then B is elementarily embeddable in A.

Immediately from the definitions, we have the following.

Lemma 8.1.1. *Suppose A is λ-big and $\kappa < \lambda$. Then A is κ-big. Similarly with '-saturated', '-homogeneous' or '-universal' in place of '-big'.* $\square$

The simplest links between these concepts run as follows:

$$\text{(1.5)} \qquad \lambda\text{-big} \Rightarrow \lambda\text{-saturated} \begin{array}{c} \nearrow \lambda\text{-homogeneous} \\ \\ \searrow \lambda\text{-universal} \end{array}$$

Let us prove these implications. (One can say more. For example, if A is saturated then A is $|A|$-big. Exercise 14 will prove this under a small assumption.)

Theorem 8.1.2. *Suppose A is λ-big. Then A is λ-saturated.*

Proof. Suppose A is a λ-big L-structure. Let $\bar{a}$ be a sequence of fewer than λ elements of A; let B be an elementary extension of A and b an element of A. We must show that $\mathrm{tp}_B(b/\bar{a})$ is realised in A. Let L^+ be the first-order language formed by adding a 1-ary relation symbol R to L, and make B into an L^+-structure B^+ by interpreting R as the singleton $\{b\}$. By λ-bigness there is a relation S on the domain of A, such that $(A, S, \bar{a}) \equiv (B^+, \bar{a})$. Now $B^+ \vDash$ 'Exactly one element satisfies Rx', and so S is a singleton $\{c\}$. Clearly c realises $\mathrm{tp}_B(b/\bar{a})$. $\square$

Our next lemma shows that λ-saturation is a strengthening of λ-homogeneity.

Lemma 8.1.3. *Let A be an L-structure and λ a cardinal. The following are equivalent.*
(a) *A is λ-saturated.*
(b) *For every L-structure B and every pair of sequences $\bar{a}$, $\bar{b}$ of elements of A, B respectively, if $\bar{a}$ and $\bar{b}$ have the same length $< \lambda$ and $(A, \bar{a}) \equiv (B, \bar{b})$, and d is any element of B, then there is an element c of A such that $(A, \bar{a}, c) \equiv (B, \bar{b}, d)$.*

Proof. (a) $\Rightarrow$ (b). Assume (a), and suppose $\bar{a}$, $\bar{b}$ are as in the hypothesis of (b). By elementary amalgamation (Theorem 5.3.1) there are an elementary extension D of A and an elementary embedding $f: B \to D$ such that $f\bar{b} = \bar{a}$. Now if d is any element of B, then $(D, \bar{a}, fd) \equiv (B, \bar{b}, d)$ since f is elementary. But $\bar{a}$ contains fewer than λ elements and A is λ-saturated, so that A contains an element c such that $\text{tp}_A(c/\bar{a}) = \text{tp}_D(fd/\bar{a})$. Then $(A, \bar{a}, c) \equiv (D, \bar{a}, fd) \equiv (B, \bar{b}, d)$ as required.

The implication (b) $\Rightarrow$ (a) is immediate from the definitions. $\square$

Theorem 8.1.4. *If A is λ-saturated then A is λ-homogeneous.*

Proof. The definition of λ-homogeneity is the special case of Lemma 8.1.3(b) where $A = B$. $\square$

Lemma 8.1.3 can be applied over and over again, to build up maps between structures, as follows.

Lemma 8.1.5. *Let L be a first-order language and A an L-structure.*

(a) *Suppose A is λ-saturated, B is an L-structure and $\bar{a}$, $\bar{b}$ are sequences of elements of A, B respectively such that $(A, \bar{a}) \equiv (B, \bar{b})$. Suppose $\bar{a}$, $\bar{b}$ have length $< \lambda$, and let $\bar{d}$ be a sequence of elements of B, of length $\leq \lambda$. Then there is a sequence $\bar{c}$ of elements of A such that $(A, \bar{a}, \bar{c}) \equiv (B, \bar{b}, \bar{d})$.*

(b) *The same holds if we replace 'λ-saturated' by 'λ-homogeneous' and add the assumption that $A = B$.*

Proof. We prove (a); the proof of (b) is similar. By induction we shall define a sequence $\bar{c} = (c_i : i < \lambda)$ of elements of A so that

$$(1.6) \qquad \text{for each } i \leq \lambda, \ (A, \bar{a}, \bar{c}|i) \equiv (B, \bar{b}, \bar{d}|i).$$

For $i = 0$, (1.6) says that $(B, \bar{b}) \equiv (A, \bar{a})$, which is given by assumption. There is nothing to do at limit ordinals, since any formula of L has only finitely many free variables. Suppose then that $\bar{c}|i$ has just been chosen and $i < \lambda$. Since A is λ-saturated and $\bar{c}|i$ has length $< \lambda$, Lemma 8.1.3 gives us an element c_i of A such that $(A, \bar{a}, \bar{c}|i, c_i) \equiv (B, \bar{b}, \bar{d}|i, d_i)$. $\square$

Theorem 8.1.6. *Let L be a first-order language and A a λ-saturated L-structure. Then A is λ^+-universal.*

Proof. We have to show that if B is an L-structure of cardinality $\leq \lambda$ and $B \equiv A$, then there is an elementary embedding $e : B \to A$. List the elements of B as $\bar{d} = (d_i : i < \lambda)$; repetitions are allowed. By the lemma there is a sequence $\bar{c}$ of elements of A such that $(B, \bar{d}) \equiv (A, \bar{c})$. It follows by the elementary diagram lemma (Lemma 2.5.3) that there is an elementary embedding of B into A taking $\bar{d}$ to $\bar{c}$. $\square$

In fact λ-saturation is exactly λ-homogeneity plus λ-universality. I leave this as an exercise (Exercise 7 below).

Lemma 8.1.5 implies that when λ is infinite, the definition of λ-saturation need not be limited to types in one variable.

Theorem 8.1.7. *Let L be a first-order language, A an L-structure, λ an infinite cardinal and $\bar{y}$ any tuple of variables. Suppose A is λ-saturated. Let $\bar{a}$ be a sequence of fewer than λ elements of A, and $\Phi(\bar{x}, \bar{y})$ a set of formulas of L such that for each finite subset Ψ of Φ, $A \vDash \exists \bar{y} \bigwedge \Psi(\bar{a}, \bar{y})$. Then there is a tuple $\bar{b}$ of elements of A such that $A \vDash \bigwedge \Phi(\bar{a}, \bar{b})$.*

Proof. By the compactness theorem (see Theorem 5.2.1(a)) there is certainly an elementary extension B of A containing a tuple $\bar{d} = (d_0, \ldots, d_{m-1})$ such that $B \vDash \bigwedge \Phi(\bar{a}, \bar{d})$. Now $(A, \bar{a}) \equiv (B, \bar{a})$, and since λ is infinite, the sequence $\bar{d}$ has fewer than λ elements. So by Lemma 8.1.5 there is a tuple $\bar{c}$ of elements of A such that $(A, \bar{a}, \bar{c}) \equiv (B, \bar{a}, \bar{d})$. Hence $A \vDash \bigwedge \Phi(\bar{a}, \bar{c})$ as required. □

The idea of Lemma 8.1.5 can be run back and forth between two structures.

Theorem 8.1.8. *Let A and B be elementarily equivalent L-structures of the same cardinality λ.*

(a) *If A and B are both saturated then $A \cong B$.*

(b) *If A and B are both homogeneous and realise the same n-types over $\varnothing$ for each $n < \omega$, then $A \cong B$.*

Proof. (a) First assume that λ is infinite. List the elements of A as $(a_i : i < \lambda)$ and those of B as $(b_i : i < \lambda)$. We claim that there are sequences $\bar{c}$, $\bar{d}$ of elements of A and B respectively, both of length λ, such that for each $i \leqslant \lambda$,

(1.7) $$(A, \bar{a}|i, \bar{c}|i) \equiv (B, \bar{d}|i, \bar{b}|i).$$

The proof is by induction on i. Again the case $i = 0$ is given in the theorem hypothesis, and there is nothing to do at limit ordinals. When (1.7) has been established for some $i < \lambda$, fewer than λ parameters have been chosen (since λ is infinite). We use the saturation of B to find d_i such that $(A, \bar{a}|i, a_i, \bar{c}|i) \equiv (B, \bar{d}|i, d_i, \bar{b}|i)$; then we use the saturation of A to find c_i such that $(A, \bar{a}|i, a_i, \bar{c}|i, c_i) \equiv (B, \bar{d}|i, d_i, \bar{b}|i, b_i)$. At the end of the construction, the diagram lemma gives us an embedding $f : A \to B$ such that $f\bar{a} = \bar{d}$ and $f\bar{c} = \bar{b}$. The embedding is onto B since $\bar{b}$ includes all the elements of B.

When λ is finite, Theorem 8.1.6 tells us that there is an elementary embedding $f : A \to B$. Then f must be an isomorphism since A and B both have cardinality λ.

(b) The argument here is similar but with a couple of extra twists. We can assume that λ is infinite. We begin by establishing

(1.8)   if $i < \lambda$ and $\bar{b}$ is a sequence in B of length i, then there is a sequence $\bar{a}$ of elements of A such that $(A, \bar{a}) \equiv (B, \bar{b})$. (And the same with A and B transposed.)

The proof is by induction on i. Since the hypotheses are symmetrical in A and B, we only need prove (1.8) one way round.

When i is finite, (1.8) is given by the theorem hypothesis. When i is infinite we distinguish two cases. First suppose that i is a cardinal. Then we build up $\bar{a}$ so that for each $j < i$,

(1.9) $(A, \bar{a}|j) \equiv (B, \bar{b}|j)$.

The theorem hypothesis gives the case $j = 0$. When j is a limit ordinal there is nothing to do. Given (1.9) for j, we find a_j as follows. By induction hypothesis, since $|j + 1| < i$, there is a sequence $\bar{c} = (c_k : k \leqslant j)$ in A such that $(A, \bar{c}) \equiv (B, \bar{b}|(j + 1))$. Then $(A, \bar{a}|j) \equiv (A, \bar{c}|j)$, and so by the homogeneity of A there is a_j such that $(A, \bar{a}|j, a_j) \equiv (A, \bar{c}) \equiv (B, \bar{b}|(j + 1))$, giving (1.9) for $j + 1$. This establishes (1.8) when i is a cardinal. Finally when i is not a cardinal, we reduce to the case where it is a cardinal by rearranging the elements of $\bar{b}$ into a sequence of order-type $|i|$.

To prove the theorem, we go back and forth as in (a). For example, to find d_i we first use (1.8) to choose a sequence $\bar{e}$ in D such that $(A, \bar{a}|i, a_i, \bar{c}|i) \equiv (B, \bar{e})$, and then we use the homogeneity of B to find d_i so that $(B, \bar{e}) \equiv (B, \bar{d}|i, d_i, \bar{b}|i)$. $\square$

Examples

Example 2: *Finite structures.* If the structure A is finite, then any structure elementarily equivalent to A is isomorphic to A. It follows that A is λ-big for all cardinals λ. In particular A is saturated and homogeneous. This is a freak case, but it explains why the word 'infinite' will keep appearing in this chapter.

Example 3: *Vector spaces.* Let K be a field, λ an infinite cardinal $\geqslant |K|$ and A a vector space of dimension λ over K. Then A is λ-big. For this, let L be the language of A, and L^+ a language got by adding one relation symbol R. Let $\bar{a}$ be a sequence of fewer than λ elements of A, and B an L^+-structure with elements $\bar{b}$ such that $(B|L, \bar{b}) \equiv (A, \bar{a})$. The language of $(B, \bar{b})$ has cardinality at most λ, and so by using compactness and the downward Löwenheim–Skolem theorem we can suppose without loss that $B|L$ is also a vector space of cardinality $|A|$ and dimension λ over K. (If $|A| > |K|$ it suffices to get $|B| = |A|$, since then B must also have dimension λ over K. If

$|A| = |K|$, we need to realise λ types of linearly independent elements in B, which we can do by compactness.) Then linear algebra shows that $(B|L, \bar{b}) \cong (A, \bar{a})$, and the isomorphism carries R^B across to A as required.

If A is infinite but has dimension less than λ, then A is no longer λ-saturated: consider a basis $\bar{a}$ of A and a set of formulas which express 'x is not linearly dependent on $\bar{a}$'. However, every vector space is ω-homogeneous, since an isomorphism between finite-dimensional subspaces always extends to an automorphism of the whole space.

Example 4: *Algebraically closed fields*. An argument like that of Example 3 shows that every algebraically closed field A of infinite transcendence degree over the prime field is $|A|$-big and hence saturated. (When the transcendence degree is finite, the field is homogeneous but not saturated.) André Weil proposed using algebraically closed fields of infinite transcendence degree as big models for field theory; he called them **universal domains**. In fact his proposal only codified what was already common practice – when we add roots of a polynomial, where are they supposed to come from?

Eample 5: *Countable ω-categorical structures*. Exercise 6.2.6 showed that every countable ω-categorical structure is saturated.

Example 6: *Dense linear orderings without endpoints*. The previous example shows that every countable dense linear ordering without endpoints is ω-saturated. Suppose A is an uncountable dense linear ordering without endpoints. Then A is certainly ω-saturated (see Exercise 8 below). Suppose that in fact A is ω_1-saturated. Then A has an element x sitting in each gap of the form

$$(1.10) \qquad a_0 \leqslant a_1 \leqslant a_2 \leqslant \ldots < x < \ldots \leqslant b_2 \leqslant b_1 \leqslant b_0$$

where a_i and b_i ($i < \omega$) are elements of A. Now it follows by Cantor's famous argument for the reals that A must have at least 2^ω elements. Closer inspection shows that A is not much like the ordering of the reals. No strictly increasing sequence ($a_i: i < \omega$) of elements of A has a supremum in A. For if b was its supremum, we could get a contradiction by taking $b_i = b$ for each $i < \omega$ in (1.10). Saturated dense linear orderings without endpoints, of cardinality ω_α, are known as η_α**-sets**.

Exercises for section 8.1

1. Suppose λ is an infinite cardinal and A is a λ-saturated L-structure. Show that if E is an equivalence relation on n-tuples of elements of A which is first-order definable with parameters, then the number of equivalence classes of E is either finite or $\geqslant \lambda$.

(In particular if X is a subset of dom(A) which is first-order definable with parameters, then $|X|$ is either finite or $\geq \lambda$.)

2. Let A be an L-structure and λ a cardinal $> |A|$. Show that the following are equivalent. (a) A is λ-big. (b) A is λ-saturated. (c) A is finite.

3. We define λ^- to be μ if λ is a successor cardinal μ^+, and λ otherwise. Show that if an L-structure A is not λ-saturated, then for all κ with $\max(|L|, \lambda^-) \leq \kappa < |A|$ there are elementary substructures of A of cardinality κ which are not λ-saturated.

A structure A is said to be **strongly λ-homogeneous** *if for every pair of sequences $\bar{a}, \bar{b}$ of fewer than λ elements of A, if $(A, \bar{a}) \equiv (A, \bar{b})$ then there is an automorphism of A which takes $\bar{a}$ to $\bar{b}$.*
4. (a) Show that if A is λ-big then A is strongly λ-homogeneous. [It suffices to find an elementary extension A' of A with an automorphism taking $\bar{a}$ to $\bar{b}$. This can be done with repeated applications of elementary amalgamation (Theorem 5.3.1) back and forth.] (b) Show that if A is strongly λ-homogeneous then A is λ-homogeneous. (c) Show that if A is $|A|$-homogeneous then A is strongly $|A|$-homogeneous.

Suppose L is a first-order language, A an L-structure, X a set of elements of A and a an element of A. We say that a is **definable over** *X in A if there is a formula $\phi(x)$ of L with parameters from X, such that $A \vDash \phi(a) \wedge \exists_{=1} x \phi(x)$.*
5. Let λ be an infinite cardinal, A a λ-saturated structure and X a set of fewer than λ elements of A. Show (a) if an element a of A is not algebraic over X, then infinitely many elements of A realise $\text{tp}_A(a/X)$, (b) if an element a of A is not definable over X, then at least two elements of A realise $\text{tp}_A(a/X)$.

6. Show that if A is a λ-big L-structure and $\bar{a}$ is a sequence of fewer than λ elements of A, then $(A, \bar{a})$ is λ-big. Show that the same holds for λ-saturation, λ-homogeneity and strong λ-homogeneity.

7. Show that the following are equivalent, where λ, μ are any cardinals with $\min(\lambda, \omega) \leq \mu \leq \lambda^+$. (a) A is λ-saturated. (b) A is λ-homogeneous and μ-universal. [For (b) $\Rightarrow$ (a), adapt the proof of Theorem 8.1.8(b).]

8. Show that if A and B are elementarily equivalent structures and A is ω-saturated, then B is ω-saturated if and only if it is back-and-forth equivalent to A.

9. Show that if $(A_i : i < \kappa)$ is an elementary chain of λ-saturated structures and $\text{cf}(\kappa) \geq \lambda$ then $\bigcup_{i<\kappa} A_i$ is λ-saturated. Show the same for λ-homogeneity.

10. Show that the result of Exercise 9 fails for λ-bigness and strong λ-homogeneity. [Take a countable structure consisting of an equivalence relation with just two equivalence classes, both infinite. Form a chain of extensions of length ω_1, by adding elements to just one of the equivalence classes.]

11. Show that if L and L^+ are languages with $L \subseteq L^+$, and A is a λ-big L^+-structure, then $A|L$ is a λ-big L-structure. Show the same for λ-saturation. On the other hand, show that if A is strongly λ-homogeneous, it need not follow that $A|L$ is λ-homogeneous. [For λ-big, use Theorem 5.5.1.]

12. Show that if L, L^+ are first-order languages with $L \subseteq L^+$, A is an L^+-structure, $A|L$ is λ-big and A is a definitional expansion of $A|L$ (see section 2.6), then A is λ-big. Show that the same holds for λ-saturation, λ-homogeneity and strong λ-homogeneity.

13. Let L be a first-order language. Show that if A is a λ-big L-structure and $\phi(x)$ is a formula of L such that $\phi(A)$ is the domain of a substructure B of A, then B is λ-big. Show that the same holds for λ-saturation, λ-homogeneity and strong λ-homogeneity.

14. Show that if A and B are λ-big structures, then the disjoint sum of A and B (see Exercise 3.2.8) is λ-big. Show the same for λ-saturation, λ-homogeneity and strong λ-homogeneity.

15. Let L be a first-order language and A a saturated L-structure of cardinality $> |L|$. Show that A is $|A|$-big. [Put $\lambda = |A|$. Since A remains saturated after adding fewer than λ parameters, it suffices to show that A is splendid. Take L^+ as in (1.1), and an L^+-structure B of cardinality λ with $B|L \equiv A$. Write B as the union of a chain $(B_i : i < \lambda)$ of elementary substructures with $|B_i| = |L| + |i|$. By induction on i, build an elementary chain $(A_i : i < \lambda)$ with $|A_i| = |B_i|$, and an elementary chain $(C_i : i < \lambda)$ of L^+-structures with a commuting family of elementary embeddings $f_i : B_i \to C_i$, so that $A_i = C_i|L$. At each successor step, first form an amalgam C_i' of C_i and B_{i+1} over B_i, then use Theorem 5.5.1 to amalgamate C_i' and A over C_i; then cut down to size.]

*A structure A is said to be λ-**compact** if every type (not necessarily complete) of cardinality $< \lambda$ over a set of elements of A is realised in A.*

16. (a) Show that if L is a first-order language and $\lambda > |L|$, then an L-structure λ is λ-saturated if and only if it is λ-compact. (b) Show that every structure is ω-compact. (c) Show that for every infinite cardinal λ, a λ-saturated structure is λ-compact. (d) Give an example of an $|L|$-compact L-structure which is not $|L|$-saturated.

17. Suppose the L-structure A is λ-compact, $\bar{a}$ is a tuple from A, $\psi(\bar{x}, \bar{y})$ is a formula of L and $\Phi(\bar{x}, \bar{y})$ is a set of fewer than λ formulas of L. Show that if $A \vDash \forall \bar{x}(\bigwedge \Phi(\bar{x}, \bar{a}) \to \psi(\bar{x}, \bar{a}))$, then there is a finite subset Φ_0 of Φ such that $A \vDash \forall \bar{x}(\bigwedge \Phi_0(\bar{x}, \bar{a}) \to \psi(\bar{x}, \bar{a}))$.

18. Let L be a countable first-order language and A a countable atomic L-structure. Show that A is homogeneous.

The final exercise shows that a natural extension of the notion of λ-big – allowing us to expand to a model of a theory with many new symbols – is not really an extension at all.

19. Let L and L^+ be first-order languages with $L \subseteq L^+$. Let λ be an infinite cardinal such that the number of symbols in the signature of L^+ but not in L is less than λ. Show that if A is an L^+-structure and B is a λ-big L-structure such that $A|L \equiv B$, then B can be expanded to a structure $B' \equiv A$. [Adding a pairing function to L^+, we can suppose that all the other new symbols of L^+ are 1-ary relation symbols. They can then be coded up by a single 2-ary relation symbol and a set of fewer than λ constants.]

8.2 Big models exist

We shall show that for every structure A and every cardinal λ, there is an elementary extension of A which is λ-big. The proof will illustrate a common model-theoretic trick: we name the elements of a structure before we build the structure. This helps us to plan the construction in advance.

The cardinal $\mu^{<\lambda}$ is the sum of all cardinals μ^κ with $\kappa < \lambda$; so for example if $\lambda = \kappa^+$ then $\mu^{<\lambda}$ is just μ^κ. (See the end of the chapter for a reference on cardinal arithmetic.)

Existence of λ-big models

Theorem 8.2.1. *Let L be a first-order language, A an L-structure and λ a regular cardinal $> |L|$. Then A has a λ-big elementary extension B such that $|B| \leqslant |A|^{<\lambda}$.*

Proof. If A is finite then A is already λ-big for any cardinal λ. So we can assume henceforth that A is infinite.

Let C and D be structures. We shall say that D is an **expanded elementary extension** of C if D is an expansion of some elementary extension of C. An **expanded elementary chain** is a chain $(C_i : i < \kappa)$ of structures such that whenever $i < j < \kappa$, C_j is an expanded elementary extension of C_i. Using the Tarski–Vaught theorem on elementary chains (Theorem 2.5.2) it's not hard to see that each expanded elementary chain has a union D which is an expanded elementary extension of every structure in the chain. We write $\bigcup_{i<\kappa} C_i$ for the union of the expanded elementary chain $(C_i : i < \kappa)$.

Put $\mu = (|A| + |L^+|)^{<\lambda}$. Then $\mu = \mu^{<\lambda} \geqslant \lambda$. The ordinal $\mu^2 \cdot \lambda$ consists of $\mu \cdot \lambda$ copies of μ laid end to end. The object will be to construct B (or rather, an expansion of B) as the union of an expanded elementary chain $(A_i : 0 < i < \mu \cdot \lambda)$ where for each $i < \mu \cdot \lambda$ the domain of A_i is the ordinal $\mu \cdot i$. Then B will have cardinality $|\mu^2 \cdot \lambda| = \mu$ as required. The ordinals $< \mu^2 \cdot \lambda$ will be called **witnesses**. We can regard them either as elements of the structure to be built, or as new constants which will be used as names of themselves.

Since A is infinite, we can suppose without loss that A has cardinality μ (by the upward Löwenheim–Skolem theorem, Corollary 5.1.4). Then we can identify $\mathrm{dom}(A)$ with the ordinal $\mu \cdot 1 = \mu$ and put $A_1 = A$. At limit ordinals $\delta < \mu$ we put $A_\delta = \bigcup_{0 < i < \delta} A_i$. It remains to define A_{i+1} when A_i has been defined; this is where the work is done.

Suppose L_0 is a first-order language and L', L'' are first-order languages got by adding new relation symbols R', R'' respectively to L_0. We say that theories T', T'' in L', L'' respectively are **conjugate** if T'' comes from T' by replacing R' by R'' throughout. We can list as $((X_i, T_i) : 0 < i < \mu \cdot \lambda)$ the set of 'all' pairs (X_i, T_i) where X_i is a set of fewer than λ witnesses, and T_i is a complete theory in the first-order language L_i formed by adding to L the witnesses in X_i and one new relation symbol R_i. Here 'all' means that for each such pair (X, T) there is a pair (X_i, T_i) with $X = X_i$ and T conjugate to T_i. To check the arithmetic, note first that for each cardinal $v < \lambda$, the number of sets X consisting of v witnesses is $\mu^v = \mu$, and the number of complete theories T (up to conjugacy) in the language got by adding X and a relation symbol R to L is at most $2^{|L|+v} \leqslant \mu^{<\lambda} = \mu$. So the total number of pairs that we need is at most $\mu \cdot \lambda = \mu$. The listing can be done so that

(2.1)   the relation symbols R_i are all distinct, and

(2.2)   up to conjugacy, each possible pair (X, T) appears as (X_i, T_i) cofinally often in the listing.

In fact $\mu \cdot \lambda$ consists of λ blocks of length μ; we can make sure that each (X, T) appears at least once – up to conjugacy – in each of these blocks.

Now we define A_{i+1} inductively, assuming A_i has been defined with domain $\mu \cdot i$. Consider the pair (X_i, T_i). If some witness $\geqslant \mu \cdot i$ appears in X_i, then we take A_{i+1} to be an arbitrary elementary extension of A_i with domain $\mu \cdot (i + 1)$ (which is possible by Exercise 5.1.11).

Suppose then that every witness in X_i is an element of A_i. If the theory T_i is inconsistent with the elementary diagram of A_i, then again we take A_{i+1} to be an arbitrary elementary extension of A_i with domain $\mu \cdot (i + 1)$.

Finally suppose that every witness in X_i is an element of A_i, and T_i is consistent with the elementary diagram of A_i. Then some expanded elementary extension D of A_i is a model of T_i. By the downward Löwenheim–Skolem theorem we can suppose that D has cardinality μ, and so again (after adding at most μ elements if necessary) we can identify the elements of D with the ordinals $< \mu \cdot (i + 1)$. This done, we take A_{i+1} to be D.

Thus the chain $(A_i : 0 < i < \mu \cdot \lambda)$ is defined, and we put $B^+ = \bigcup_{0 < i < \mu \cdot \lambda} A_i$ and $B = B^+ | L$. The structure B^+ is an expanded elementary extension of A, so B is an elementary extension of A. Since B is the union of a chain of length μ in which every structure has cardinality μ, B has cardinality μ.

It remains to show that B is λ-big. Suppose $\bar{a}$ is a sequence of fewer than λ elements of B, and C is a structure with a new relation symbol R, such that $(C|L, \bar{c}) \equiv (B, \bar{a})$ for some sequence $\bar{c}$ in C. Adjusting C, we can suppose without loss that $\bar{c}$ is $\bar{a}$. Since $\mu \cdot \lambda$ is an ordinal of cofinality λ and λ is regular, there is some $j < \mu \cdot \lambda$ such that all the witnesses in $\bar{a}$ are less than j, and thus $(C|L, \bar{a}) \equiv (A_j, \bar{a})$. By (2.2) there is some $i \geqslant j$ such that T_i is conjugate to $\text{Th}(C, \bar{a})$. Then $\text{Th}(A_i|L, \bar{a}) \cup T_i$ is consistent, and so by (2.1) and Theorem 5.5.1, T_i is consistent with the elementary diagram of A_i. So by construction A_{i+1} is a model of T_i, and hence so is B^+. Thus B expands to a model of T_i as required.   □

We deduce the following.

Corollary 8.2.2. *Let A be an L-structure and λ a cardinal $\geqslant |L|$. Then A has a λ^+-big (and hence λ^+-saturated) elementary extension of cardinality $\leqslant |A|^\lambda$.*

Proof. Direct from the theorem.   □

In section 5.3 there was advance notice of a result saying that one can realise many complete types together. Corollary 8.2.2 establishes this. It's worth remarking that if we only want λ^+-saturation, not λ^+-bigness, then the elementary amalgamation theorem can be used in place of Theorem 5.5.1.

Corollary 8.2.3. *Let λ be any cardinal. Then every structure is elementarily equivalent to a λ-big structure.*

Proof. This follows from Corollary 8.2.2 and Lemma 8.1.1.   □

Corollary 8.2.3 is important. It tells us that if we want to classify the models of a first-order theory T up to elementary equivalence, it's enough to choose a cardinal λ and classify the λ-big models up to elementary equivalence. Since the λ-big models of T may be a much better behaved collection than the models of T in general, this is real progress. For example if R is an infinite ring, $|R|^+$-saturated left R-modules are algebraically compact. A good structure theory is known for such modules.

If the generalised continuum hypothesis (GCH) holds, then $\lambda^{<\lambda} = \lambda$ for every regular cardinal λ. So by Theorem 8.2.1, if λ is a regular cardinal $> |L|$ and A is an L-structure of cardinality $\leqslant \lambda$, then A has a λ-big elementary extension B of cardinality at most λ. In particular, we have the following.

Corollary 8.2.4. *If the GCH holds, every structure has a saturated elementary extension.*   □

Existence of λ-homogeneous models

Since every λ-big structure is λ-homogeneous, Theorem 8.2.1 creates λ-homogeneous elementary extensions too. But if all we want is λ-homogeneity, we can generally get it with a smaller structure than Theorem 8.2.1 offers. This is useful.

Theorem 8.2.5. *Let L be a first-order language, A an L-structure and λ a regular cardinal. Then A has a λ-homogeneous elementary extension C such that $|C| \leq (|A| + |L|)^{<\lambda}$.*

Proof. By Theorem 8.2.1 we have a λ-big elementary extension B of A; never mind its cardinality. Write ν for $(|A| + |L|)^{<\lambda}$, noting that $\nu \geq \lambda$. (Otherwise $\nu = \nu^{<\lambda} = (\nu^{<\lambda})^\nu \geq 2^\nu > \nu$.) If D is any elementary substructure of B with cardinality at most ν, we can find a structure D^* with $D \leq D^* \leq B$ so that

(2.3)   if $\bar{a}$ and $\bar{b}$ are two sequences of elements of D, both of length $< \lambda$, and $(D, \bar{a}) \equiv (D, \bar{b})$, then for every element c of D there is an element d of D^* such that $(D, \bar{a}, c) \equiv (D^*, \bar{b}, d)$.

We can find D^* as the union of a chain of elementary substructures of B, taking one such substructure for each triple $(\bar{a}, \bar{b}, c)$ such that $(D, \bar{a}) \equiv (D, \bar{b})$ and c is in D. Such a chain is automatically elementary. As we move one step up the chain, we choose the next structure so that it contains some d with $(D, \bar{a}, c) \equiv (B, \bar{b}, d)$. This is possible since B is λ-homogeneous. The number of triples $(\bar{a}, \bar{b}, c)$ is at most $\nu^{<\lambda} = \nu$, and each structure in the chain can be chosen of cardinality at most ν; so the union D^* can be found with cardinality at most ν.

Now we build a chain $(A_i: i < \lambda)$ of elementary substructures of B, so that for each $i < \lambda$, A_{i+1} is A_i^*. At limit ordinals we take unions. Let C be $\bigcup_{i<\lambda} A_i$. Then C has cardinality at most $\nu \cdot \lambda = \nu$. If $(C, \bar{a}) \equiv (C, \bar{b})$ where $\bar{a}$ and $\bar{b}$ are sequences of length $< \lambda$ in C, and c is an element of C, then since λ is regular, all of $\bar{a}$, $\bar{b}$ and c must lie within some A_i, so that A_{i+1} contains d with $(C, \bar{a}, c) \equiv (A_{i+1}, \bar{a}, c) \equiv (A_{i+1}, \bar{b}, d) \equiv (C, \bar{b}, d)$. Thus C is λ-homogeneous as required. $\square$

This theorem most often appears as (a) of the corollary below. Strictly (b) is not a corollary but an analogue. We start with Exercise 8.1.4(a) and form a chain as in the proof of Theorem 8.2.5, adding enough elements to allow the required automorphisms. I leave the details as Exercise 5.

Corollary 8.2.6. *Let A be an infinite L-structure and μ a cardinal $\geqslant |A| + |L|$.*

(a) *A has an ω-homogeneous elementary extension of cardinality μ. In particular every complete and countable first-order theory with infinite models has a countable homogeneous model.*

(b) *A has an elementary extension B of cardinality μ which is strongly ω-homogeneous (i.e. if $\bar{a}$, $\bar{b}$ are tuples in B such that $(B, \bar{a}) \equiv (B, \bar{b})$, then there is an automorphism of B taking $\bar{a}$ to $\bar{b}$).*   ☐

Exercises for section 8.2

The first three exercises give consequences of the fact that countable complete theories have countable homogeneous models.

1. (a) Let T be a countable first-order theory with infinite models. Show that T has a countable strongly ω-homogeneous model. (b) Show that if the continuum hypothesis fails, then there is a countable first-order theory with infinite models but with no strongly ω_1-homogeneous model of cardinality ω_1. [Consider a very symmetrical tree with 2^ω branches.]

2. Let T be a complete theory in a countable first-order language. Suppose T has infinite models, and there is a finite set of types of T, such that all countable models of T realising these types are isomorphic. Show that T is ω-categorical. [There is a countable strongly ω-homogeneous model in which all these types are realised; so we have an ω-categorical theory got by adding finitely many parameters to T.]

3. Show that if T is a countable complete first-order theory, then the number of countable models of T, counted up to isomorphism, is not 2. [If it isn't 1, then by the theorem of Engeler, Ryll-Nardzewski and Svenonius (Theorem 6.3.1), T has a non-principal type $p(\bar{x})$ over the empty set. If it is 2, then apply the previous exercise.]

4. Show that if $2 < n < \omega$ then there is a countable complete first-order theory T such that up to isomorphism, T has exactly n countable models. [Start with the theory of a dense linear ordering without endpoints. To get three models, add constants for some sequence of elements of order-type ω. All known solutions to this exercise are variants of this.]

5. Show that Theorem 8.2.5 holds with 'strongly λ-homogeneous' in place of 'λ-homogeneous'. (This is perceptibly harder than Theorem 8.2.5, though the extra difficulty is mostly book-keeping.)

6. Let L be a first-order language, T an $\forall_2$ theory in L, A an L-structure which is a model of T and λ a regular cardinal $> |L|$. Show that there is an e.c. model B of T such that $A \subseteq B$, $|B| \leqslant |A|^{<\lambda}$ and

(2.4)   for every sequence $\bar{b}$ of $< \lambda$ elements of B, every e.c. model C of T
extending B, and every element c of C, there is an element d of B
such that $(B, \bar{b}, d) \equiv_1 (C, \bar{b}, c)$.

[Mingle the proofs of Theorem 8.2.1 and Theorem 7.2.1.] *In Abraham Robinson's*
terminology, an e.c. model of T which satisfies (2.4) when $\lambda = \omega$ is said to be
existentially universal. *An* **infinite-generic** *model of T is a model which is an elementary*
substructure of an existentially universal model.

7. Let T be an $\forall_2$ theory in a first-order language L. (a) Show that every model of T
can be embedded in some infinite-generic model of T. (b) Show that if A and B are
infinite-generic models of T with $A \subseteq B$ then $A \preccurlyeq B$. [In fact if $\bar{a}$ is any tuple of
elements of A, then $(A, \bar{a})$ and $(B, \bar{a})$ are back-and-forth equivalent.] (c) Show that if
T is companionable, then the infinite-generic models of T are exactly the e.c. models
of T.

8.3 Syntactic characterisations

Lemma 8.1.5 is handy for setting up maps between λ-saturated structures.
Sometimes these maps give intuitive proofs of preservation theorems. Two
examples will serve (with more in the Exercises). After these examples we
shall extract some general principles for getting syntactic characterisations
out of saturated models. The principles take the form of a game of infinite length.

Our first example shows how to embed a structure in a λ-saturated
structure. Recall that if A and B are L-structures, then '$A \Rightarrow_1 B$' means that
for every $\exists_1$ first-order sentence ϕ of L, if $A \vDash \phi$ then $B \vDash \phi$.

Theorem 8.3.1. *Let L be a first-order language. Let A and B be L-structures,
and suppose B is $|A|$-saturated and $A \Rightarrow_1 B$. Then A is embeddable in B.*

Proof. List the elements of A as $\bar{a} = (a_i : i < \lambda)$ where $\lambda = |A|$. We claim that
there is a sequence $\bar{b} = (b_i : i < \lambda)$ of elements of B such that

(3.1)   for each $i \leqslant \lambda$, $(A, \bar{a}|i) \Rightarrow_1 (B, \bar{b}|i)$.

The proof is by induction on i. When $i = 0$, $A \Rightarrow_1 B$ by assumption. When i
is a limit ordinal, (3.1) holds at i provided it holds at all smaller ordinals.

This leaves the case where i is a successor ordinal $j + 1$. Let $\bar{x}$ be the
sequence of variables $(x_\alpha : \alpha < i)$, and let $\Phi(\bar{x}, y)$ be the set of all $\exists_1$ formulas
$\phi(\bar{x}, y)$ of L such that $A \vDash \phi(\bar{a}|i, a_i)$. For each finite set of formulas $\phi_0, \ldots,$
ϕ_{n-1} from Φ, we have $A \vDash \exists y \bigwedge_{k<n} \phi_k(\bar{a}|j, y)$. But $\exists y \bigwedge_{k<n} \phi_k$ is equivalent
to an $\exists_1$ formula, and so $B \vDash \exists y \phi(\bar{b}|j, y)$ by inductive assumption. It follows
by Theorem 5.2.1(a) that $\Phi(\bar{b}|j, y)$ is a type with respect to B. Since $j < \lambda$
and B is λ-saturated, this type is realised in B, say by an element b_j. Then
$(A, \bar{a}|i) \Rightarrow_1 (B, \bar{b}|i)$ as required. This proves the claim.

Hence $(A, \bar{a}) \Rightarrow_1 (B, \bar{b})$. It follows by the diagram lemma that there is an embedding $f\colon A \to B$ such that $f\bar{a} = \bar{b}$. $\quad\square$

This theorem leads at once to a new proof of the dual of the Łoś–Tarski theorem (cf. Exercise 5.4.1).

Corollary 8.3.2. *Let L be a first-order language, T a theory in L and $\Phi(\bar{x})$ a set of formulas of L (where the sequence $\bar{x}$ may be infinite). Suppose that whenever A and B are models of T with $A \subseteq B$, and $\bar{a}$ is a sequence of elements of A such that $A \vDash \bigwedge\Phi(\bar{a})$, we have $B \vDash \bigwedge\Phi(\bar{a})$. Then Φ is equivalent modulo T to a set $\Psi(\bar{x})$ of $\exists_1$ formulas of L.*

Proof. Putting new constants for the variables $\bar{x}$, we can suppose that the formulas in Φ are sentences. Let Ψ be the set of all $\exists_1$ sentences ψ of L such that $T \cup \Phi \vdash \psi$. It suffices to show that $T \cup \Psi \vdash \bigwedge\Phi$. If $T \cup \Psi$ has no models then this holds trivially. If $T \cup \Psi$ has models, let B' be one. By Corollary 8.2.2, B' is elementarily equivalent to a λ-saturated structure B where $\lambda \geqslant |L|$. Write U for the set of all $\forall_1$ sentences of L which are true in B. Then $T \cup \Phi \cup U$ has a model. (For otherwise by the compactness theorem there is a finite subset $\{\theta_0, \ldots, \theta_{m-1}\}$ of U such that $T \cup \Phi \vdash \neg\theta_0 \vee \ldots \vee \neg\theta_{m-1}$. Then $\neg\theta_0 \vee \ldots \vee \neg\theta_{m-1}$ is equivalent to a sentence in Ψ, and hence it is true in B' and B; contradiction.) Let A be a model of $T \cup \Phi \cup U$ of cardinality $\leqslant |L|$. By the choice of U, $A \Rightarrow_1 B$. So A is embeddable in B by Theorem 8.3.1, and it follows that $B \vDash \bigwedge\Phi$ since Φ is a set of $\exists_1$ sentences. Thus $B' \vDash \bigwedge\Phi$ as required. $\quad\square$

Lyndon's theorem

Our second application is a useful preservation result about positive occurrences of relation symbols in formulas.

Let $f\colon A \to B$ be a homomorphism of L-structures and R a relation symbol of L. We shall say that f **fixes** R if for every tuple $\bar{a}$ of elements of A, $A \vDash R\bar{a}$ if and only if $B \vDash Rf\bar{a}$. In this definition we allow R to be the equality symbol $=$. Thus f fixes $=$ if and only if f is injective; f fixes all relation symbols if and only if f is an embedding. What formulas are preserved by a homomorphism which fixes certain relations but not others?

Let ϕ be a formula and R a relation symbol. We say that R is **positive in** ϕ if ϕ can be brought to negation normal form (see Exercise 2.6.2) in such a way that there are no subformulas of the form $\neg R\bar{t}$.

Note that up to logical equivalence, a formula ϕ is positive (in the sense we defined in section 2.4) if and only if every relation symbol, including $=$, is positive in ϕ. (See Exercises 2.4.10 and 2.6.3.)

We aim to prove the following.

Theorem 8.3.3. *Let L be a first-order language, Σ a set of relation symbols of L (possibly including $=$) and $\phi(\bar{x})$ a formula of L in which every relation symbol in Σ is positive.*

(a) *If $f: A \to B$ is a surjective homomorphism of L-structures, and f fixes all relation symbols (including possibly $=$) which are not in Σ, then f preserves ϕ.*

(b) *Suppose that every surjective homomorphism between models of T which fixes all relation symbols not in Σ preserves ϕ. Then ϕ is equivalent modulo T to a formula $\psi(\bar{x})$ of L in which every relation symbol in Σ is positive.*

Proof. (a) is a straightforward variant of Theorem 2.4.3. To prove (b), we start along the same track as the proof of Corollary 8.3.2. Replacing the variables $\bar{x}$ by distinct new constants, we can assume that ϕ is a sentence. Let Θ be the set of all formulas of L in which every relation symbol in Σ is positive.

We shall use Θ in the same way as we used $\exists_1$ in Theorem 8.3.1. Thus if C and D are any L-structures, we write $(C, \bar{c}) \Rrightarrow_\Theta (D, \bar{d})$ to mean that if $\theta(\bar{x})$ is any formula in Θ such that $C \vDash \theta(\bar{c})$, then $D \vDash \theta(\bar{d})$. In particular $C \Rrightarrow_\Theta D$ means that every sentence in Θ which is true in C is also true in D. In place of Theorem 8.3.1, we shall show the following.

Lemma 8.3.4. *Let L, Σ and Θ be as in the theorem. Let λ be a cardinal $\geq |L|$, and suppose A and B are λ-saturated structures such that $A \Rrightarrow_\Theta B$. Then there are elementary substructures A', B' of A, B respectively, and a surjective homomorphism $f: A' \to B'$ which fixes all relation symbols not in Σ.*

Proof of lemma. We shall build up sequences $\bar{a}$, $\bar{b}$ of elements of A, B respectively, both of length λ, in such a way that

(3.2)   for every $i \leq \lambda$, $(A, \bar{a}|i) \Rrightarrow_\Theta (B, \bar{b}|i)$, and

(3.3) $\bar{a}$ (resp. $\bar{b}$) is the domain of an elementary substructure of A
          (resp. B).

The construction will be by induction on i, as in the proof of Theorem 8.3.1.

There is one main difference from the proof of Theorem 8.3.1. In that proof, each a_j was given and we had to find an element b_j to match. Here we shall sometimes choose the b_j first and then look for an answering a_j. One can think of the process as a back-and-forth game of length λ between A and B: player $\forall$ chooses an element a_j (or b_j), and player $\exists$ has to find a corresponding element b_j (or a_j). Player $\exists$ wins iff (3.2) holds after λ steps.

We shall show that player $\exists$ can always win this game. At the beginning of the game, $A \Rrightarrow_\Theta B$ by assumption. If i is a limit ordinal and $(A, \bar{a}|j) \Rrightarrow_\Theta (B, \bar{b}|j)$ for all $j < i$, then $(A, \bar{a}|i) \Rrightarrow_\Theta (B, \bar{b}|i)$ since all formulas are finite. So again we are led to the case where i is a successor ordinal $j + 1$. There are two situations, according as player $\forall$ chooses from A or from B.

Suppose first that player $\forall$ has just chosen a_j from A. Let $\Phi(\bar{x}, y)$ be the set of all formulas $\phi(\bar{x}, y)$ in Θ such that $A \vDash \phi(\bar{a}|j, a_j)$. Since Φ is closed under conjunctions and existential quantification, exactly the same argument as in the proof of Theorem 8.3.1 shows that $\Phi(\bar{b}|j, y)$ is a type over $\bar{b}|j$ with respect to B, and so there is an element b in B such that $(A, \bar{a}|j, a_j) \Rightarrow_\Theta (B, \bar{b}|i, b)$ as required. Let player $\exists$ choose b_j to be this element b.

Second, suppose player $\forall$ chose b_j from B, so that player $\exists$ must find a suitable a_j. The argument is just the same but from right to left, using the set $\{\neg\theta: \theta \in \Theta\}$ in place of Θ.

So player $\exists$ can be sure of winning. This takes care of (3.2). To make (3.3) true too, we issue some instructions to player $\forall$. As the play proceeds, he must keep a note of all the formulas of the form $\phi(\bar{a}|i, y)$, with ϕ in L, such that $A \vDash \exists y\, \phi(\bar{a}|i, y)$. For each such formula he must make sure that at some stage j later than i, he chooses a_j so that $A \vDash \phi(\bar{a}|i, a_j)$. He must do the same with B. At the end of the play, (3.3) will hold by the Tarski–Vaught criterion, Theorem 2.5.1.

Finally suppose the game is played, and sequences $\bar{a}$, $\bar{b}$ satisfying (3.2) and (3.3) have been found. Let A', B' be the substructures of A, B with domains listed by $\bar{a}$, $\bar{b}$ respectively. Since all atomic formulas of L are in Θ, the diagram lemma gives us a homomorphism $f: A' \to B'$ such that $f\bar{a} = \bar{b}$. Clearly f is surjective. If R is a relation symbol not in Σ, then the formula $\neg R\bar{z}$ is in Θ, and so (3.2) implies that f fixes R. $\qquad\qquad$ $\square$ Lemma.

The rest of the argument is much as in the proof of Corollary 8.3.2, and I leave it to the reader. $\qquad\qquad\qquad\qquad\qquad\qquad\qquad\qquad$ $\square$ Theorem.

Probably the best known corollary of Theorem 8.3.3 is the following converse of Theorem 2.4.3(b).

Corollary 8.3.5 *(Lyndon's preservation theorem). Let T be a theory in a first-order language L and $\phi(\bar{x})$ a formula of L which is preserved by all surjective homomorphism between models of T. Then ϕ is equivalent modulo T to a positive formula $\psi(\bar{x})$ of L.*

Proof. Let Σ in the theorem be the set of all relation symbols of L, including the symbol $=$. $\qquad\qquad\qquad\qquad\qquad\qquad\qquad\qquad\qquad\qquad$ $\square$

Using the same argument, we can replace ϕ and ψ in this corollary by sets Φ, Ψ of formulas; see Exercise 1 below.

Keisler games

In several of the arguments of this section and section 8.1, we inductively constructed a sequence $\bar{a}$ of elements of a structure A, using some saturation property of A. The sequence $\bar{a}$ had to satisfy certain conditions. Games are a handy way of organising arguments of this type.

In a Keisler game on a structure A, the two players $\forall$ and $\exists$ take turns to pick elements of A. Player $\exists$'s aim is to define some new relations on A, and these new relations must satisfy some conditions which depend on the choices of player $\forall$.

Let L be a first-order language and λ an infinite cardinal. A **Keisler sentence** of length λ in L is an infinitary expression of the form

$$(3.4) \qquad Q_0 x_0 Q_1 x_1 \ldots Q_i x_i \, (i < \lambda) \ldots \bigwedge \Phi$$

where each Q_i is either $\forall$ or $\exists$, and Φ is a set of formulas $\phi(x_0, x_1, \ldots)$ of L. If χ is the Keisler sentence (3.4) and A is an L-structure, then the **Keisler game** $G(\chi, A)$ is played as follows. There are λ steps. At the ith step, one of the players chooses an element a_i of A; player $\forall$ makes the choice if Q_i is $\forall$, and player $\exists$ makes it otherwise. At the end of the play, player $\exists$ wins if $A \vDash \bigwedge \Phi(a_0, a_1, \ldots)$. We define

$$(3.5) \qquad\qquad A \vDash \chi$$

to mean that player $\exists$ has a winning strategy for the game $G(\chi, A)$.

A **finite approximation** to the Keisler sentence (3.4) is a sentence $\bar{Q} \bigwedge \Psi$, where Ψ is a finite subset of Φ and $\bar{Q}$ is a finite subsequence of the quantifier prefix in (3.4), containing quantifiers to bind all the free variables of Ψ. We write app(χ) for the set of all finite approximations to the Keisler sentence χ.

These definitions adapt in an obvious way to give **Keisler formulas** $\chi(\bar{w})$ and games $G(\chi(\bar{w}), A, \bar{c})$. Thus $A \vDash \chi(\bar{c})$ holds if player $\exists$ has a winning strategy for $G(\chi(\bar{w}), A, \bar{c})$. In particular, let χ be the Keisler sentence (3.4) and let α be an ordinal $< \lambda$. Then we write $\chi^\alpha(x_i : i < \alpha)$ for the Keisler formula got from χ by removing the quantifiers $Q_i x_i \, (i < \alpha)$.

The following lemma tells us that we can detach the leftmost quantifier $Q_0 x_0$ of a Keisler sentence and treat it exactly like an ordinary quantifier. Of course the lemma generalises to Keisler formulas $\chi(\bar{w})$ too.

Lemma 8.3.6. *With the notation above, we have*

$$(3.6) \qquad A \vDash \chi \qquad iff \qquad A \vDash Q_0 x_0 \chi^1(x_0).$$

Proof. Suppose first that Q_0 is $\forall$. If $A \vDash \chi$, then the initial position in $G(\chi, A)$ is winning for player $\exists$, so that every choice a of player $\forall$ puts player $\exists$ into winning position in $G(\chi^1, A, a)$, whence $A \vDash \chi^1(a)$; so $A \vDash \forall x_0 \chi^1(x_0)$. The converse, and the corresponding arguments for the case $Q_0 = \exists$, are similar.

$\square$

The next theorem makes the crucial connection between Keisler games and saturation.

Theorem 8.3.7. *Let A be a non-empty L-structure, λ an infinite cardinal and χ a Keisler sentence of L of length λ.*
(a) *If $A \vDash \chi$ then $A \vDash \bigwedge \mathrm{app}(\chi)$.*
(b) *If $A \vDash \bigwedge \mathrm{app}(\chi)$ and A is λ-saturated then $A \vDash \chi$.*

Proof. (a) We show that if $\alpha < \lambda$ and $\theta(x_i: i < \alpha)$ is a finite approximation to χ^α, and $\bar{a}$ is a sequence of elements of A such that $A \vDash \chi^\alpha(\bar{a})$, then $A \vDash \theta(\bar{a})$. The proof is by induction on the number n of quantifiers in the quantifier prefix of θ. We write χ as in (3.4).

If $n = 0$ then θ is a conjunction of formulas $\phi(x_i: i < \alpha)$ from Φ. If $A \vDash \chi^\alpha(\bar{a})$ then player $\exists$ has a winning strategy for $G(\chi^\alpha, A, \bar{a})$, and it follows that $A \vDash \theta(\bar{a})$.

Suppose $n > 0$ and the quantifier prefix of θ begins with a universal quantifier $\forall x_\beta$. Then $\beta \geqslant \alpha$ and we can write θ as $\forall x_\beta \theta'(x_i: i \leqslant \beta)$. (Note that none of the variables x_i with $i > \alpha$ are free in θ.) If $A \vDash \chi^\alpha(\bar{a})$, then player $\exists$ has a winning strategy for $G(\chi^\alpha, A, \bar{a})$; let the players play this game through the steps $Q_i x_i$ ($\alpha \leqslant i < \beta$), with player $\exists$ using her winning strategy, and let $\bar{b}$ be the sequence of elements chosen. (This is possible because A is not empty.) Then $A \vDash \chi^\beta(\bar{a}, \bar{b})$, and hence $A \vDash \forall x_\beta \chi^{\beta+1}(\bar{a}, \bar{b}, x_\beta)$ by Lemma 8.3.6. So for every element c of A, $A \vDash \chi^{\beta+1}(\bar{a}, \bar{b}, c)$, which by induction hypothesis implies $A \vDash \theta'(\bar{a}, \bar{b}, c)$. Thus $A \vDash \theta(\bar{a})$. The argument when θ begins with an existential quantifier is similar.

Finally putting $\alpha = 0$, we have (a) of the theorem.

(b) Assume A is λ-saturated and $A \vDash \bigwedge \mathrm{app}(\chi)$. Then player $\exists$ should adopt the following rule for playing $G(\chi, A)$: always choose so that for each $\alpha < \lambda$, if $\bar{b}$ is the sequence of elements chosen before the αth step, then $A \vDash \bigwedge \mathrm{app}(\chi^\alpha)(\bar{b})$. If she succeeds in following this rule until the end of the game, when a sequence $\bar{a}$ of length λ has been chosen, then $A \vDash \bigwedge \Phi(\bar{a})$, so that she wins. We only have to show that she can follow the rule.

Suppose then that she has followed this rule up to the choice of $\bar{b} = (b_i: i < \alpha)$, so that $A \vDash \bigwedge \mathrm{app}(\chi^\alpha)(\bar{b})$. First suppose that Q_α is $\exists$. Without loss write any finite approximation θ to χ^α as $\exists x_\alpha \, \theta'(x_i: i \leqslant \alpha)$. To maintain the rule, player $\exists$ has to choose an element b_α so that $A \vDash \theta'(\bar{b}, b_\alpha)$ for each $\theta \in \mathrm{app}(\chi^\alpha)$. Now A is λ-saturated and $\bar{b}$ has length less than λ. Hence we only need show that if $\{\theta_0, \dots \theta_{n-1}\}$ is a finite set of formulas in $\mathrm{app}(\chi^\alpha)$, then $A \vDash \exists x_\alpha (\theta_0' \wedge \dots \wedge \theta_{n-1}')(\bar{b}, x_\alpha)$. But clearly there is some finite approximation θ to χ^α which begins with $\exists x_\alpha$ and is such that θ' implies $\theta_0', \dots, \theta_{n-1}'$. By assumption $A \vDash \theta(\bar{b})$, in other words $A \vDash \exists x_\alpha \, \theta'(\bar{b}, x_\alpha)$. This completes the argument when Q_α is $\exists$.

Next suppose that Q_α is $\forall$, and let $\theta'(x_i : i \leqslant \alpha)$ be a finite approximation to $\chi^{\alpha+1}$. Then $\forall x_\alpha \, \theta'$ is a finite approximation to χ^α, and so by assumption $A \vDash \forall x_\alpha \, \theta'(\bar{b}, x_\alpha)$. Hence $A \vDash \theta'(\bar{b}, b_\alpha)$ regardless of the choice of b_α. So player $\forall$ can never break player $\exists$'s rule.

The reader can check that limit ordinals are no threat to player $\exists$'s rule. So she can follow the rule and win. Therefore $A \vDash \chi$. $\square$

Theorem 8.3.7 generalises straightforwardly to Keisler formulas $\chi(\bar{w})$ with fewer than λ free variables $\bar{w}$. In this setting, part (a) of the theorem reads

$$(3.7) \qquad \text{If } A \vDash \chi(\bar{b}) \text{ then } A \vDash (\textstyle\bigwedge \mathrm{app}(\chi))(\bar{b}),$$

and likewise with part (b).

We shall put Keisler games to good use in the next section. But for the moment, here is how Theorem 8.3.7 relates to Theorem 8.3.1. Under the assumptions of Theorem 8.3.1, list the elements of A as $\bar{a} = (a_i : i < \lambda)$. Let $\bar{x}$ be the sequence of variables $(x_i : i < \lambda)$, and write Θ for the set of $\exists_1$ formulas $\theta(\bar{x})$ of L such that $A \vDash \theta(\bar{a})$. Let χ be the sentence

$$(3.8) \qquad \exists_0 x_0 \exists_1 x_1 \dots \textstyle\bigwedge \Theta.$$

Then $\bigwedge \mathrm{app}(\chi)$ is a conjunction of $\exists_1$ sentences true in A, and hence $B \vDash \bigwedge \mathrm{app}(\chi)$ since $A \Rightarrow_1 B$. By Theorem 8.3.7(b) it follows that $B \vDash \chi$, and we see at once that A is embeddable in B.

Exercises for section 8.3

1. Let T be a theory in a first-order language L and $\Phi(\bar{x})$ a set of formulas of L such that if $f : A \to B$ is any surjective homomorphism between models A, B of T and $A \vDash \bigwedge \Phi(\bar{a})$ then $B \vDash \bigwedge \Phi(\bar{b})$. Show that Φ is equivalent modulo T to a set $\Psi(\bar{x})$ of positive formulas of L.

The next exercise extracts the crucial facts about saturation used in the constructions in Theorem 8.3.1 and Lemma 8.3.4.

2. Let L be a first-order language and Θ a set of formulas of L. If A, B are L-structures, we write $(A, \bar{a}) \Rightarrow_\Theta (B, \bar{b})$ to mean that for every formula θ in Θ, if $A \vDash \theta(\bar{a})$ then $B \vDash \theta(\bar{b})$. Show (a) if Θ is closed under conjunction, disjunction and both existential and universal quantification, and A and B are λ-saturated L-structures with $\lambda \geqslant |L|$ such that $A \Rightarrow_\Theta B$, then there are sequences $\bar{a}$, $\bar{b}$ in A, B respectively, both of length λ, such that $\bar{a}$, $\bar{b}$ list the domains of elementary substructures A', B' of A, B respectively, and $(A, \bar{a}) \Rightarrow_\Theta (B, \bar{b})$, (b) if Θ is closed under conjunction and existential quantification, and A and B are L-structures such that $A \Rightarrow_\Theta B$ and B is $|A|$-saturated, then there are sequences $\bar{a}$, $\bar{b}$ in A, B respectively such that $\bar{a}$ lists the domain of A and $(A, \bar{a}) \Rightarrow_\Theta (B, \bar{b})$.

3. Let L be a first-order language containing a 1-ary relation symbol P. Let A and B be elementarily equivalent L-structures of cardinality λ. Suppose that B is saturated,

but make the following weaker assumption on A: if X is any set of fewer than λ elements of A, and $\Phi(x)$ is a type over X with respect to A, which contains the formula $P(x)$, then Φ is realised in A. Show that there is an elementary embedding $f: A \to B$ which is a bijection from P^A to P^B.

4. Let L be a first-order language, let A and B be $|L|$-saturated L-structures, and suppose that every sentence of Th(A) which is either positive or $\forall_1$ is in Th(B). Show that there are elementary substructures A', B' of A, B respectively, a surjective homomorphism $f: A' \to B'$ and an embedding $e: B' \to A$. [Build up sequences $\bar{a}$, $\bar{b}$, $\bar{c}$ so that the right relationships hold between $(A, \bar{a})$, $(B, \bar{b})$ and $(A, \bar{c})$.]

5. Let L be a first-order language and T a theory in L. Show that the following are equivalent, for every sentence ϕ of L. (a) For every model A of T and endomorphism $e: A \to A$, if ϕ is true in A then ϕ is true in the image of A. (b) ϕ is equivalent modulo T to a positive boolean combination of positive sentences of L and $\forall_1$ sentences of L. [Use the previous exercise.]

6. Let L be a first-order language whose symbols include a 2-ary relation symbol $<$, and let T be a theory in L and $\phi(\bar{x})$ a formula of L. In the terminology of Exercise 2.4.5, show that the following are equivalent. (a) If A and B are models of T and A is an end-extension of B, then for every tuple $\bar{b}$ of elements of B, $B \vDash \phi(\bar{b})$ implies $A \vDash \phi(\bar{a})$. (b) ϕ is equivalent modulo T to a Σ_1^0 formula $\psi(\bar{x})$ of L.

7. Let L be a first-order language whose symbols include a 2-ary relation symbol $<$, and let T be a theory in L which implies '$<$ linearly orders the set of all elements, with no last element'. Recall from Exercise 2.4.9 the notion of a **cofinal substructure**; let Φ be defined as in that exercise, except that the formulas in Φ are required to be first-order. Show (a) if A and B are models of T, A and B are $|L|$-saturated and $A \Rightarrow_\Phi B$, then there is an embedding of an elementary substructure A' of A onto a cofinal substructure of an elementary substructure B' of B, (b) if ϕ is a formula of L, and f preserves ϕ whenever f is an embedding of a model of T onto a cofinal substructure of a model of T, then ϕ is equivalent modulo T to a formula in Φ.

8. Let L be a first-order language whose symbols include a 2-ary relation symbol R, and let T be a theory in L. Suppose that T implies that R expresses a reflexive symmetric relation. If A is a model of T, we define a relation $\sim$ on dom(A) by '$\sim$ is the smallest equivalence relation containing R^A'. A **closed** substructure of A is a substructure whose domain is a union of equivalence classes of $\sim$. We define Θ to be the least set of formulas of L such that (i) Θ contains all quantifier-free formulas, (ii) Θ is closed under disjunction, conjunction and existential quantification, and (iii) if $\phi(\bar{x}yz)$ is in Θ, and y occurs free in ϕ, then the formula $\forall z(Ryz \to \phi)$ is in Θ. Show (a) if A and B are models of T, A and B are $|L|$-saturated and $A \Rightarrow_\Theta B$, then there is an embedding of an elementary substructure A' of A onto a closed substructure of an elementary substructure B' of B, (b) if ϕ is a formula of L, and f preserves ϕ whenever f is an embedding of a model of T onto a closed substructure of a model of T, then ϕ is equivalent modulo T to a formula in Θ.

Contrast the next exercise with Corollary 5.4.3.

9. Let ϕ be the sentence $\exists xy(Rxy \wedge \forall z(Rxz \rightarrow \exists t(Rxt \wedge Rzt)))$. Show (a) if a structure A is a homomorphic image of a model of ϕ, then A contains elements a_i $(i < \omega)$, not necessarily distinct, such that $A \vDash R(a_0, a_i) \wedge R(a_i, a_{i+1})$ whenever $0 < i < \omega$, (b) if a structure B contains arbitrarily long finite sequences like the sequence of length ω in (a), then some elementary extension of B is a homomorphic image of a model of ϕ, (c) by (a) and (b), the class of homomorphic images of models of a first-order sentence need not be closed under elementary equivalence.

8.4 One-cardinal and two-cardinal theorems

A **two-cardinal theorem** is a theorem which tells us that under certain conditions, a theory has a model A in which the $\varnothing$-definable subsets $\phi(A)$, $\psi(A)$ defined by two given formulas $\phi(x)$, $\psi(x)$ have different cardinalities. A **one-cardinal theorem** is a theorem with the opposite conclusion. For example if κ and λ are any two infinite cardinals with $\kappa \leqslant \lambda$, then there is a group G of cardinality λ whose centre has cardinality κ; this is a two-cardinal theorem. But if G is a finite-dimensional general linear group over an infinite field, then G has the same cardinality as its centre; this is a one-cardinal theorem.

The following theorem, a memorable result of Vaught, was the first model-theoretic two-cardinal theorem to be proved. Its proof uses the notion of homogeneous structures as in section 8.1 above.

Theorem 8.4.1 (*Vaught's two-cardinal theorem*). *Let L be a countable first-order language, $\phi(x)$ and $\psi(x)$ two formulas of L and T a theory in L. Then the following are equivalent.*
(a) *T has a model A in which $|\phi(A)| \leqslant \omega$ but $|\psi(A)| = \omega_1$.*
(b) *T has a model A in which $|\phi(A)| < |\psi(A)| \geqslant \omega$.*
(c) *T has models A, B such that $B \leqslant A$ and $\phi(A) = \phi(B)$ but $\psi(A) \neq \psi(B)$.*

Proof. (a) $\Rightarrow$ (b) is trivial. (b) $\Rightarrow$ (c): assuming (b), when $|\phi(A)|$ is infinite, let B be any elementary substructure of A which contains $\phi(A)$ and has cardinality $|\phi(A)|$.

(c) $\Rightarrow$ (a). Assume (c). Form a first-order language L^+ by adding to L a 1-ary relation symbol P, and let A^+ be the L^+-structure got from A by interpreting P as dom(B). This puts us into the setting of section 4.2; in the notation of that section $(A^+)_P = B$. Now A is infinite since $\psi(A) \neq \psi(B)$ and $B \leqslant A$. Hence A^+ has an elementary substructure C of cardinality ω. By the relativisation theorem (Theorem 4.2.1), C_P is defined, $C_P \leqslant C|L$, $\phi(C_P) = \phi(C|L)$ and $\psi(C_P) \neq \psi(C|L)$, since all these facts are recorded in Th(C).

We claim that C has a countable elementary extension D such that $D|L$ and D_P are homogeneous, and for every $n < \omega$, each n-type over $\varnothing$ which is realised in $D|L$ is also realised in D_P. (This makes sense since $D_P \preccurlyeq D|L$).

To construct D we build a countable chain $C = C_0 \preccurlyeq C_1 \preccurlyeq \ldots$ as follows. When C_i has been constructed, we find a countable elementary extension C_{i+1} of C_i such that

(4.1) if $\bar{a}$, $\bar{b}$ are tuples of elements of C_i with $(C_i|L, \bar{a}) \equiv (C_i|L, \bar{b})$, and d is any element of C_i, then there is an element c of C_{i+1} such that $(C_{i+1}|L, \bar{a}, c) \equiv (C_{i+1}|L, \bar{b}, d)$,

(4.2) if $\bar{a}$, $\bar{b}$ are tuples of elements of $(C_i)_P$ with $(C_i|L, \bar{a}) \equiv (C_i|L, \bar{b})$, and d is any element of C_i, then there is an element c of $(C_{i+1})_P$ such that $(C_{i+1}|L, \bar{a}, c) \equiv (C_{i+1}|L, \bar{b}, d)$.

Thus, let E be an ω-saturated elementary extension of C_i; then $E|L$ and E_P are ω-saturated elementary extensions of $C_i|L$ and $(C_i)_P$ respectively, by Exercises 8.1.11 and 8.1.13. So we can choose C_{i+1} to be a suitable elementary substructure of E. Finally we put $D = \bigcup_{i<\omega} C_i$, and it readily follows that $D_P = \bigcup_{i<\omega}(C_i)_P$. Both D and D_P are countable. By (4.1) and (4.2) respectively, $D|L$ and D_P are both homogeneous. By (4.2) and induction on n, every n-tuple over $\varnothing$ which is realised in $D|L$ is also realised in D_P. The claim is proved.

If D is as in the claim, then Theorem 8.1.8(b) tells us that $D|L$ and D_P are isomorphic. Changing notation, write A for $D|L$ and B for D_P. Since D was an elementary extension of C, clause (c) of the theorem now holds with A and B countable, ω-homogeneous and isomorphic.

We build an elementary chain $(A_i : i < \omega_1)$ as follows, so that each structure A_i is isomorphic to A. We put $A_0 = B$. When A_i has been defined, we choose A_{i+1} and an isomorphism $f: A \to A_{i+1}$ so that the image of B under f is A_i. At a limit ordinal δ we put $A_\delta = \bigcup_{i<\delta} A_i$. The definition of homogeneity implies at once that the union of a countable elementary chain of countable homogeneous structures is homogeneous; so A_δ is homogeneous, and clearly it realises the same types over $\varnothing$ as do the structures A_i ($i < \delta$). So by Theorem 8.1.8(b) again, $A_\delta \cong A$.

Put $A_{\omega_1} = \bigcup_{i<\omega_1} A_i$. By the construction, $\phi(A_{\omega_1}) = \phi(A_i)$ for each $i < \omega_1$, so that $\phi(A_{\omega_1})$ is at most countable. But for each $i < \omega_1$, $\psi(A_i) \subset \psi(A_{i+1})$, and so $\psi(A_{\omega_1})$ is uncountable. This proves (a) of the theorem. $\qquad\square$

Suppose T is a theory in a language L. A formula $\phi(x)$ of L is said to be **two-cardinal** for T if there is a model A of T such that $|A|$ and $|\phi(A)|$ are distinct; otherwise it is **one-cardinal** for T. We say the theory T is **two-cardinal** if there is a two-cardinal formula ϕ for T such that $\phi(A)$ is infinite in every model A of T; otherwise T is **one-cardinal**. A **Vaught pair** for the formula ϕ is a pair of structures A, B such that $B \leqslant A$, $B \neq A$ and $\phi(A)$, $\phi(B)$ are infinite and equal. For a countable complete first-order theory T, the implication (c) $\Rightarrow$ (b) in Vaught's two-cardinal theorem tells us that if some formula $\phi(x)$ has a Vaught pair of models of T, then ϕ is two-cardinal for T and T is a two-cardinal theory.

Layerings

To the end of the section, L is a first-order language and T is a theory in L. If $\phi(x)$ and $\psi(x)$ are formulas of L, we write $\psi \leqslant \phi$ to mean that for every pair of models A, B of T with $B \leqslant A$, if $\phi(A) = \phi(B)$ then $\psi(A) = \psi(B)$. We shall give a syntactic characterisation of the relation $\leqslant$. (Recall that we met the relation $\leqslant$ in Theorem 8.4.1 above. But what follows is independent of that theorem, and unlike that theorem it is not limited to countable languages.)

If $\phi(x)$ is a formula of L, we write $(\forall x \in \phi)\chi$ and $(\exists x \in \phi)\chi$ for $\forall x(\phi \to \chi)$ and $\exists x(\phi \wedge \chi)$ respectively. A **layering by** $\phi(x)$ is a formula $\theta(x)$ of the following form, for some positive integer n:

$$(4.3) \quad (\exists y_0 \in \phi)\forall z_0((\exists z_0 \eta_0 \to \eta_0) \to$$
$$(\exists y_1 \in \phi)\forall z_1((\exists z_1 \eta_1 \to \eta_1) \to$$
$$\cdots$$
$$(\exists y_{n-1} \in \phi)\forall z_{n-1}((\exists z_{n-1} \eta_{n-1} \to \eta_{n-1}) \to$$
$$x = z_0 \vee \ldots \vee x = z_{n-1}) \ldots),$$

where each η_i is a formula $\eta_i(y_0, z_0, y_1, \ldots, y_i, z_i)$ of L.

Theorem 8.4.2. *Let L and T be as above, and let $\phi(x)$, $\psi(x)$ be formulas of L such that $T \vdash \exists x \, \phi$. Then $\psi \leqslant \phi$ if and only if there exists a layering $\theta(x)$ by ϕ such that $T \vdash \forall x(\psi \to \theta)$.*

Proof. The easy direction is from right to left. Suppose A, B are models of T with $B \leqslant A$, and θ is a layering by ϕ such that $T \vdash \forall x(\psi \to \theta)$. Suppose also that $\phi(A) = \phi(B)$. Let a be any element of $\psi(A)$; we must show that a is in $\psi(B)$. Now by induction on $i < n$, where n is as in (4.3), we can choose elements b_i in $\phi(A) = \phi(B)$ and c_i in B so that

$$(4.4) \qquad\qquad B \vDash (\exists z_i \, \eta_i \to \eta_i)(b_0, c_0, \ldots, b_i, c_i).$$

(This is trivial. Since $T \vdash \exists x \, \phi$, we can find elements b_i in $\phi(A)$. When b_0, $\ldots$, b_i have been chosen, if $B \vdash \exists z_i \, \eta_i(b_0, \ldots, b_i)$ we have the required c_i; otherwise any element c_i in B will do.) Since $A \vDash \theta(a)$, it follows that a is one of $c_0, \ldots, c_{n-1}$, so that a is in B. Since $\psi(B) = \operatorname{dom}(B) \cap \psi(A)$, we have $\psi(A) = \psi(B)$.

For the converse we use Keisler games, slightly adapted. Assume $\psi \leqslant \phi$, and let λ be a strong limit number $> |L|$. In (3.4) of section 8.3, let each of the quantifiers $Q_i x_i$ have one of the forms $\forall x_i$, $\exists x_i$, $(\forall x_i \in \phi)$, $(\exists x_i \in \psi)$. We already know what the first two kinds of quantifier mean in the corresponding game on a structure A. When $Q_i x_i$ is $(\forall x_i \in \phi)$ (resp. $(\exists x_i \in \psi)$), player $\forall$ (resp. $\exists$) chooses the element a_i at step i, but now a_i must be an element of $\phi(A)$ (resp. $\psi(A)$). The rest is as before.

We shall need distinct variables y_i, z_i $(i < \lambda)$. We shall also need a list $(\eta_i : i < \lambda)$ of formulas of L such that

(4.5)   each η_i is of the form $\chi_i(y_0, z_0, y_1, z_1, \ldots, y_i, z_i)$,

(4.6)   if $\bar{y}$ (resp. $\bar{z}$) is a tuple of variables y_j (resp. z_j) and $\eta(x, \bar{y}, \bar{z})$ is any formula of L, then there is $i < \lambda$ such that $\eta(z_i, \bar{y}, \bar{z})$ is η_i and z_i is not in $\bar{z}$.

It is clear that such a list can be made.

Let χ be the Keisler sentence

(4.7) $(\exists x \in \psi)(\forall y_0 \in \phi)\exists z_0 \ldots (\forall y_i \in \phi)\exists z_i \ldots$

$$\left(\bigwedge_{i<\lambda} (x \neq z_i) \wedge \bigwedge_{i<\lambda} (\exists z_i\, \eta_i \rightarrow \eta_i) \right).$$

We claim that $T \cup \operatorname{app}(\chi)$ has no model.

Assume for contradiction that $T \cup \operatorname{app}(\chi)$ has a model. To make life easier, assume the GCH too, so that by Corollary 8.2.4, $T \cup \operatorname{app}(\chi)$ has a saturated model A. (See the end of this chapter on ways to eliminate the GCH.) Then by Theorem 8.3.7(b), player $\exists$ has a winning strategy in the game $G(\chi, A)$. Let the two players play this game; let player $\exists$ play to win, and let player $\forall$ choose so that his moves exhaust $\phi(A)$. Suppose the resulting play is

(4.8) $a, b_0, c_0, \ldots, b_i, c_i, \ldots$

and let X be the set $\{c_i : i < \lambda\}$. By (4.6) and the second conjunction in χ, if $\eta(z)$ is any formula with parameters from among the b_i and the c_i, and $A \vDash \exists z\, \eta$, then $A \vDash \eta(c)$ for some $c \in X$. Hence X is the domain of an elementary substructure B of A (using the Tarski–Vaught criterion, Exercise 2.5.1). Hence also every b_i is in X, and so $\phi(B) = \phi(A)$ in view of player $\forall$'s moves. But by the first conjunction in χ, $a \notin X$ and hence $\psi(B) \neq \psi(A)$. Thus $\psi \not\leqslant \phi$; contradiction. The claim is proved.

So by compactness there are sentences $\xi_0, \ldots, \xi_{m-1}$ in $\operatorname{app}(\chi)$ such that $T \vdash \neg(\xi_0 \wedge \ldots \wedge \xi_{m-1})$. But by the definition of $\operatorname{app}(\chi)$, there is some single sentence ξ in $\operatorname{app}(\chi)$ which implies all of $\xi_0, \ldots, \xi_{m-1}$, so that

$T \vdash \neg \xi$. An easy rearrangement of $\neg \xi$ brings it to the form $\forall x(\psi \rightarrow \theta)$ where θ is a layering by ϕ. $\qquad\qquad\qquad\qquad\qquad\qquad\qquad\qquad\square$

Exercises for section 8.4

1. Let L be a countable first-order language, T a theory in L and $\phi(x)$, $\psi(x)$ formulas of L. Suppose that for every model A of T, $|\psi(A)| \leq |\phi(A)| + \omega$. Show that there is a polynomial $p(x)$ with integer coefficients, such that for every model A of T, if $|\phi(A)| = m < \omega$ then $|\psi(A)| \leq p(m)$. [Use Theorem 8.4.2.]

2. Let L be a first-order language, T a complete theory in L with infinite models, and $\phi(x)$, $\psi(x)$ formulas of L. By a **stratification** of ψ over ϕ in a model A of T we mean a formula $\sigma(x, y)$ of L with parameters from A, such that $A \vDash \forall x(\psi(x) \leftrightarrow \exists y(\sigma(x, y) \wedge \phi(y)))$; we call the stratification σ **algebraic** if for every element b of $\phi(A)$, the set $\{a: A \vDash \sigma(a, b)\}$ is finite. Show (a) even when $\psi \leq \phi$, there need not be an algebraic stratification of ψ over ϕ in any model of T, (b) if A is a model of T, $\psi \leq \phi$ and $\psi(A)$ is infinite, then there are a formula $\rho(x)$ of L with parameters from A such that $\rho(A)$ is an infinite subset of $\psi(A)$, and an algebraic stratification of ρ over ϕ in A. [Referring to the formula $\theta(x)$ of (4.3), write $\theta_k(x, y_0, z_0, y_1, \ldots, z_{k-1}, y_k)$ for the formula which results if we delete everything up to $(\exists y_k \in \phi)$ inclusive. *Case 1.* For each $b \in \phi(A)$ there are only finitely many $a \in \psi(A)$ such that $A \vDash \theta_0(a, b)$. Then put $\rho = \psi$ and $\sigma(x, y) = \theta_0(x, y)$. *Case 2.* There is $b_0 \in \phi(A)$ such that the set $X_0 = \{a: A \vDash \theta_0(a, b_0)\}$ is infinite, but for each $b \in \phi(A)$ there are only finitely many $a \in X_0$ such that $A \vDash \theta_1(a, b_0, c_0, b)$ (where c_0 is some fixed element of A such that $A \vDash \eta_0(b_0, c_0)$ if there is such an element). Then put $\rho(x) = \theta_0(x, b_0)$ and $\sigma(x, y) = \theta_1(x, b_0, c_0, y)$. Etc. up to case n.]

8.5 Ultraproducts and ultrapowers

Ultraproducts are a method for constructing new structures out of old ones. The ingredients are a family $(A_i: i \in I)$ of L-structures, for some signature L, and a set $\mathcal{U}$ known as an ultrafilter on I. The outcome is a new L-structure which is written $\prod_{i \in I} A_i / \mathcal{U}$, and known as the ultraproduct of $(A_i: i \in I)$ over $\mathcal{U}$.

The ultraproduct construction has four distinctive features which make it unlike anything else in the book. The first is that the ultraproduct is defined outright in terms of the structures A_i and the ultrafilter; the definition uses no induction on the ordinals and no logical formulas. Algebraists tend to like ultraproducts for this reason. But one should not be misled: the messier features of model theory are still there, hidden in the choice of the ultrafilter. In practice we often find that the construction of the ultrafilter uses either logical formulas or induction on ordinals, or both.

Second, there is a simple rule which determines the complete first-order theory of $\prod_{i \in I} A_i / \mathcal{U}$ in terms of $\mathcal{U}$ and the complete first-order theories $(\mathrm{Th}(A_i): i \in I)$. This is Łos's Theorem, Theorem 8.5.3 below.

Third, ultraproducts generally realise a large number of types. In fact there are ways of choosing the ultrafilter $\mathcal{U}$ so that $\prod_{i \in I} A_i / \mathcal{U}$ is guaranteed to be saturated – though we shall not consider this here.

Fourth, 'forming ultraproducts commutes with forming reducts'. What this means is that the interpretation of each symbol of L in $\prod_{i \in I} A_i / \mathcal{U}$ is independent of the interpretations of other symbols, so that we can add or remove symbols in the signature without affecting the rest of the ultraproduct. This fact has many applications. One of them is that we can build elementary extensions in which the cardinalities of certain definable sets are controlled independently of each other. In fact this leads to some results, like Corollaries 8.5.8 and 8.5.9 below, which have no known proof except by ultraproducts.

Apart from these corollaries and a few results like them, ultraproducts are hardly more than a way of using the compactness theorem – they tell us nothing that we couldn't prove just as easily in other ways. This is why I have given them just one section.

Ultraproducts defined

The definition of the ultraproduct $\prod_{i \in I} A_i / \mathcal{U}$ is in two steps. First we define the direct product $\prod_{i \in I} A_i$, and then we form a homomorphic image by factoring out $\mathcal{U}$. In order to avoid some bad behaviour in the direct products, we assume for the rest of this section that *the index set I is not empty, and none of the structures A_i are empty*.

Let L be a signature and I a non-empty set. Suppose that for each $i \in I$ an L-structure A_i is given. We define the **direct product** $\prod_{i \in I} A_i$ (or $\prod_I A_i$ for short) to be the L-structure B defined in the next paragraph. Direct products are also known as **Cartesian products**; for brevity we usually call them just **products**.

Write X for the set of all maps $a: I \to \bigcup_{i \in I} \mathrm{dom}(A_i)$ such that for each $i \in I$, $a(i) \in \mathrm{dom}(A_i)$. We put $\mathrm{dom}(B) = X$. For each constant c of L we take c^B to be the element a of X such that $a(i) = c^{A_i}$ for each $i \in I$. For each n-ary function symbol F of L and n-tuple $\bar{a} = (a_0, \ldots, a_{n-1})$ from X, we define $F^B(\bar{a})$ to be the element b of X such that for each $i \in I$, $b(i) = F^{A_i}(a_0(i), \ldots, a_{n-1}(i))$. For each n-ary relation symbol R of L and n-tuple $\bar{a}$ from X, we put $\bar{a}$ in R^B iff for every $i \in I$, $(a_0(i), \ldots, a_{n-1}(i)) \in R^{A_i}$. Then B is an L-structure, and we define $\prod_I A_i$ to be B. The structure A_i is called the ith **factor** of the product. If $I = \{0, \ldots, n-1\}$ we write $A_0 \times \ldots \times A_{n-1}$ for $\prod_I A_i$.

If $\bar{a}$ is a sequence of elements $(a_0, a_1, \ldots)$ of the product $\prod_I A_i$, then $\bar{a}(i)$ will always mean $(a_0(i), a_1(i), \ldots)$, and never a_i.

In algebra one defines various kinds of product, and these often turn out to be identical with direct products as we have defined them. For example a product of groups is exactly a direct product in our sense. Model theory has a good deal to say about direct products; but it is not our concern here. Instead we turn to the second (longer) part of the definition of ultraproducts.

By a **filter** over the set I we mean a non-empty set $\mathcal{F}$ of subsets of I such that

(5.1) $X, Y \in \mathcal{F} \Rightarrow X \cap Y \in \mathcal{F}$; $X \in \mathcal{F}, X \subseteq Y \subseteq I \Rightarrow Y \in \mathcal{F}$;
        and $\varnothing \notin \mathcal{F}$.

In particular $I \in \mathcal{F}$ by the second part of (5.1) and the fact that $\mathcal{F}$ is not empty. A filter $\mathcal{F}$ is called an **ultrafilter** if it has the further property

(5.2)   For every set $X \subseteq I$, exactly one of X, $I \backslash X$ is in $\mathcal{F}$.

In a moment we shall find a way of constructing interesting ultrafilters. But it is very easy to find uninteresting ones. Let i be any element of I, and let $\mathcal{U}$ be the set of all subsets X of I such that $i \in X$. Then $\mathcal{U}$ is an ultrafilter over I. Ultrafilters of this form are said to be **principal**.

Let $\phi(\bar{x})$ be a formula of L and $\bar{a}$ a tuple of elements of the product $\prod_I A_i$. We define the **boolean value** of $\phi(\bar{a})$, in symbols $\|\phi(\bar{a})\|$, to be the set $\{i \in I: A_i \vDash \phi(\bar{a}(i))\}$.

This definition is lifted almost directly from George Boole's first logical monograph, published in 1847. In essence he pointed out the laws

(5.3) $\|\phi \wedge \psi\| = \|\phi\| \cap \|\psi\|$, $\|\phi \vee \psi\| = \|\phi\| \cup \|\psi\|$, $\|\neg \phi\| = I \backslash \|\phi\|$,

and used them to show that both logic and set theory provide interpretations of his boolean calculus.

The analogue of (5.3) for the existential quantifier should say that $\|\exists x\, \phi(x)\|$ is the union of the sets $\|\phi(a)\|$ with a in $\prod_I A_i$. But in fact something stronger is true, both for $\prod_I A_i$ and for some of its substructures C. We say that C **respects** $\exists$ if for every formula $\phi(x)$ of L with parameters from C.

(5.4) $\|\exists x\, \phi(x)\| = \|\phi(a)\|$ for some element a of C.

It is clear that $\prod_I A_i$ respects $\exists$: for each $i \in \|\exists x\, \phi(x)\|$, choose an element a_i such that $A_i \vDash \phi(a_i)$, and take the element a of $\prod_I A_i$ such that $a(i) = a_i$ for each $i \in \|\exists x\, \phi(x)\|$. (Here we invoke the axiom of choice.)

Let I be a non-empty set, $(A_i: i \in I)$ a family of non-empty L-structures and $\mathcal{F}$ a filter over I. We form the product $\prod_I A_i$, and using $\mathcal{F}$ we define an equivalence relation $\sim$ on dom $\prod_I A_i$ by

(5.5) $a \sim b$ iff $\|a = b\| \in \mathcal{F}$.

We verify that $\sim$ is an equivalence relation. Reflexive: for each element a of $\prod_I A_i$, $\|a = a\| = I \in \mathscr{F}$. Symmetry is clear. Transitive: $\|a = b\| \cap \|b = c\| \subseteq \|a = c\|$, so that if $\|a = b\|$, $\|b = c\| \in \mathscr{F}$, then $\|a = c\| \in \mathscr{F}$ by (5.1). Thus $\sim$ is an equivalence relation. We write $a/\mathscr{F}$ for the equivalence class of the element a.

We define an L-structure D as follows. The domain $\mathrm{dom}(D)$ is the set of equivalence classes $a/\mathscr{F}$ with $a \in \mathrm{dom} \prod_I A_i$. For each constant symbol c of L we put

(5.6) $c^D = a/\mathscr{F}$ where $a(i) = c^{A_i}$ for each $i \in I$.

Next let F be an n-ary function symbol of L, and $a_0, \ldots, a_{n-1}$ elements of $\prod_I A_i$. We define

(5.7) $F^D(a_0/\mathscr{F}, \ldots, a_{n-1}/\mathscr{F}) = b/\mathscr{F}$
        where $b(i) = F^{A_i}(a_0(i), \ldots, a_{n-1}(i))$ for each $i \in I$.

It has to be checked that (5.7) is a sound definition. Suppose $a_j \sim a_j'$ for each $j < n$. Then by (5.1) there is a set $X \in \mathscr{F}$ such that $X \subseteq \|a_j = a_j'\|$ for each $j < n$. It follows that $X \subseteq \|F(a_0, \ldots, a_{n-1}) = F(a_0', \ldots, a_{n-1}')\|$, which justifies the definition. Finally if R is an n-ary relation symbol of L and $a_0, \ldots, a_{n-1}$ are elements of $\prod_I A_i$, then we put

(5.8) $(a_0/\mathscr{F}, \ldots, a_{n-1}/\mathscr{F}) \in R^D$ iff $\|R(a_0, \ldots, a_{n-1})\| \in \mathscr{F}$.

Again (5.1) shows that this definition is sound.

We have defined an L-structure D. This structure is called the **reduced product** of $(A_i : i \in I)$ over $\mathscr{F}$, in symbols $\prod_I A_i/\mathscr{F}$. When $\mathscr{F}$ is an ultrafilter, the structure is called the **ultraproduct** of $(A_i : i \in I)$ over $\mathscr{F}$. The effect of definitions (5.6)–(5.8) is that for every unnested atomic formula $\phi(\bar{x})$ of L and every tuple $\bar{a}$ of elements of $\prod_I A_i$.

(5.9) $\prod_I A_i/\mathscr{F} \vDash \phi(\bar{a}/\mathscr{F})$ iff $\|\phi(\bar{a})\| \in \mathscr{F}$.

Note that $\prod_I A_i$ itself is just the reduced product $\prod_I A_i/\{I\}$, so that every direct product is a reduced product.

Our first result says that taking reduced products commutes with taking relativised reducts.

Theorem 8.5.1. *Let L and L^+ be signatures and P a 1-ary relation symbol of L^+. Let $(A_i : i \in I)$ be a non-empty family of non-empty L^+-structures such that $(A_i)_P$ is defined (see section 4.2) and $\mathscr{F}$ a filter over I. Then $(\prod_I A_i/\mathscr{F})_P = \prod_I ((A_i)_P)/\mathscr{F}$.*

Proof. Define $f: \prod_I((A_i)_P)/\mathscr{F} \to \prod_I A_i/\mathscr{F}$ by taking any element $a/\mathscr{F}$ of $\prod_I((A_i)_P)/\mathscr{F}$ to the corresponding element $a/\mathscr{F}$ of $\prod_I A_i/\mathscr{F}$. One can check from the definition of reduced products that this definition is sound, and that f is an embedding with image $(\prod_I A_i/\mathcal{F})_P$. □

When all the structures A_i are equal to a fixed structure A, we call $\prod_I A/\mathcal{F}$ the **reduced power** $A^I/\mathcal{F}$; if $\mathcal{F}$ is an ultrafilter, we call the structure the **ultrapower** of A over $\mathcal{F}$. There is an embedding $e: A \to A^I/\mathcal{F}$ defined by $e(b) = a/\mathcal{F}$ where $a(i) = b$ for all $i \in I$. The fact that e is an embedding follows from the next lemma, but it's easy to check directly. We call e the **diagonal embedding**.

Recall that a **positive primitive (p.p.)** formula is a first-order formula of the form $\exists \bar{y} \bigwedge \Phi$ where Φ is a set of atomic formulas.

Lemma 8.5.2. *Let L be a signature and $\phi(\bar{x})$ a p.p. formula of L. Let $(A_i: i \in I)$ be a non-empty family of non-empty L-structures and $\bar{a}$ a tuple of elements of $\prod_I A_i$. Let $\mathcal{F}$ be a filter over I. Then*

$$(5.10) \qquad \prod_I A_i/\mathcal{F} \vDash \phi(\bar{a}/\mathcal{F}) \qquad \text{if and only if} \qquad \|\phi(\bar{a})\| \in \mathcal{F}.$$

Proof. We go by induction on the complexity of ϕ. Since $\|\psi\| = \|\chi\|$ whenever ψ and χ are logically equivalent, we can quote Corollary 2.6.2 and assume that ϕ is unnested. Then (5.10) for atomic formulas is just (5.9).

If (5.10) holds for formulas $\phi(\bar{x})$, $\psi(\bar{x})$ then it holds for their conjunction. From left to right, suppose $\prod_I A_i/\mathcal{F} \vDash (\phi \wedge \psi)(\bar{a}/\mathcal{F})$. Then by assumption $\|\phi(\bar{a})\|$ and $\|\psi(\bar{a})\|$ are both in $\mathcal{F}$. It follows that $\|(\phi \wedge \psi)(\bar{a})\| \in \mathcal{F}$ by the first parts of (5.3) and (5.1). From right to left, if $\|(\phi \wedge \psi)(\bar{a})\| \in \mathcal{F}$ then $\|\phi(\bar{a})\| \in \mathcal{F}$ by the second part of (5.1), since $\|(\phi \wedge \psi)(\bar{a})\| \subseteq \|\phi(\bar{a})\|$. The rest is clear.

If (5.10) holds for $\psi(\bar{x}, \bar{y})$ then it holds for $\exists \bar{y}\, \psi(\bar{x}, \bar{y})$. From left to right, suppose $\prod_I A_i/\mathcal{F} \vDash \exists \bar{y}\, \psi(\bar{a}/\mathcal{F}, \bar{y})$. Then there are elements $\bar{b}$ of $\prod_I A_i$ such that $\prod_I A_i/\mathcal{F} \vDash \psi(\bar{a}/\mathcal{F}, \bar{b}/\mathcal{F})$, so that $\|\psi(\bar{a}, \bar{b})\| \in \mathcal{F}$ by assumption. Since $\|\psi(\bar{a}, \bar{b})\| \subseteq \|\exists \bar{y}\, \psi(\bar{a}, \bar{y})\|$, it follows that $\|\exists \bar{y}\, \psi(\bar{a}, \bar{y})\| \in \mathcal{F}$ by (5.1). Conversely suppose $\|\exists \bar{y}\, \psi(\bar{a}, \bar{y})\| \in \mathcal{F}$. Since $\prod_I A_i$ respects $\exists$, there are elements $\bar{b}$ of $\prod_I A_i$ such that $\|\psi(\bar{a}, \bar{b})\| = \|\exists \bar{y}\, \psi(\bar{a}, \bar{y})\|$; whence $\prod_I A_i/\mathcal{F} \vDash \psi(\bar{a}/\mathcal{F}, \bar{b}/\mathcal{F})$ by assumption. Hence $\prod_I A_i \vDash \exists \bar{y}\, \psi(\bar{a}/\mathcal{F}, \bar{y})$. □

Theorem 8.5.3. (*Łoś's theorem*). *Let L be a first-order language, $(A_i: i \in I)$ a non-empty family of non-empty L-structures and $\mathcal{U}$ an ultrafilter over I. Then for any formula $\phi(\bar{x})$ of L and tuple $\bar{a}$ of elements of $\prod_I A_i$,*

$$(5.11) \qquad \prod_I A_i/\mathcal{U} \vDash \phi(\bar{a}/\mathcal{U}) \qquad \text{if and only if} \qquad \|\phi(\bar{a})\| \in \mathcal{U}.$$

Proof. We go by induction on the complexity of ϕ. Comparing with the proof of Lemma 8.5.2, we see that only one more thing is needed: assuming that (5.11) holds for ϕ, we have to deduce it for $\neg \phi$ too. But this is easy by (5.2):

$$(5.12) \qquad \prod_I A_i/\mathcal{U} \vDash \neg \phi(\bar{a}/\mathcal{U}) \Leftrightarrow \|\phi(\bar{a})\| \notin \mathcal{U} \Leftrightarrow \|\neg \phi(\bar{a})\| \in \mathcal{U}. \qquad \square$$

Corollary 8.5.4. *If $A^I/\mathcal{U}$ is an ultrapower of A, then the diagonal map $e: A \to A^I/\mathcal{U}$ is an elementary embedding.*

Proof. Immediate. $\qquad \square$

By the usual manipulation (see Exercise 1.2.4(b)), Corollary 8.5.4 allows us to regard A as an elementary substructure of $A^I/\mathcal{U}$. So ultrapowers give elementary extensions. By Exercise 1 below, this is useful only when the ultrafilter is non-principal; it's high time we found some non-principal ultrafilters.

Finding ultrafilters

Let I be a non-empty set and W a set of subsets of I. We say that I has the **finite intersection property** if for every finite set $X_0, \ldots, X_{m-1}$ of elements of W, the intersection $X_0 \cap \ldots \cap X_{m-1}$ is not empty. Note that every filter over I has the finite intersection property.

Lemma 8.5.5. *Let I be a non-empty set and W a set of subsets of I with the finite intersection property. Then there is an ultrafilter $\mathcal{U}$ over I such that $W \subseteq \mathcal{U}$.*

Proof. Let L be the first-order language with the following signature: each subset of I is a constant, and there is one 1-ary relation symbol P. Let T be the theory

$$(5.13) \quad \{P(a) \to P(b): a \subseteq b\}$$
$$\cup \{P(a) \wedge P(b) \to P(c): a \cap b = c\}$$
$$\cup \{P(a) \leftrightarrow \neg P(b): b = I \backslash a\} \cup \{P(a): a \in W\}.$$

We claim that T has a model. For suppose not. Then by the compactness theorem (Theorem 5.1.1), some finite subset U of T has no model. Let $X_0, \ldots, X_{m-1}$ be the elements a of W such that '$P(a)$' $\in U$. Since W has the finite intersection property, there is some element $i \in I$ such that $i \in X_0 \cap \ldots \cap X_{m-1}$. Let $\mathcal{V}$ be the principal ultrafilter consisting of all the subsets of I that contain i. Then we form a model of U by interpreting each subset of I as a name of itself, and reading '$P(c)$' as '$c \in \mathcal{V}$'. This proves the claim.

Now let B be a model of T. Define a set $\mathcal{U}$ of subsets of I by putting $b \in \mathcal{U}$ if and only if $B \vDash P(b)$. Then we can read off from (5.13) that $\mathcal{U}$ is an ultrafilter containing all of W. $\qquad\square$

Assuming that I is infinite, this lemma gives us a non-principal ultrafilter if we take W to be the set of all cofinite subsets of I (i.e. sets $I \backslash X$ where X is finite). But a much stronger statement is within our grasp.

Lemma 8.5.6. *Let I be an infinite set. Then there is an ultrafilter $\mathcal{U}$ over I containing sets X_j $(j \in I)$ such that for each $i \in I$ the set $\{j : i \in X_j\}$ is finite.*

Proof. Clearly it is enough if we prove the lemma for some set J of the same cardinality as I. Let J be the set of all finite subsets of I. For each $i \in I$, let $X(i)$ be $\{j \in J : i \in j\}$, so that $j \in X(i) \Leftrightarrow i \in j$. If $i_0, \ldots, i_{n-1}$ are in I then $X(i_0) \cap \ldots \cap X(i_{n-1})$ is not empty, since it is the set of all $j \in J$ with $i_0, \ldots, i_{n-1} \in j$. So by Lemma 8.5.5 there is an ultrafilter $\mathcal{U}$ over J which contains $X(i)$ for each $i \in I$. Use a bijection between I and J to relabel the sets $X(i)$ as X_j $(j \in J)$. $\qquad\square$

An ultrafilter $\mathcal{U}$ with the property of Lemma 8.5.6 is said to be **regular**. It is obviously not principal.

Theorem 8.5.7. *Let L be a first-order language, A an L-structure, I an infinite set and $\mathcal{U}$ a regular ultrafilter over I.*

(a) *If $\phi(x)$ is a formula of L such that $|\phi(A)|$ is infinite, then $|\phi(A^I/\mathcal{U})| = |\phi(A)|^{|I|}$.*

(b) *If $\Phi(\bar{x})$ is a type over $\mathrm{dom}(A)$ with respect to A, and $|\Phi| \leqslant |I|$, then some tuple $\bar{a}$ in $A^I/\mathcal{U}$ realises Φ.*

Proof. (a) We first prove $\leqslant$. By Łoś's theorem (Theorem 8.5.3) each element of $\phi(A^I/\mathcal{U})$ is of the form $b/\mathcal{U}$ for some b such that $\|\phi(b)\| \in \mathcal{U}$. Since we can change b anywhere outside a set in $\mathcal{U}$ without affecting $b/\mathcal{U}$, we can choose b so that $\|\phi(b)\| = I$. This sets up an injection from $\phi(A^I/\mathcal{U})$ to the set $\phi(A)^I$ of all maps from I to $\phi(A)$.

Next we prove $\geqslant$. Since $\mathcal{U}$ is regular, there are sets X_i $(i \in I)$ in $\mathcal{U}$ such that for each $j \in I$ the set $Z_j = \{i \in I : j \in X_i\}$ is finite. For each $j \in I$, let μ_j be a bijection taking the set $\phi(A)^{Z_j}$ (of all maps from Z_j to $\phi(A)$) to $\phi(A)$. Such a μ_j exists since $\phi(A)$ is infinite. For each function $f : I \to \phi(A)$, define f^μ to be the map from I to $\phi(A)$ such that

(5.14) $\qquad$ for each $j \in I$, $f^\mu(j) = \mu_j(f \restriction Z_j)$.

Each function $f^\mu: I \to \phi(A)$ is an element of A^I, and by Łoś's theorem $f^\mu/\mathcal{U} \in \phi(A^I/\mathcal{U})$. So it remains only to show that if f, g are distinct maps from I to $\phi(A)$ then $f^\mu/\mathcal{U} \neq g^\mu/\mathcal{U}$. Suppose then that $f(i) \neq g(i)$ for some $i \in I$. It follows that $f|Z_j \neq g|Z_j$ whenever $i \in Z_j$, i.e. whenever $j \in X_i$. Hence $X_i \subseteq \|f^\mu \neq g^\mu\|$, and so $f^\mu/\mathcal{U} \neq g^\mu/\mathcal{U}$ since $X_i \in \mathcal{U}$.

(b) Since $\mathcal{U}$ is regular, there is a family $\{X_\phi: \phi \in \Phi\}$ of sets in $\mathcal{U}$, such that for each $i \in I$ the set $Z_i = \{\phi \in \Phi: i \in X_\phi\}$ is finite. Since Φ is a type over $\text{dom}(A)$, for each $i \in I$ there is a tuple $\bar{a}_i$ in A which satisfies Z_i. Let $\bar{a}$ be the tuple in A^I such that $\bar{a}(i) = \bar{a}_i$ for each i. Then for each formula ϕ in Φ, if $i \in X_\phi$ then $\phi \in Z_i$ and so $A \vDash \phi(\bar{a}_i)$. Thus $X_\phi \subseteq \|\phi(\bar{a})\|$, and by Łoś's theorem (Theorem 8.5.3) we deduce that $A^I/\mathcal{U} \vDash \phi(\bar{a})$. $\qquad\square$

Corollary 8.5.4 and Theorem 8.5.7(a) give us arbitrarily large elementary extensions of any infinite structure A. They also give us the following stronger statement, for which no other proof is known.

Corollary 8.5.8. *Let L be a first-order language, A an L-structure and κ an infinite cardinal. Then A has an elementary extension B such that for every formula $\phi(\bar{x})$ of L, $|\phi(B)|$ is either finite or equal to $|\phi(A)|^\kappa$.* $\qquad\square$

The next application also has no other known proof. It is more complicated than Corollary 8.5.8, but also more useful. The finite cover property was defined in section 4.4 (and will be used in section 9.5).

Corollary 8.5.9. *Let L be a first-order language and T a complete theory in L which has infinite models. Suppose T is λ-categorical for some $\lambda \geq \max((2^\omega)^+, |L|)$. Then T doesn't have the finite cover property.*

Proof. I use Shelah's definition of the finite cover property, as in section 4.4. The proof adapts at once to Keisler's definition (given at Exercise 4.4.3).

Let A be a model of T of cardinality at least λ. If T has the finite cover property, then there is a formula $\phi(\bar{x}, \bar{y}, \bar{z})$ of L such that for each tuple $\bar{c}$ in A, $\phi(\bar{x}, \bar{y}, \bar{c})$ defines an equivalence relation $E_{\bar{c}}$, and for each $n < \omega$ there is a tuple $\bar{c}_n$ such that the number of equivalence classes of $E_{\bar{c}_n}$ is finite and at least n. In some suitable finite slice B of A^{eq}, the equivalence classes of each $E_{\bar{c}}$ form a set of elements $X_{\bar{c}}$; $X_{\bar{c}}$ is definable in terms of $\bar{c}$.

Let $\mathcal{U}$ be a regular ultrafilter over ω and $(X_m: m < \omega)$ a descending chain of sets in $\mathcal{U}$ with empty intersection. Choose a tuple $\bar{c}/\mathcal{U}$ in $B^\omega/\mathcal{U}$ so that for each $i < \omega$, if $i \in X_n \backslash X_{n+1}$ then $\bar{c}(i) = \bar{c}_n$. For every $n < \omega$, the set

(5.15) $\|\phi(\bar{x}, \bar{y}, \bar{c})$ defines an equivalence relation whose set $X_{\bar{c}}$ of

equivalence classes has more than n elements$\|$

contains all but finitely many $i \in \omega$. So by Łoś's theorem, $X_{\bar{c}}$ is an infinite set consisting of the equivalence classes of $\phi(\bar{x}, \bar{y}, \bar{c})$ in $B^\omega/\mathcal{U}$. But $X_{\bar{c}}$ is also a relativised reduct of $B^\omega/\mathcal{U}$ (with parameters $\bar{c}$). Hence we can apply Theorem 8.5.1 to deduce that $|X_{\bar{c}}| = |\prod_\omega X_{\bar{c}(i)}/\mathcal{U}| \leqslant \prod_\omega |X_{\bar{c}(i)}| \leqslant 2^\omega$. So in $A^\omega/\mathcal{U}$, the number of equivalence classes of $\phi(\bar{x}, \bar{y}, \bar{c})$ is infinite but $\leqslant 2^\omega$.

Now by the downward Löwenheim–Skolem theorem, $A^\omega/\mathcal{U}$ has an elementary substructure of cardinality λ in which the number of equivalence classes of $\phi(\bar{x}, \bar{y}, \bar{c})$ is infinite but $\leqslant 2^\omega$. But one easily applies the compactness theorem to construct a model of T of cardinality λ in which for every tuple $\bar{d}$, the number of equivalence classes of $\phi(\bar{x}, \bar{y}, \bar{d})$ is either finite or at least λ (see Exercise 6.1.10). This contradicts the assumption that T is λ-categorical. $\qquad\square$

Elementary equivalence

Theorem 8.5.10 (Keisler–Shelah theorem). *Let L be a signature and let A, B be L-structures. The following are equivalent.*
(a) *$A \equiv B$.*
(b) *There are a set I and an ultrafilter $\mathcal{U}$ over I such that $A^I/\mathcal{U} \cong B^I/\mathcal{U}$.*

Proof. The proof uses some quite difficult combinatorics. It can be found in the proof (but not the statement) of Theorem 6.1.15 in Chang and Keisler, *Model theory* (see the references at the end of this chapter). $\qquad\square$

The Keisler–Shelah theorem was an impressive solution of a natural problem. But it hasn't led to much new information. The following application is typical in two ways: it uses Theorem 8.5.1, and there is also a straightforward proof by elementary means (see Exercise 5.5.1).

Corollary 8.5.11 (Robinson's joint consistency lemma). *Let L_1 and L_2 be first-order languages and $L = L_1 \cap L_2$. Let T_1 and T_2 be consistent theories in L_1 and L_2 respectively, such that $T_1 \cap T_2$ is a complete theory in L. Then $T_1 \cup T_2$ is consistent.*

Proof. Let A_1, A_2 be models of T_1, T_2 respectively. Then since $T_1 \cap T_2$ is complete, $A_1|L \equiv A_2|L$. By the Keisler–Shelah theorem there is an ultrafilter $\mathcal{U}$ over a set I such that $(A_1|L)^I/\mathcal{U} \cong (A_2|L)^I/\mathcal{U}$. Corollary 8.5.4 tells us that $A_1^I/\mathcal{U} \vDash T_1$ and $A_2^I/\mathcal{U} \vDash T_2$. By Theorem 8.5.1, $A_1^I/\mathcal{U}$ is an expansion of $(A_1|L)^I/\mathcal{U}$. But also Theorem 8.5.1 tells us that $A_2^I/\mathcal{U}$ is an expansion of an isomorphic copy of $(A_1|L)^I/\mathcal{U}$. So we can use $A_2^I/\mathcal{U}$ as a template to expand $A_1^I/\mathcal{U}$ to a model of T_2. $\qquad\square$

Let L be a first-order language, and let S be the set of all theories in L which are of the form $\text{Th}(A)$ for some L-structure A. Let X be a subset of S and T a set of sentences of L. Let us call T a **limit point of** X if

(5.16)   for every sentence ϕ of L, exactly one of ϕ, $\neg\phi$ is in T, and

(5.17)   for every finite $T_0 \subseteq T$ there is $T' \in X$ with $T_0 \subseteq T'$.

The next theorem is one way of showing that such a set T is in fact an element of S. Readers who know the Stone topology on S can read (5.17) as saying that T is a limit point of X in that topology.

Theorem 8.5.12. *Let L be a first-order language, $\mathbf{K}$ a class of L-structures and T a limit point of $\{\text{Th}(A): A \in \mathbf{K}\}$. Then T is $\text{Th}(B)$ for some ultraproduct B of structures in $\mathbf{K}$.*

Proof. The proof is a variant of that of Theorem 8.5.7(b). Let $\mathcal{U}$ be a regular ultrafilter over the set T. Then there is a family $\{X_\phi: \phi \in T\}$ of sets in $\mathcal{U}$, such that for each $i \in T$ the set $Z_i = \{\phi \in T: i \in X_\phi\}$ is finite. Since T is a limit point of $\{\text{Th}(A): A \in \mathbf{K}\}$, for each $i \in T$ there is a structure $A_i \in \mathbf{K}$ such that $A_i \vDash Z_i$. Put $B = \prod_T A_i/\mathcal{U}$. If $i \in X_\phi$ then $\phi \in Z_i$ and so $A_i \vDash \phi$; hence $X_\phi \subseteq \|\phi\|$ for each sentence ϕ in T. It follows by Łoś's theorem (Theorem 8.5.3) that $B \vDash T$, and so $T = \text{Th}(B)$ by (5.16). $\square$

Readers who enjoy going round in circles should spare a minute to deduce the compactness theorem from Theorem 8.5.12. The rest of us will move on to deduce a criterion for first-order axiomatisability.

Corollary 8.5.13. *Let L be a first-order language and $\mathbf{K}$ a class of L-structures. Then the following are equivalent.*
(a) $\mathbf{K}$ is axiomatisable by a set of sentences of L.
(b) $\mathbf{K}$ is closed under ultraproducts and isomorphic copies, and if A is an L-structure such that some ultrapower of A lies in $\mathbf{K}$, then A is in $\mathbf{K}$.

Proof. (a) $\Rightarrow$ (b) follows at once from Theorem 8.5.3 and Corollary 8.5.4.

Conversely suppose (b) holds, and let T be the set of all sentences of L which are true in every structure in $\mathbf{K}$. To prove (a) it suffices to show that any model A of T lies in $\mathbf{K}$.

We begin by showing that $\text{Th}(A)$ is a limit point of $\{\text{Th}(C): C \in \mathbf{K}\}$. For this, let U be a finite set of sentences of L which are true in A. Then $\bigwedge U$ is a sentence ϕ which is true in A, and so $\neg\phi \notin T$ since A is a model of T. It follows by the definition of T that some structure in $\mathbf{K}$ is a model of ϕ. Thus $\text{Th}(A)$ is a limit point of $\{\text{Th}(C): C \in \mathbf{K}\}$. By Theorem 8.5.12 we deduce that A is elementarily equivalent to some ultraproduct of structures in $\mathbf{K}$, and hence (by (b)) to some structure B in **K**. By the Keisler–Shelah theorem (Theorem 8.5.10) it follows that some ultrapower of A is isomorphic to an ultrapower of B, and so by (b) again, A is in **K**.   □

Exercises for section 8.5

Throughout these exercises we assume that all structures are non-empty.

1. Show that if $\mathcal{U}$ is a principal ultrafilter then the ultraproduct $\prod_I A_i/\mathcal{U}$ is isomorphic to one of the A_i.

2. Show that if $|A_i| \leq |B_i|$ for all $i \in I$ then $|\prod_I A_i/\mathcal{U}| \leq |\prod_I B_i/\mathcal{U}|$.

*An ultrafilter $\mathcal{U}$ over a set I is λ-**complete** if for every set X of fewer than λ sets in $\mathcal{U}$, the intersection $\bigcap X$ is also in $\mathcal{U}$. Every ultrafilter is ω-complete. If there are any non-principal ω_1-complete ultrafilters, then a measurable cardinal exists.*
3. Show that no regular ultrafilter over an infinite set is ω_1-complete.

4. Show that if the ultrafilter $\mathcal{U}$ is not ω_1-complete, then every ultraproduct $\prod_I A_i/\mathcal{U}$ has cardinality $< \omega$ or $\geq 2^\omega$.

5. Show that if B is an ultraproduct of finite structures then $|B|$ is either finite or $\geq 2^\omega$.

6. Show that the following conditions on an ultrafilter $\mathcal{U}$ over I are equivalent. (a) $\mathcal{U}$ is not ω_1-complete. (b) There are disjoint non-empty sets $X_i \subseteq I$ ($i < \omega$) such that for each $n < \omega$, $\bigcup_{i \geq n} X_i \in \mathcal{U}$. (c) The ultrapower $(\omega, <)^I/\mathcal{U}$ is not well-ordered.

7. An ultrafilter over a set I is said to be **uniform** if every set in the ultrafilter has cardinality $|I|$. Show that every regular ultrafilter over an infinite set is uniform.

8. (Frayne's theorem) Show that two L-structures A and B are elementarily equivalent if and only if A is elementarily embeddable in some ultrapower of B. (Give a direct proof without quoting the Keisler–Shelah theorem.)

9. Let **K** be a class of L-structures. Show (a) **K** is first-order axiomatisable if and only if **K** is closed under ultraproducts and under elementary equivalence, (b) **K** is first-order definable if and only if both **K** and its complement in the class of L-structures are closed under ultraproducts and elementary equivalence.

10. Use the compactness theorem to deduce each of Robinson's joint consistency lemma (Corollary 8.5.11) and Craig's interpolation theorem (Theorem 5.5.3) from the other.

11. Let L be a signature, **K** a class of L-structures and A an L-structure. Show that the following are equivalent. (a) Every $\forall_1$ sentence in Th($\mathbf{K}$) is true in A. (b) A is embeddable in an ultraproduct of structures in **K**.

12. Show that if $\mathcal{U}$ is a regular ultrafilter over a set I of cardinality κ then the structure $(\kappa^+, <)$ is embeddable in $(\omega, <)^I/\mathcal{U}$.

13. Let $\mathcal{U}$ be an ultrafilter over a set I of cardinality κ. Show that $\mathcal{U}$ is regular if and only if for every signature L with $|L| \leq \kappa$ and every L-structure A, $A^I/\mathcal{U}$ is κ^+-universal. [For right to left, consider a structure A and a type $\Phi(x)$ over $\varnothing$ with respect to A, such that Φ has cardinality κ and is not realised in A. If b realises Φ in $A^I/\mathcal{U}$, for each formula ϕ in Φ consider $\{i \in I : A \vDash \phi(b(i))\}$.]

By a **basic Horn formula** *we mean a formula of the form* $\bigwedge \Phi \to \psi$ *where* Φ *is a set of atomic formulas and* ψ *is either an atomic formula or* $\bot$. *We allow* Φ *to be empty; in this case the basic Horn formula is just* ψ. *A* **Horn formula** *is a formula consisting of a finite (possibly empty) string of quantifiers, followed by a conjunction of basic Horn formulas. A theory consisting of Horn formulas is said to be a* **Horn theory**.

14. Show that if T is a Horn theory in the first-order language L, $\mathcal{F}$ is a filter over I and $(A_i : i \in I)$ is a family of L-structures which are models of T, then the reduced product $\prod_I A_i/\mathcal{F}$ is also a model of T. [Start from Lemma 8.5.2.] *So for example a reduced product of groups is always a group, and likewise with torsion-free abelian groups.*

Further reading

The text

Chang, C. C. and Keisler, H. J. *Model theory*, third edition. Amsterdam: North-Holland 1990.

has very thorough treatments both of saturated structures and of ultraproducts. It also discusses ways of eliminating the generalised continuum hypothesis from proofs which assume the existence of saturated models. Many features of saturation depend on the properties of infinite cardinal numbers; a reference for these properties is

Levy, A. *Basic set theory*. Berlin: Springer-Verlag 1979.

Saturated models are a useful tool for studying definability; see for example

Kueker, D. W. Generalized interpolation and definability. *Ann. Math. Logic*, **1** (1970), 423–468.

Kueker's paper was a precursor of the theory of recursive saturation, which one can study in:

Kaye, R. *Models of Peano arithmetic*. Oxford: Oxford University Press 1991.

The next paper is a highly readable survey on applications of ultraproducts, though most of Eklof's examples don't really need ultraproducts:

Eklof, P. C. Ultraproducts for algebraists. In *Handbook of mathematical logic*, ed. K. J. Barwise, pp. 105–137. Amsterdam: North-Holland 1977.

Reduced products and the related 'boolean products' are of interest in universal algebra:

Burris, S. and Sankappanavar, H. P. *A course in universal algebra*. New York: Springer-Verlag 1981.

To see how model theory forms a natural setting for many questions on the structure of modules, a good reference is:

Prest, M. *Model theory and modules*. Cambridge: Cambridge University Press 1988.

9

Structure and categoricity

William Byrd, *Non vos relinquam.*

Some of the best introductory courses in model theory set it as their goal to reach Morley's theorem on uncountably categorical theories. The statement of the theorem is no big deal; the value lies in two other things. These are, first, the techniques which one develops in order to prove the theorem, and second, the elegant structure theory which emerges from the proof.

The body of the proof is in section 9.5. I have noted all the places where the proof rests on some prior theory. Most of these prerequisites are gathered in the first four sections of the chapter. There is probably too much in them for a beginning model theory course, but I hope an instructor will not find it hard to make a selection. Certainly a first course can't go deep into structure theory; some relevant books are listed at the end of the chapter.

The passage of Byrd's motet quoted above is – with allowances for artistry – an indiscernible sequence of length four. Stretched to ω, it represents the legend that Jesus ascended into heaven. Sir Michael Tippett put a similar indiscernible sequence of length three at the climax of the closing spiritual in his oratorio *A child of our time*, set to the words 'Walk into heaven'.

9.1 Ehrenfeucht–Mostowski models

Andrzej Ehrenfeucht and Andrzej Mostowski had the ingenious idea of building structures around linearly ordered sets, so that properties of the linearly ordered sets would control the properties of the resulting structures. For readers familiar with categories, the natural way to describe the Ehrenfeucht–Mostowski construction is as a functor from the category of linear orderings to the category of L-structures for some signature L. But the discussion below will not assume anything from category theory.

Definition of EM functors

A linearly ordered set, or a **linear ordering** as we shall say for brevity, is an ordered pair $(\eta, <^{\eta})$ where η is a set and $<^{\eta}$ is an irreflexive linear ordering of η. Often we refer to the pair simply as η. We use variables η, ζ, ξ to stand for linear orderings. As in section 5.6, we write $[\eta]^k$ for the set of all increasing k-tuples from η (i.e. all finite sequences $(a_0, \ldots, a_{k-1})$ of elements of η such that $a_0 <^{\eta} \ldots <^{\eta} a_{k-1}$). An **embedding** of the linear ordering η in the linear ordering ξ is a map f from η to ξ such that $x <^{\eta} y$ implies $fx <^{\xi} fy$.

We shall say that the structure A **contains** the linear ordering η if every element of η is an element of A; there need not be any other connection between the ordering relation $<^{\eta}$ and the structure A.

Let L be a language (for example a first-order language, or one of the form $L_{\infty\omega}$). An **Ehrenfeucht–Mostowski functor** in L, or **EM functor** for short, is defined to be a function F which takes each linear ordering η to an L-structure $F(\eta)$ so that the following three axioms are satisfied.

(1.1)   For each linear ordering η, the structure $F(\eta)$ contains η as a set of generators.

(1.2)   For each embedding $f: \eta \to \xi$ there is an embedding $F(f): F(\eta) \to F(\xi)$ which extends f.

(1.3) F is functorial; i.e. for all embeddings $f: \eta \to \xi$ and $g: \xi \to \zeta$, $F(gf) = F(g) \cdot F(f)$, and for every linear ordering η, $F(1_{\eta}) = 1_{F(\eta)}$.

For example, if f is an automorphism of the linear ordering η then $F(f): F(\eta) \to F(\eta)$ is an automorphism extending f. The automorphism group of η is embedded in the automorphism group of $F(\eta)$.

By (1.1), $F(\eta)$ contains η. We call η the **spine** of $F(\eta)$. Since the spine generates $F(\eta)$, every element of $F(\eta)$ is of the form $t^{F(\eta)}(\bar{a})$ for some term $t(x_0, \ldots, x_{k-1})$ of L and some increasing tuple $\bar{a} \in [\eta]^k$.

The two central properties of EM functors are known in the trade as

sliding (i.e. we can slide elements up and down the spine without noticing) and **stretching** (i.e. we can pull out the spine into as long an ordering as we like). Theorems 9.1.1 and 9.1.4 will make this precise.

Theorem 9.1.1 (Sliding). *Let F be an EM functor in L, and let $\bar{a}$, $\bar{b}$ be increasing k-tuples from linear orderings η, ξ respectively. Then for every quantifier-free formula $\phi(x_0, \ldots, x_{k-1})$ of L we have $F(\eta) \models \phi(\bar{a}) \Leftrightarrow F(\xi) \models \phi(\bar{b})$.*

Proof. Find a linear ordering ζ and embeddings $f\colon \eta \to \zeta$ and $g\colon \xi \to \zeta$ such that $f\bar{a} = g\bar{b}$. Consider the diagram

$$(1.4) \qquad\qquad F(\eta) \xrightarrow{F(f)} F(\zeta) \xleftarrow{F(g)} F(\xi).$$

Assuming that $F(\eta) \models \phi(\bar{a})$ and recalling that embeddings preserve quantifier-free formulas (Theorem 2.4.1), we have $F(\zeta) \models \phi(f\bar{a})$ by $F(f)$. So $F(\zeta) \models \phi(g\bar{b})$, and hence $F(\xi) \models \phi(\bar{b})$ by $F(g)$. $\qquad\square$

In section 5.6 we defined 'ϕ-indiscernible sequence'.

Corollary 9.1.2. *If F is an EM functor and ζ an ordering, then ζ is a ϕ-indiscernible sequence in $F(\zeta)$ for every quantifier-free formula ϕ.* $\qquad\square$

Suppose A is an L-structure and η is a linear ordering contained in A. We define the **theory** of η in A, $\mathrm{Th}(A, \eta)$, to be the set of all first-order formulas $\phi(x_0, \ldots, x_{n-1})$ of L such that $A \models \phi(\bar{a})$ for every increasing n-tuple $\bar{a}$ from η. (If formulas of $L_{\omega_1\omega}$ were allowed, we would write $\mathrm{Th}_{\omega_1\omega}(A, \eta)$, and so on.) The **theory** of the EM functor F in L, $\mathrm{Th}(F)$, is defined to be the set of all first-order formulas $\phi(x_0, \ldots, x_{n-1})$ of L such that for *every* linear ordering η and every increasing n-tuple $\bar{a}$ from η, $F(\eta) \models \phi(\bar{a})$.

Since every first-order formula has just finitely many free variables, Theorem 9.1.1 tells us that $\mathrm{Th}(F)$ contains exactly the same quantifier-free formulas as $\mathrm{Th}(F(\eta), \eta)$ for any infinite linear ordering η. We can say more.

Lemma 9.1.3. *Let F be an EM functor in the first-order language L; suppose η is an infinite linear ordering and ϕ is an $\forall_1$ sentence of L which is true in $F(\eta)$. Then $\phi \in \mathrm{Th}(F)$.*

Proof. Suppose ϕ is $\forall \bar{x}\, \psi(\bar{x})$ with ψ quantifier-free. Let ζ be any linear ordering and $\bar{a}$ a tuple of elements of $F(\zeta)$; we must show $F(\zeta) \models \psi(\bar{a})$. Since ζ generates $F(\zeta)$, there is some finite subordering ζ_0 of ζ with $\bar{a}$ in $F(\zeta_0)$. Since η is infinite, there is an embedding $f\colon \zeta_0 \to \eta$. By assumption $F(f)\bar{a}$ satisfies ψ in $F(\eta)$, so $F(\zeta_0) \models \psi(\bar{a})$ since ψ is quantifier-free; then $F(\zeta) \models \psi(\bar{a})$ likewise. $\qquad\square$

Lemma 9.1.3 implies that much of $\text{Th}(F)$ is recoverable from $\text{Th}(F(\omega), \omega)$. But in fact the whole of F is recoverable up to isomorphism from $F(\omega)$.

Theorem 9.1.4 (Stretching). *Let L be a signature; let A be any L-structure containing the linear ordering ω as a set of generators. If ω is a ϕ-indiscernible sequence in A for all atomic formulas ϕ of L, then there is an EM functor F in L such that $A = F(\omega)$. This functor F is unique up to natural isomorphism of functors (i.e. if G is any other EM functor with this property, then for each linear ordering η there is an isomorphism $i_\eta: F(\eta) \to G(\eta)$ which is the identity on η).*

Proof. To construct F, take any ordering η and write $L(\eta)$ for L with the elements of η added as new constants. We shall define a set $S(\eta)$ of atomic sentences of $L(\eta)$. Let ϕ be any atomic sentence of $L(\eta)$. Then ϕ can be written as $\psi(\bar{c})$ for some atomic formula $\psi(\bar{x})$ of L and some increasing tuple $\bar{c}$ from η. We put ϕ into $S(\eta)$ if $\psi(\bar{x}) \in \text{Th}(A, \omega)$. The choice of ψ here is not unique (there could be redundant variables in $\bar{x}$), but an easy sliding argument shows that the definition is sound.

We claim that $S(\eta)$ is $=$-closed in $L(\eta)$ (see section 1.5). Clearly $S(\eta)$ contains $t = t$ for each closed term t, since $x_0 = x_0 \in \text{Th}(A, \omega)$. Suppose $S(\eta)$ contains both $\psi(s(\bar{c}), \bar{c})$ and $s(\bar{c}) = t(\bar{c})$, where $\psi(s(\bar{x}), \bar{x})$ is an atomic formula of L and $\bar{c}$ is increasing in η. Then for every increasing tuple $\bar{a}$ from ω, $A \vDash \psi(s(\bar{a}), \bar{a}) \wedge s(\bar{a}) = t(\bar{a})$, so that $\psi(t(\bar{x}), \bar{x}) \in \text{Th}(A, \omega)$ and hence $\psi(t(\bar{c}), \bar{c}) \in S(\eta)$. This proves the claim.

Now define $F(\eta)$ to be the L-reduct of the canonical model of $S(\eta)$. Since $x_0 = x_1 \notin \text{Th}(A, \omega)$, the elements $a^{F(\eta)}$ with a in η are pairwise distinct, and hence we can identify each $a^{F(\eta)}$ with a. Then $F(\eta)$ contains η as a set of generators. Let $f: \eta \to \xi$ be an embedding of linear orderings. Then for each atomic formula $\psi(\bar{x})$ of L and each increasing tuple $\bar{a}$ from η,

(1.5) $\qquad F(\eta) \vDash \psi(\bar{a}) \Leftrightarrow \psi(\bar{x}) \in \text{Th}(A, \omega) \Leftrightarrow F(\xi) \vDash \psi(f\bar{a}).$

It follows by the diagram lemma (Lemma 1.4.2) that we can define an embedding $F(f): F(\eta) \to F(\xi)$ by putting $F(f)(t^{F(\eta)}\bar{a}) = t^{F(\xi)}f\bar{a}$, for each term $t(\bar{x})$ of L and each increasing tuple $\bar{a}$ from η. This definition satisfies (1.3), so that F is an EM functor.

We constructed F so that $\text{Th}(F)$ agrees with $\text{Th}(A, \omega)$ in all atomic formulas of L. Let G be any other EM functor with this property. Then for every linear ordering η, every atomic formula $\psi(\bar{x})$ of L and every increasing tuple $\bar{a}$ from η, $F(\eta) \vDash \psi(\bar{a})$ if and only if $G(\eta) \vDash \psi(\bar{a})$. Since η generates both $F(\eta)$ and $G(\eta)$, it follows that we can define an isomorphism $i_\eta: F(\eta) \to G(\eta)$ by putting $i_\eta(t^{F(\eta)}\bar{a}) = t^{G(\eta)}\bar{a}$. Taking t to be x_0, i_η is the identity on η.

By the same argument, $F(\omega)$ can be identified with A. $\qquad \square$

Finding Ehrenfeucht–Mostowski models

If T is a theory, an **Ehrenfeucht–Mostowski model** of T is a structure of the form $F(\eta)$ which is a model of T, where F is a EM functor. (In practice, reducts of $F(\eta)$ are also known as Ehrenfeucht–Mostowski models.) How do we find Ehrenfeucht–Mostowski models of a given first-order theory?

We do it in two steps: first skolemise, then use Ramsey's theorem. The next two lemmas give the details.

Lemma 9.1.5. *Let F be an EM functor and suppose $\mathrm{Th}(F(\omega))$ is a Skolem theory. Then for every first-order formula $\phi(\bar{x})$, either ϕ or $\neg\phi$ is in $\mathrm{Th}(F)$. In particular all the structures $F(\eta)$ are elementarily equivalent, and in each structure $F(\eta)$, η is an indiscernible sequence.*

Proof. A Skolem theory is axiomatised by a set of $\forall_1$ sentences, and modulo the theory, every formula is equivalent to a quantifier-free formula. Now quote Lemma 9.1.3 and Theorem 9.1.1. □

Theorem 9.1.6 (Ehrenfeucht–Mostowski theorem). *Let L be a first-order language and A an L-structure such that $\mathrm{Th}(A)$ is a Skolem theory. Suppose A contains an infinite linear ordering η. (The ordering relation $<^\eta$ need not have anything to do with A.) Then there is an EM functor F in L whose theory contains $\mathrm{Th}(A, \eta)$.*

Proof. Let $\bar{c}$ be a sequence $(c_i: i < \omega)$ of pairwise distinct constants not in L, and write $L(\bar{c})$ for the language got by adding the constants c_i to L. Let T be the following set of sentences of $L(\bar{c})$:

(1.6) $\phi(\bar{a}) \leftrightarrow \phi(\bar{b})$ for each first-order formula $\phi(x_0, \ldots, x_{k-1})$ of L and all $\bar{a}, \bar{b} \in [\bar{c}]^k$;

(1.7) $\phi(c_0, \ldots, c_{k-1})$ for each formula $\phi(x_0, \ldots, x_{k-1})$ $\in \mathrm{Th}(A, \eta)$.

We claim that T has a model.

The claim follows by the compactness theorem if we show that every finite subset of T has a model. Let U be a finite subset of T. The formulas $\phi(\bar{x})$ in (1.6), (1.7) which occur in U can be listed as $\phi_0, \ldots, \phi_{m-1}$ for some finite m, and for some finite k the new constants which occur in U are all among $c_0, \ldots, c_{k-1}$. By adding redundant variables at the right-hand end, we can write each of the formulas ϕ_i as $\phi(x_0, \ldots, x_{k-1})$. Now if $\bar{a}, \bar{b} \in [\eta]^k$, write $\bar{a} \sim \bar{b}$ if

(1.8) $A \vDash \phi_j(\bar{a}) \Leftrightarrow A \vDash \phi_j(\bar{b})$ for every $j < m$.

Then $\sim$ is an equivalence relation on $[\eta]^k$ with a finite number of equivalence classes. Hence by Ramsey's theorem (Theorem 5.6.1) there is an increasing sequence $\bar{e} = (e_j : j < 2k)$ in η such that any two increasing k-tuples from $\bar{e}$ lie in the same equivalence class of $\sim$. Interpreting each c_j as e_j ($j < k$), we make A into a model of U. (We chose $\bar{e}$ of length $2k$ to allow space for any redundant variables in the formulas ϕ_i.) The claim is proved.

Now let B be any model of T. Since the formula $x_0 \neq x_1$ is in $\mathrm{Th}(A, \eta)$, the elements c_i^B are pairwise distinct. Hence we can identify each c_i^B with the number i, so that B contains ω. Let $B|L$ be the L-reduct of B and let C be the substructure of $B|L$ generated by ω. By (1.7), $\mathrm{Th}(A, \eta) \subseteq \mathrm{Th}(B|L, \omega)$. In particular $\mathrm{Th}(B|L)$ is a Skolem theory, so that $C \preccurlyeq B|L$. It follows that $\mathrm{Th}(A, \eta) \subseteq \mathrm{Th}(C, \omega)$. By (1.6), ω is an indiscernible sequence in C. The theorem follows by Theorem 9.1.4 and Lemma 9.1.5. $\qquad\square$

Theorem 9.1.6 tells us that every theory with infinite models has Ehrenfeucht–Mostowski models with spines of any order types we care to choose – though we may have to skolemise the theory before we construct the models.

Features of Ehrenfeucht–Mostowski models

Ehrenfeucht–Mostowski models of a theory T form the 'thinnest possible' models of T – a kind of opposite to saturated models. The next result is one way of making this precise.

Theorem 9.1.7. *Let L be a first-order language, T a Skolem theory in L and A an Ehrenfeucht–Mostowski model of T.*

(a) *For every $n < \omega$, the number of complete types $\in S_n(T)$ which are realised in A is at most $|L|$.*

(b) *If the spine of A is well-ordered and X is a set of elements of A, then the number of complete 1-types over X which are realised in A is at most $|L| + |X|$.*

Proof. Let F be an EM functor, and suppose $A = F(\eta)$.

(a) Taking a fixed $n < \omega$, let $\bar{a} = (a_0, \dots, a_{n-1})$ be an n-tuple of elements of A. Since η generates A, for each $i < n$ we can choose a term $t_i(\bar{y}_i)$ of L and an increasing tuple $\bar{b}_i$ of elements of η, such that a_i is $t_i^A(\bar{b}_i)$. By adding redundant variables to the terms t_i, we can suppose without loss that the tuples $\bar{b}_i$ are all equal; write them $\bar{b}$.

Suppose now that $\bar{b}'$ is an increasing tuple of elements of η, of the same length as $\bar{b}$. Write $\bar{a}' = (a_0', \dots, a_{n-1}')$ for the n-tuple where each a_i' is $t_i^A(\bar{b}')$. Let $\phi(x_0, \dots, x_{n-1})$ be any formula of L. By the indiscernibility of η we have

(1.9) $\qquad A \vDash \phi(t_0(\bar{b}), \dots, t_{n-1}(\bar{b})) \Leftrightarrow A \vDash \phi(t_0(\bar{b}'), \dots, t_{n-1}(\bar{b}')).$

So $A \vDash \phi(\bar{a}) \Leftrightarrow A \vDash \phi(\bar{a}')$. It follows that the complete type of $\bar{a}$ is determined once we know the terms $t_0, \ldots, t_{n-1}$. But there are only $|L|$ ways of choosing these terms.

(b) Assume η is a cardinal κ. Let X be any set of elements of $F(\kappa)$. For each element a of X we can choose a representation of the form $t_a^{F(\kappa)}(\bar{b}_a)$ where $t_a(\bar{x})$ is a term of L and $\bar{b}_a$ is an increasing tuple from κ. Let W be the smallest subset of κ such that all $\bar{b}_a$ ($a \in X$) lie in W. Then $|W| \leqslant |X| + \omega$. Let $s(\bar{y})$ be any term of L. By indiscernibility, for each increasing tuple $\bar{c}$ from κ the type of the element $s^{F(\kappa)}(\bar{c})$ over X is completely determined by the positions of the elements of $\bar{c}$ relative to the elements of W in κ. Since κ is well-ordered, there are at most $|W| + \omega$ ways that $\bar{c}$ can lie relative to W. So the elements $s^{F(\kappa)}(\bar{c})$ with $\bar{c}$ increasing in κ account for at most $|W| + \omega$ complete types over X. There are at most $|L|$ terms $s(\bar{y})$, yielding a total of $|W| + |L| = |X| + |L|$ types of elements over X. $\square$

Exercises for section 9.1

1. Give an example to show that if F is an EM functor and $f: \omega \to \omega$ an order-preserving map, $F(f): F(\omega) \to F(\omega)$ need not preserve $\forall_1$ first-order formulas.

2. Show that if T is any first-order theory with infinite models and G is a group of automorphisms of a linear ordering η, then there is a model A of T which contains η, such that G is the restriction to η of a subgroup of $\text{Aut}(A)$. In particular show that T has a model on which the automorphism group of the ordering $(\mathbb{Q}, <)$ of the rationals acts faithfully.

3. Let F be an EM functor in the first-order language L, with skolemised theory. Show that if η is any infinite linear ordering and X is any set which is first-order definable in $F(\eta)$ without parameters, then X has cardinality either $|\eta|$ or $\leqslant |L|$.

4. Let L and L^+ be first-order languages with $L \subseteq L^+$, and suppose every symbol in L^+ but not in L is a relation symbol. Let F be an EM functor in L and T an $\forall_1$ theory in L^+ which is consistent with $\text{Th}(F(\omega))$. Show that there is an EM functor F^+ in L^+ such that for each linear ordering η, $F^+(\eta)$ is a model of T and $F(\eta) = F^+(\eta)|L$.

5. Let L be a first-order language containing a 2-ary relation symbol $<$, and A an L-structure such that $<^A$ linearly orders the elements of A in order-type κ for some infinite cardinal κ. Writing η for the ordering $(\text{dom } A, <^A)$, show that $\text{Th}(A, \eta)$ contains the following formulas: (i) '$<$ linearly orders the universe' and $x_0 < x_1$; (ii) for each term $t(x_0, \ldots, x_{n-1})$ of L, the formula $t(x_0, \ldots, x_{n-1}) < x_n$; (iii) for each term $t(x_0, \ldots, x_{n-1})$ of L and each $i < n$, the formula $t(x_0, \ldots, x_{n-1}) \leqslant x_i \to t(x_0, \ldots, x_{n-1}) = t(x_0, \ldots, x_i, x_n, x_{n+1}, \ldots, x_{2n-i-2})$.

6. Let L be a first-order language containing a 2-ary relation symbol $<$, and F an EM functor for L which contains all the formulas (i)–(iii) of the preceding exercise. Show that if λ is any infinite cardinal, then in $F(\lambda)$ there are no $<$-descending sequences of length $|L|^+$, the spine is cofinal and no element α of the spine has more than $|\alpha| + |L|$ predecessors in the $<$-ordering.

9.2 Minimal sets

For the remainder of this chapter we shall assume that a complete theory T is given, in a first-order language L. We shall be interested in building up models of T. It will be helpful to take Michelangelo's perspective and imagine that we are carving out our models inside a large model M which is already given. For this, let M be a μ-big model of T, where μ is some conveniently large infinite cardinal. There is such a model M by Theorem 8.2.1. When we say 'model' we shall normally mean 'elementary substructure of M'. If we wanted to consider a model B of T which is not elementarily embeddable in M, we could quietly use Theorem 8.2.1 again to replace M by an elementary extension M' of M which is $|B|^+$-big and hence $|B|^+$-universal; B is elementarily embeddable in M' for sure. So there is no real loss of generality in working with a μ-big model in this way. We call M simply **the big model**.

If A is a model, we can think of A as a subset of the domain of M; the relations and functions of A are just the restrictions of those of M. Thus if $\phi(x)$ is a formula of L, we have $\phi(A) = A \cap \phi(M)$.

As in section 2.1 above, we say that a definable set X of elements of A is **minimal** if X is infinite but for every formula $\phi(x)$ of L, maybe with parameters from A, one of the sets $X \cap \phi(A)$ and $X \backslash \phi(A)$ is finite. We say that the structure A itself is a **minimal structure** if $\mathrm{dom}(A)$ is a minimal set.

In section 2.1 we lacked the equipment to find any interesting examples. But the quantifier elimination of section 2.7 puts several in our hands.

Example 1: *Algebraically closed fields.* In an algebraically closed field A, every formula $\phi(x)$ of L is equivalent to a boolean combination of polynomial equations $p(x) = 0$, where p has coefficients from A. (See Theorem 2.7.3.) The solution set of $p(x)$ is either the whole of A (when p is the zero polynomial) or a finite set (otherwise). It follows at once that A is minimal.

Example 2: *Vector spaces.* Let V be an infinite vector space over a field K. Every formula $\phi(x)$ of L is equivalent to a boolean combination of linear equations $rx = a$ with r in K and a in V. (See Exercise 2.7.9 above or Example 1 in section 7.4.) The set of vectors satisfying the equation $rx = a$ is either the whole of V or a singleton or the empty set. Thus again A is minimal.

Example 3: *Affine spaces.* Once again let V be an infinite vector space over a finite field K. This time we put a different structure on V, to create a structure A. The elements of A are the vectors in V. For each field element α we introduce a 2-ary function symbol F_α with the following interpretation:

$$(2.1) \qquad\qquad F_\alpha(a, b) = \alpha a + (1 - \alpha)b.$$

We also introduce a 3-ary function symbol G with the interpretation

$$(2.2) \qquad\qquad G(a, b, c) = a - b + c.$$

The signature of A consists of these symbols F_α ($\alpha \in K$) and G. In particular there is no symbol for 0. A structure A formed in this way is called an **affine space**. Its substructures are the cosets (under the additive group of V) of the subspaces of V; these substructures are known as the **affine flats**. The structure A is certainly minimal, since each of its definable relations is already a definable relation in the vector space V.

We say that a formula $\psi(x)$ of L, maybe with parameters in the structure A, is **minimal** (for A) if $\psi(A)$ is minimal. We say that ψ and $\psi(A)$ are **strongly minimal** (for A) if for every elementary extension B of A, $\psi(B)$ is minimal in B. In particular the structure A itself is **strongly minimal** if every elementary extension of A is a minimal structure. The classes of structures mentioned in Examples 1–3 above are all closed under taking elementary extensions, and so these structures are all strongly minimal.

Fact 9.2.1. *If $\psi(x)$ is a formula of L, then the following are equivalent.*
(a) *$\psi(A)$ is strongly minimal.*
(b) *For every structure B which is elementarily equivalent to A, $\psi(B)$ is minimal in B.*
(c) *$\psi(M)$ is minimal in M.*

Proof. (b) implies (a) by the definitions, and (a) implies (c) since we can regard A as an elementary substructure of M. To complete the circle it suffices to show that if M is ω-saturated and $\psi(M)$ is minimal in M, then $\psi(M)$ is strongly minimal in M. (We know that M is ω-saturated since it is big.) Suppose to the contrary that B is an elementary extension of M and there is a formula $\phi(x, \bar{b})$ with parameters $\bar{b}$ in B, such that $\psi(B) \cap \phi(B, \bar{b})$ and $\psi(B) \backslash \phi(B, \bar{b})$ are both infinite. (Here we are temporarily dropping the assumption that every model is an elementary substructure of M.) Let $\bar{a}$ be the parameters of ψ. Then since M is ω-saturated, there is a tuple $\bar{c}$ in M such that $(M, \bar{a}, \bar{c}) \equiv (B, \bar{a}, \bar{b})$. This implies that both $\psi(M) \cap \phi(M, \bar{c})$ and $\psi(M) \backslash \phi(M, \bar{c})$ are infinite, contradicting the assumption that $\psi(M)$ is minimal in M. $\qquad\square$

Let us say that a formula ψ of L is **strongly minimal for** a theory T in L if ψ defines a strongly minimal set in every model of T. Then $\psi(A)$ is strongly minimal in A if and only if ψ is strongly minimal for $\text{Th}(A)$. As a matter of style, it can be helpful to hide the parameters used to define a strongly minimal set, by adding them to the language as new constants. I shall often do this without comment.

Algebraic dependence and minimal sets

In section 5.3 we met the notion of algebraic dependence. This notion is particularly well behaved in minimal sets. For example, we have the following result.

Lemma 9.2.2. *(Exchange lemma) Let X be a set of elements of A and Ω an X-definable minimal set in A. Suppose a is an element of A and b is an element of Ω. If $a \in \text{acl}(X \cup \{b\}) \backslash \text{acl}(X)$, then $b \in \text{acl}(X \cup \{a\})$.*

Proof. We deny this and aim for a contradiction. Adding the elements of X as parameters to L, we can replace X by $\varnothing$ in the lemma. Since $a \in \text{acl}(b)$, there are a formula $\phi(x, y)$ of L and a positive integer n such that

$$(2.3) \qquad A \vDash \phi(a, b) \wedge \exists_{=n} x\, \phi(x, b).$$

Hence, since Ω is minimal and $b \in \Omega \backslash \text{acl}(a)$, there is a finite subset Y of Ω such that

$$(2.4) \qquad \text{for all } b' \in \Omega \backslash Y,\ A \vDash \phi(a, b') \wedge \exists_{=n} x\, \phi(x, b').$$

Since $a \notin \text{acl}(\varnothing)$, there is an infinite set Z of elements a' in A such that

$$(2.5) \quad \text{for all but } |Y| \text{ elements } b' \text{ of } \Omega,\ A \vDash \phi(a', b') \wedge \exists_{=n} x\, \phi(x, b').$$

If $a_0, \ldots, a_n$ are distinct elements of Z, then by (2.5) there is some element b' of Ω such that $A \vDash \bigwedge_{i \leqslant n} \phi(a_i, b') \wedge \exists_{=n} x\, \phi(x, b')$; this is a contradiction. $\square$

We put this result together with the facts that we proved about algebraic closure in section 5.3.

Theorem 9.2.3. *Let Ω be a minimal set in A, and let U be a set of elements of A such that Ω is U-definable. Then for all subsets X and Y of Ω we have*

(2.6) $X \subseteq \text{acl}(X)$

(2.7) $X \subseteq \text{acl}(Y) \Rightarrow \text{acl}(X) \subseteq \text{acl}(Y),$

(2.8) $a \in \text{acl}(X) \Rightarrow a \in \text{acl}(Z)$ *for some finite* $Z \subseteq X,$

(2.9)   (Exchange law) *For any set $W \supseteq U$ of elements of A, and any two elements a, b of Ω, if a is algebraic over $W \cup \{b\}$ but not over W, then b is algebraic over $W \cup \{a\}$.*

Proof. (2.6) and (2.7) were in Exercise 5.3.6, (2.8) is trivial and (2.9) is from Lemma 9.2.2.   ☐

Together these four laws show that algebraic dependence in a minimal set Ω behaves very much like linear dependence in a vector space. As it happens, infinite vector spaces are a special case of Theorem 9.2.3, since in an infinite vector space the algebraic closure of a set X of vectors is exactly the subspace spanned by X.

This allows us to commit piracy on the language of linear algebra, and steal terms like 'independent' and 'basis' for use in any minimal set Ω. We call a subset Y of Ω **closed** if it contains every element of Ω which is dependent on it; this is the analogue of a subspace. The set Ω itself is closed. Every subset Z of Ω lies in a smallest closed set, $\Omega \cap \mathrm{acl}(Z)$; we call this set the **closure** of Z, and we say that Z **spans** this set. If Y is a closed set, then a **basis** of Y is a maximal independent set of elements of Y. The usual vector space arguments go over without any alteration at all, to give the following facts.

Theorem 9.2.4. *Let Ω be a $\varnothing$-definable minimal set in A.*

(a) *Every set $X \subseteq \Omega$ contains a maximal independent set W of elements, and any such set W is a basis of the closure of X.*

(b) *A subset W of a closed set X is a basis of X if and only if it is minimal with the property that $X \subseteq \mathrm{acl}(W)$.*

(c) *Any two bases of a closed set have the same cardinality.*

(d) *If X and Y are closed sets with $X \subseteq Y$, then any basis of X can be extended to a basis of Y.*   ☐

The **dimension** of a closed set X in Ω is defined to be the cardinality of any basis of X. This is well-defined by (c).

Example 1 *continued.* In an algebraically closed field, 'dependence' means algebraic dependence. A basis in our sense is the same as a transcendence basis, and the dimension is the transcendence degree over the prime field.

Example 3 *continued.* In an infinite affine space the closed sets are the affine flats. **Warning.** Geometers count the dimension of an affine flat differently from us. Their dimension, the **geometric dimension**, is one less than ours. For example an affine line has dimension 2 for us, because it needs two independent points to specify it. But the geometers give it geometric dimension one, because lines in vector spaces have dimension 1.

Automorphisms of minimal sets

Suppose that A and B are L-structures – and recall that L is first-order throughout this section. By an **elementary map** between A and B we mean a map $f: X \to \mathrm{dom}(B)$ where $X \subseteq \mathrm{dom}(A)$, such that for every tuple $\bar{a}$ of elements of X and every formula $\phi(\bar{x})$ of L, $A \vDash \phi(\bar{a})$ iff $B \vDash \phi(f\bar{a})$. (This includes the empty tuple: an empty map from A to B is elementary if and only if $A \equiv B$.) An **elementary bijection** $f: X \to Y$ between A and B is an elementary map f between A and B which is a bijection from X to Y.

The next two lemmas show how one can build up elementary maps between minimal sets.

Lemma 9.2.5. *Let $f: X \to Y$ be an elementary bijection between the L-structures A, B. Then f can be extended to an elementary bijection $g: \mathrm{acl}_A(X) \to \mathrm{acl}_B(Y)$.*

Proof. By Zorn's lemma there is a maximal elementary bijection $f': X' \to Y'$ with $X \subseteq X' \subseteq \mathrm{acl}_A(X)$ and extending f. Since every element of X' is algebraic over X, it's easily checked that $Y' \subseteq \mathrm{acl}_B(Y)$ too. We show first that $X' = \mathrm{acl}_A(X)$.

Suppose $X' \neq \mathrm{acl}_A(X)$, and let a be an element of $\mathrm{acl}_A(X) \backslash X'$. Since $a \in \mathrm{acl}_A(X')$, we can choose a formula $\phi(x)$ with parameters in X' so that $A \vDash \phi(a) \wedge \exists_{=n} x\, \phi(x)$ for some n; we choose ϕ to make n as small as possible. Since f' is elementary, $B \vDash \exists_{=n} x\, \phi(x)$, so we can find an element b in $\mathrm{acl}(Y')$ such that $B \vDash \phi(b)$. If $\bar{c}$ is any tuple of elements on X' and $\psi(x, \bar{z})$ any formula of L, we claim that

$$(2.10) \qquad A \vDash \psi(a, \bar{c}) \Rightarrow B \vDash \psi(b, f'\bar{c}).$$

For suppose $A \vDash \psi(a, \bar{d})$. Then by choice of n as minimal, $A \vDash \forall x(\phi(x) \to \psi(x, \bar{c}))$, so $B \vDash \forall x(\phi(x) \to \psi(x, f'\bar{c}))$ by applying f', and hence $B \vDash \psi(b, f'\bar{c})$ as required. The claim shows that we can extend f' by taking a to b, contradicting the maximality of X'.

Thus $X' = \mathrm{acl}_A(X)$, and so $\mathrm{acl}_A(X)$ is the domain of f'. We must still show that the image of f' contains $\mathrm{acl}_B(Y)$. But every element of $\mathrm{acl}_B(Y)$ is one of a finite set of elements satisfying some formula over Y, and the domain of f' must contain the same number of elements satisfying the corresponding formula over X. $\qquad \square$

Now let us suppose that A contains a minimal set Ω. After adding parameters we can assume that Ω is $\varnothing$-definable, and so without loss we can suppose that $\Omega = P^A$ for some 1-ary relation symbol $P(x)$ of L. An **independent sequence in Ω over** a set of elements X is a sequence $(a_i: i < \gamma)$ of distinct elements of Ω, such that for each $i < \gamma$, $a_i \notin \mathrm{acl}(X \cup \{a_j: j \neq i\})$. The sequence is simply **independent** if it is independent over the empty set.

Lemma 9.2.6. *Suppose A and B are elementarily equivalent L-structures, and P^A, P^B are minimal sets in A, B respectively. Let X, Y be sets of elements of A, B respectively, and $f: X \to Y$ an elementary bijection. Suppose $(a_i: i < \gamma)$ is an independent sequence in P^A over X, and $(b_i: i < \gamma)$ is an independent sequence in P^B over Y. Then f can be extended to an elementary map g which takes each a_i to b_i.*

Proof. Write g_i for the extension of f which takes a_j to b_j whenever $j < i$. We show by induction on i that each map g_i is elementary. First, g_0 is f, which is elementary by assumption. Next, if δ is a limit ordinal and g_i is elementary for each $i < \delta$, then g_δ is elementary too, since first-order formulas mention only finitely many elements.

Finally suppose $i = k + 1$ and g_k is elementary. Let $\phi(\bar{c}, \bar{d}, x)$ be a formula of L with $\bar{c}$ in X and $\bar{d}$ in $\{a_j: j < k\}$. Then one of $P^A \cap \phi(\bar{c}, \bar{d}, A)$ and $P^A \backslash \phi(\bar{c}, \bar{d}, A)$ is finite. But a_k is independent of X and $\bar{d}$, and so a_k satisfies whichever one of $\phi(\bar{c}, \bar{d}, x)$, $\neg \phi(\bar{c}, \bar{d}, x)$ picks out an infinite subset of P^A. If it be $\phi(\bar{c}, \bar{d}, x)$, then since g_k is elementary, $\phi(g_k\bar{c}, g_k\bar{d}, B)$ is infinite too, and so $B \vDash \phi(g_k\bar{c}, g_k\bar{d}, g_i a_k)$. This shows that g_i is elementary.   □

These two lemmas give us a way of building up automorphisms of a minimal set: we choose two maximal independent sets, use Lemma 9.2.6 to find an elementary bijection from one to the other, and then cover the whole minimal set by means of Lemma 9.2.5.

Thus the following grand old theorem of Steinitz is really a result in the model theory of minimal structures.

Corollary 9.2.7. *Let A and B be algebraically closed fields of the same characteristic and the same uncountable cardinality λ. Then A is isomorphic to B.*

Proof. Since they have the same characteristic, A and B satisfy the same quantifier-free sentences. Hence quantifier elimination (Theorem 2.7.3) tells us that A and B are elementarily equivalent. By Example 1, A and B are minimal structures. Let X and Y be bases of A and B respectively. Since the language is countable, the cardinality of $\text{acl}_A(X)$ is at most $|X| + \omega$; but A is uncountable and equal to $\text{acl}_A(X)$, so that $|X| = \lambda$. Similarly $|Y| = \lambda$. The empty map between A and B is elementary since $A \equiv B$; so by Lemma 9.2.6 there is an elementary bijection from X to Y. By Lemma 9.2.5 this map extends to an elementary map from A onto B, this is clearly an isomorphism.

          □

Corollary 9.2.7 suggests – correctly – that there should be some close connection between λ-categoricity for uncountable cardinals λ and minimal sets. But the reader should meditate on the following example before making any rash conjectures.

Example 4: A structure which is λ-categorical for every infinite cardinal λ but not minimal. Let κ be an infinite cardinal and p a prime, and let A be the abelian group $\oplus_{i<\kappa}\mathbb{Z}(p^2)b_i$, i.e. the direct sum of κ cyclic groups $\mathbb{Z}(p^2)b_i$ of order p^2. Then A carries a strongly minimal set Ω, namely the socle (the set of elements a such that $pa = 0$). It is not hard to see that A is λ-categorical for every infinite λ. (We considered the case $\lambda = \omega$ in Example 1 of section 6.3.) But the elements of A are not algebraic over Ω. To see this, take any element c of the socle, and let α_c be the automorphism of A which carries b_0 to $b_0 + c$ and b_i ($i \neq 0$) to b_i. Then α_c fixes the socle pointwise, but it can carry b_0 to infinitely many different elements according to the choice of c. The only other $\varnothing$-definable minimal set in A is the socle less 0.

One moral that we might draw from this example is that instead of cutting a set by just one formula (as we do in the definition of minimal sets), we should consider repeated cuts by different formulas. This thought leads directly to Morley rank, which we shall consider in the next section.

Exercises for section 9.2

1. Show that a minimal set remains minimal if we add parameters; likewise with 'strongly minimal' for 'minimal'.

2. Give an example to show that the notion 'algebraic over' need not obey the exchange law. More precisely, find a structure A with elements a, b such that a is algebraic over $\{b\}$ but not over the empty set, while b is not algebraic over $\{a\}$.

The next exercise can be handled tediously by the method of quantifier elimination, or swiftly by Lindström's test (Theorem 7.3.4) and Theorem 7.4.1.
3. Let G be a group. We define a **G-set** to be an L-structure A as follows. The signature is a family of 1-ary functions $(F_g: g \in G)$; the laws $\forall x\, F_g F_h(x) = F_{gh}(x)$, $\forall x\, F_1(x) = x$ hold in A. We say A is a **faithful** G-set if for all $g \neq h$ in G and all elements a, $F_g^A(a) \neq F_h^A(a)$. (a) Show that the class of faithful G-sets is first-order axiomatisable, in fact by $\forall_1$ sentences. (b) Show that if A is an infinite faithful G-set, then A decomposes in a natural way into a set of connected components, and each component is isomorphic to a Cayley graph of the group G (see Exercise 4.1.1 above). (c) Deduce that every infinite faithful G-set is strongly minimal, and its dimension is the number of components.

4. Let A be an L-structure, Ω a $\varnothing$-definable minimal set in A, $(a_i: i < \gamma)$ a sequence of elements of Ω, X a set of elements of A and $\bar{b}$ a sequence listing the elements of X. Show that the following are equivalent: (a) $(a_i: i < \gamma)$ is independent over X; (b) for each $i < \gamma$, $a_i \notin \operatorname{acl}(X \cup \{a_j: j < i\})$; (c) $a_i \neq a_j$ whenever $i < j$, and $\{a_i: i < \gamma\}$ is an independent set in $(A, \bar{b})$.

5. Suppose $\phi(x)$ is a formula of L and $\phi(A)$ is a minimal set in A with dimension κ. If $|L| \leq \lambda \leq \kappa$, show that A has an elementary substructure B in which $\phi(B)$ has dimension λ.

6. Show that if $\psi(A)$ is a minimal set in A, then $\psi(A)$ is strongly minimal if and only if A has an elementary extension B in which $\psi(B)$ is minimal and of infinite dimension. In particular, every minimal structure of infinite dimension is strongly minimal.

7. Show that if A is an L-structure and P^A is a minimal set of uncountable cardinality, then A is strongly minimal. [If L is countable, use the previous exercise. If L is uncountable, apply that exercise to each countable sublanguage.]

8. Show that the structure $(\omega, <)$ is minimal but not strongly minimal.

9. Let A be a structure (Ω, E) where E is an equivalence relation whose equivalence classes are all finite, and E has just one class of cardinality n for each positive integer n. Show that A is minimal but not strongly minimal.

9.3 Totally transcendental structures

Throughout this section, L is a first-order language, T is a complete theory in L with infinite models, and M is a big model of T, as explained at the beginning of section 9.2. 'Model' will mean 'model of T', and the letters A, B etc. will range over models. Unless we say otherwise, all models A are assumed to be elementary substructures of M. We identify A with $\operatorname{dom} A$ as a subset of $\operatorname{dom} M$. Thus if A and B are models, '$A \subseteq B$' means that A is an elementary substructure of B (since both are elementary substructures of M). 'Definable' means 'definable in L with parameters from M'.

We are about to set up a dimension theory for definable relations. The dimension – or as we shall say, the Morley rank – of a relation will be a measure of the number of times we can chop the relation into separate non-empty definable pieces. Readers with some knowledge of algebraic geometry might like to know that when M is an algebraically closed field and the relation $X \subseteq M^n$ is an irreducible algebraic set, then the Morley rank of X turns out to be exactly its Krull dimension. This is not at all obvious from the definition of Morley rank, which is purely model-theoretic.

For any formula $\psi(\bar{x})$ of L with parameters from the big model M, the

Morley rank $RM(\psi)$ of the formula ψ is either -1 or an ordinal or ∞, and is defined as follows (where $\bar{x}$ is an n-tuple of variables).

(3.1) $RM(\psi) \geq 0$ iff $\psi(M^n)$ is not empty.

(3.2) $RM(\psi) \geq \alpha + 1$ iff there are formulas $\psi_i(\bar{x})$ $(i < \omega)$ of L with parameters from M, such that the sets $\psi(M^n) \cap \psi_i(M^n)$ $(i < \omega)$ are pairwise disjoint and $RM(\psi \wedge \psi_i) \geq \alpha$ for each $i < \omega$.

(3.3) $RM(\psi) \geq \delta$ (limit) iff for all $\alpha < \delta$, $RM(\psi) \geq \alpha$.

This determines $RM(\psi)$ uniquely, if we regard -1 as coming before all ordinals and ∞ as coming after all ordinals.

Lemma 9.3.1. *Let* $\phi(\bar{x})$ *be a formula of L with parameters from M. Then the following are equivalent.*
(a) $RM(\phi) = \infty$.
(b) *For every finite sequence $\bar{s}$ of 0's and 1's, there is a formula $\psi_{\bar{s}}(\bar{x})$ of L with parameters from M, such that*
  (i) *for each $\bar{s}$, $M \vDash \exists \bar{x}\, \psi_{\bar{s}}(\bar{x})$,*
  (ii) *for each $\bar{s}$, $M \vDash \forall \bar{x}(\psi_{\bar{s}\frown 0} \rightarrow \psi_{\bar{s}}) \wedge \forall \bar{x}(\psi_{\bar{s}\frown 1} \rightarrow \psi_{\bar{s}})$,*
  (iii) *for each $\bar{s}$, $M \vDash \neg \exists \bar{x}(\psi_{\bar{s}\frown 0} \wedge \psi_{\bar{s}\frown 1})$,*
  (iv) *$\psi_{\langle\rangle}$ is ϕ.*
This can be illustrated thus:

(3.4)

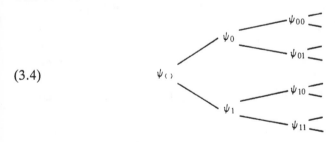

Proof. (a) $\Rightarrow$ (b). There is an ordinal α such that every formula of L with parameters in M has rank either $< \alpha$ or $= \infty$. So if $\psi_{\bar{s}}$ has been chosen and has rank ∞ by (3.2), there is an infinite set of formulas $\psi_{\bar{s}\frown i}(\bar{x})$ $(i < \omega)$ of L with parameters in M, such that the sets $\psi_{\bar{s}\frown i}(M^n)$ are pairwise disjoint subsets of $\psi_{\bar{s}}(M^n)$ and all have rank $\geq \alpha$, i.e. rank ∞. We get the tree in (b) by discarding the formulas $\psi_{\bar{s}\frown i}$ with $i \geq 2$.

(b) $\Rightarrow$ (a). Suppose there is a tree of formulas as in (b). We show by induction on the ordinal β that every formula $\psi_{\bar{s}}$ in the tree has Morley rank $\geq \beta$. For $\beta = 0$ this holds by (i), and at limit ordinals there is nothing to prove. Finally suppose $\beta = \gamma + 1$. Write ϕ_i for the formula $\psi_{\bar{s}}{}^\smallfrown 0 \ldots 01$ with i 0's after $\bar{s}$. By (ii) and (iii) the sets $\phi_i(M^n)$ are pairwise disjoint subsets of $\psi_{\bar{s}}(M^n)$, and by induction hypothesis each ϕ_i has rank $\geq \gamma$: so $\psi_{\bar{s}}$ has rank $\geq \beta$. □

The next lemma gathers together some fundamental properties of Morley rank.

Lemma 9.3.2. *Suppose $\phi(\bar{x})$, $\psi(\bar{x})$ are formulas of L with parameters in M, and $\chi(\bar{x}, \bar{y})$ is a formula of L without parameters.*

(a) *If $M \vDash \forall \bar{x}(\phi \to \psi)$ then $RM(\phi) \leq RM(\psi)$. More generally if $\psi'(\bar{x}')$ is a formula of L with parameters in M, and there is a formula of L with parameters in M which defines an injective map from $\phi(M^n)$ to $\psi'(M^m)$ for some $m < \omega$, then $RM(\phi) \leq RM(\psi')$.*

(b) *$RM(\phi \vee \psi) = \max(RM(\phi), RM(\psi))$.*

(c) *If $(M, \bar{a}) \equiv (M, \bar{b})$ then $RM(\chi(\bar{x}, \bar{a})) = RM(\chi(\bar{x}, \bar{b}))$.*

(d) *If A is an ω-saturated elementary substructure of M, and RM_A is Morley rank calculated in A instead of in M, then $RM_A(\phi) = RM(\phi)$ whenever the parameters of ϕ lie in A.*

Proof. (a) By induction on α, prove that if $RM(\phi) \geq \alpha$ then $RM(\psi) \geq \alpha$.

(b) We have $\geq$ by (a). For the converse we use induction on $\max(RM(\phi), RM(\psi))$. Suppose $RM(\phi \vee \psi) = \alpha$ but $RM(\phi)$, $RM(\psi)$ are both $\leq \beta < \alpha$. Then (putting $n = $ length of $\bar{x}$) there are infinitely many disjoint definable subsets X_i ($i < \omega$) of $(\phi \vee \psi)(M^n)$, each of Morley rank $\geq \beta$. Since $RM(\phi) \leq \beta$, there is $k_\phi < \omega$ such that $RM(X_i \cap \phi(M^n)) < \beta$ whenever $i \geq k_\phi$, and likewise there is $k_\psi < \omega$ such that $RM(X_i \cap \psi(M^n)) < \beta$ whenever $i \geq k_\psi$. Putting $k = \max(k_\phi, k_\psi)$, both $RM(X_k \cap \phi(M^n))$ and $RM(X_k \cap \psi(M^n))$ are $< \beta$ though X_k has Morley rank $\geq \beta$. This contradicts the induction hypothesis.

(c) Since M is big, it has an automorphism g which takes $\bar{a}$ to $\bar{b}$. Now show by induction on α that for any formula $\theta(\bar{x}, \bar{y})$ of L and any tuple $\bar{c}$ in M, $RM(\theta(\bar{x}, \bar{c})) \geq \alpha$ if and only if $RM(\theta(\bar{x}, g\bar{c})) \geq \alpha$.

(d) The harder direction is $\geq$. We go by induction on α, and the nontrivial case is where $\alpha = \beta + 1$. Suppose that $RM(\phi) \geq \alpha$. Then there are formulas $\phi_i(\bar{x}, \bar{b}_i)$ ($i < \omega$) of Morley rank $\geq \beta$, with parameters $\bar{b}_i$ in M, such that $\phi_i(M^n, \bar{b}_i)$ are pairwise disjoint subsets of $\phi(M^n)$. If $\bar{a}$ are the parameters of ϕ, which by assumption lie in A, then we can use the ω-saturation of A and induction on i to find tuples $\bar{c}_i$ in A so that

(3.5) $(A, \bar{a}, \bar{c}_0, \ldots, \bar{c}_i) \equiv (M, \bar{a}, \bar{b}_0, \ldots, \bar{b}_i).$

Then by (c) and induction hypothesis, the formulas $\phi_i(\bar{x}, \bar{c}_i)$ have Morley rank $\geq \beta$ and the relations $\phi_i(A^n, \bar{c}_i)$ are pairwise disjoint subsets of $\phi(A^n)$. Hence $\mathrm{RM}_A(\phi) \geq \alpha$. $\square$

This lemma allows us some new notation. First, if $\phi(\bar{x})$ and $\psi(\bar{x})$ are equivalent in M, then by (a) they have the same Morley rank. So the Morley rank really belongs to the defined relation rather than the formula, and we can write $\mathrm{RM}(X)$ for $\mathrm{RM}(\phi)$ when $X = \phi(M^n)$.

Second, if M_1 and M_2 are two big models of the same complete theory T, then by elementary amalgamation (Theorem 5.3.1) and Corollary 8.2.2, both M_1 and M_2 can be elementarily embedded in a third big model N of T. Part (d) of the lemma tells us that for any formula ϕ of L, $\mathrm{RM}(\phi)$ has the same value whether it is calculated in M_1, N or M_2, since each of these structures is ω-saturated.

Hence we can define the **Morley rank** of a complete theory T, $\mathrm{RM}(T)$, to be the Morley rank of the formula $x = x$ in any ω-saturated model of T. We say that T is **totally transcendental** if $\mathrm{RM}(T) < \infty$. And of course we say that T **has finite Morley rank** if $\mathrm{RM}(T)$ is finite. Since T is complete, it is determined by any one of its models A, and so we can say that A **has Morley rank** α, or is **totally transcendental**, when the same is true of T. There is no need for A to be ω-saturated here; but if it is not, then we should pass to an ω-saturated elementary extension of A before we try to calculate the Morley ranks of any formulas.

Morley degree

Morley rank is about splitting up into infinitely many pieces. How about splitting up into finitely many? There are two things to be said here. First, splitting up into finitely many pieces *infinitely often* is just as bad as splitting up into infinitely many. Lemma 9.3.1 told us so.

And second, there is a finite upper bound to the number of times we can split a formula with Morley rank $\alpha < \infty$ into formulas of the same Morley rank α. More precisely, we have the following.

Lemma 9.3.3. *Let $\phi(\bar{x})$ be a formula with parameters in M, of Morley rank $\alpha < \infty$. Then there is some greatest positive integer d such that $\phi(M^n)$ can be written as the union of disjoint sets $\phi_0(M^n), \ldots \phi_{d-1}(M^n)$ all of Morley rank α. This integer d is not altered if we replace M by any ω-saturated elementary substructure of M containing the parameters of ϕ.*

Proof. Suppose there is no such d. Then for every positive integer d, $\phi(M^n)$ can be written as a union of disjoint sets $\phi_{d,0}(M^n), \ldots, \phi_{d,d-1}(M^n)$ of

Morley rank α, forming a partition π_d of $\phi(M^n)$ into d sets. We shall adjust these formulas a little, by induction on d, starting at $d = 1$. For each $i \leqslant d$, the set $\phi_{d+1,i}(M^n)$ is partitioned into d sets $(\phi_{d+1,i} \wedge \phi_{d,j})(M^n)$ $(j < d)$, and by Lemma 9.3.2(b), at least one of these partition sets has Morley rank α. Using this and Lemma 9.3.2(a), we can rewrite the formulas $\phi_{d+1,j}$, still of Morley rank α, so that the partition π_{d+1} refines the partition π_d.

Suppose this adjustment has been made for all d. By König's tree lemma (Exercise 5.6.3) we can choose integers $d_0 < d_1 < \ldots$ and integers $i_0, i_1, \ldots$ so that, writing ψ_m for ϕ_{d_m, i_m}, each set $\psi_{m+1}(M^n)$ is a proper subset of $\psi_m(M^n)$. Then the sets $(\psi_m \wedge \neg \psi_{m+1})(M^n)$ $(m < \omega)$ are disjoint subsets of $\phi(M^n)$, all of Morley rank α, so that $\phi(M^n)$ has Morley rank $\geqslant \alpha + 1$; contradiction. This proves the existence of d.

Suppose A is an ω-saturated elementary substructure of M containing the parameters $\bar{a}$ of ϕ. By Lemma 9.3.2(d), any finite partition of $\phi(A^n)$ into pairwise disjoint sets $\phi_i(A^n)$ $(i < d')$ of Morley rank α lifts to a similar partition of $\phi(M^n)$; so $d' \leqslant d$. We must show that there is such a partition of $\phi(A^n)$ with $d' = d$. By assumption there are d pairwise disjoint subsets $\psi_i(M^n, \bar{b}_i)$ $(i < d)$ of $\phi(M^n)$, each of Morley rank α. Since A is ω-saturated, there are $a_0, \ldots, \bar{a}_{d-1}$ in A such that $(A, \bar{a}, \bar{a}_0, \ldots, \bar{a}_{d-1}) \equiv (M, \bar{a}, \bar{b}_0, \ldots, \bar{b}_{d-1})$. So by Lemma 9.3.2(c, d), the sets $\psi_i(A^n, \bar{a}_i)$ are pairwise disjoint subsets of $\phi(A^n)$ of Morley rank α. $\qquad \square$

The positive integer d as in Lemma 9.3.3 is called the **Morley degree** of ϕ (or of $\phi(M^n)$). (Algebraic geometers might guess that when M is an algebraically closed field and $\phi(M^n)$ is an algebraic set of dimension α, then its Morley degree is the number of irreducible components of dimension α. This is correct.)

We have already met the bottom levels of Morley's hierarchy in a different guise.

Theorem 9.3.4. *Let X be a non-empty definable set in M.*
(a) X is finite if and only if X has Morley rank 0.
(b) X is strongly minimal if and only if its Morley rank and degree are both equal to 1. In particular a strongly minimal structure has Morley rank 1.

Proof. (a) To say that X has Morley rank 0 is to say that X doesn't contain infinitely many pairwise disjoint non-empty definable subsets. Since singletons are definable, this is to say that X is finite.

(b) To say that X has Morley rank and degree 1 is to say that X is infinite (by (a)), but there is no definable set Y such that both $X \cap Y$ and $X \setminus Y$ are infinite. This is the definition of minimality. By Fact 9.2.1, minimal sets in M are strongly minimal. $\qquad \square$

From one to several variables

Lemma 9.3.2(c) tells us that if M is totally transcendental, then every formula $\phi(x)$ with one free variable has Morley rank $< \infty$. What about formulas with several free variables? The next lemma answers this question.

Lemma 9.3.5 (*Erimbetov's inequality*). *Let* $\phi(\bar{x}, \bar{y})$ *be a formula of L with parameters in M. Let α and β be ordinals such that*
(a) $\mathrm{RM}(\exists \bar{y}\, \phi(\bar{x}, \bar{y})) = \alpha$,
(b) *for every tuple $\bar{a}$ in M,* $\mathrm{RM}(\phi(\bar{a}, \bar{y})) \leqslant \beta$.
Then $\mathrm{RM}(\phi(\bar{x}, \bar{y}))$ *and* $\mathrm{RM}(\exists \bar{x}\, \phi(\bar{x}, \bar{y}))$ *are both* $\leqslant \beta \cdot (\alpha + 1)$ *when* $\beta \neq 0$, *and* $= \alpha$ *when* $\beta = 0$.

Proof. I omit the proof for $\exists \bar{x}\, \phi(\bar{x}, \bar{y})$. Assuming it, we prove the result for $\phi(\bar{x}, \bar{y})$ as follows. Let $\chi(\bar{x}, \bar{y}, \bar{z})$ be the formula $\phi(\bar{x}, \bar{y}) \wedge \bar{x} = \bar{z}$. The inequality tells us that if $\mathrm{RM}(\exists \bar{x}\bar{y}\, \chi(\bar{x}, \bar{y}, \bar{z})) = \alpha$ and for every tuple $\bar{a}$ in M, $\mathrm{RM}(\chi(\bar{x}, \bar{y}, \bar{a})) \leqslant \beta$, then

$$\mathrm{RM}(\exists \bar{z}\, \chi(\bar{x}, \bar{y}, \bar{z})) \leqslant \beta \cdot (\alpha + 1) \text{ (or } = \alpha \text{ when } \beta = 0).$$

But $\exists \bar{x}\bar{y}\, \chi(\bar{x}, \bar{y}, \bar{z})$ and $\exists \bar{z}\, \chi(\bar{x}, \bar{y}, \bar{z})$ are logically equivalent to $\exists \bar{y}\, \phi(\bar{z}, \bar{y})$ and $\phi(\bar{x}, \bar{y})$ respectively, while $\chi(\bar{x}, \bar{y}, \bar{a})$ and $\phi(\bar{a}, \bar{y})$ have the same Morley rank by Lemma 9.3.2(a). $\qquad \square$

We can apply this lemma to any formula ϕ with more than one free variable, by splitting up the variables into two groups $\bar{x}$ and $\bar{y}$. The formulas $\exists \bar{y}\, \phi$ and $\phi(\bar{a}, \bar{y})$ both have fewer free variables than ϕ. So by induction and simple arithmetic we have the following useful consequences.

Lemma 9.3.6. *Let* $\phi(\bar{x})$ *be any formula of L with parameters in M.*
(a) *If M has Morley rank α and $\bar{x}$ has length n, then ϕ has Morley rank at most $(\alpha + 1)^n$.*
(b) *If M is totally transcendental then ϕ has Morley rank $< \infty$.*
(c) *If M has finite Morley rank, then so does ϕ.* $\qquad \square$

Warning. It makes no sense to speak of the Morley rank of a formula ϕ unless we specify the variables of ϕ. By writing $x = x$ as $\phi(x, y)$, we add an extra degree of freedom and so raise the Morley rank.

One further corollary of Erimbetov's inequality will help us in section 9.5. Recall from section 9.1 that a formula $\phi(x)$ is **one-cardinal** for the theory T if in every model A of T, $\phi(A)$ has the same cardinality as A.

Corollary 9.3.7. *Suppose the formula $\phi(x)$ of L is one-cardinal for T.*
(a) *If ϕ has finite Morley rank then so does T.*
(b) *If ϕ has Morley rank $< \infty$ then so does T.*

Proof. We prove (a); the proof of (b) is much the same.

Since ϕ is a one-cardinal formula, Theorem 8.4.2 tells us that there is a layering $\theta(x)$ by ϕ such that $T \vdash \forall x\, \theta$. Referring to (4.3) in section 8.4, write $\theta_i(x, y_0, z_0, \ldots, y_{i-1}, z_{i-1})$ for θ with the first i lines removed. Thus θ_0 is θ. We claim that for every $i \leq n$, there is $m_i < \omega$ such that for all $b_0, \ldots, b_{i-1}$ in $\phi(M)$ and all $c_0, \ldots, c_{i-1}$ in M, the Morley rank of $\theta_i(x, b_0, c_0, \ldots, b_{i-1}, c_{i-1})$ is at most m_i.

The claim is proved by downwards induction from n to 0. The formula $\theta_n(x, b_0, \ldots, c_{n-1})$ is simply $x = c_0 \vee \ldots \vee x = c_{n-1}$, which has Morley rank ≤ 0. Assuming the claim holds for $i + 1$, we consider the formula $\psi(x, y_i, z_i)$, viz.

$$(3.6) \qquad (\exists z_i\, \eta_i(b_0, \ldots, c_{i-1}, y_i, z_i) \to \eta_i(b_0, \ldots, c_{i-1}, y_i, z_i))$$

$$\to \theta_{i+1}(x, b_0, \ldots, c_{i-1}, y_i, z_i),$$

noting that $\theta_i(x, b_0, \ldots, c_{i-1})$ is $(\exists y_i \in \phi)\forall z_i\, \psi(x, y_i, z_i)$. Let b be any element of $\phi(M)$. We shall put a bound on the Morley rank of $\forall z_i\, \psi(x, b, z_i)$. Take any element c of M, such that $M \vDash \eta_i(b_0, \ldots, c_{i-1}, b, c)$ if possible. Then $\psi(M, b, c)$ is $\theta_{i+1}(M, b_0, \ldots, c_{i-1}, b, c)$, which has Morley rank $\leq m_{i+1}$ by induction hypothesis. Hence

$$(3.7) \qquad\qquad \forall z_i\, \psi(x, b, z_i) \text{ has Morley rank } \leq m_{i+1}.$$

Now $\exists x(\forall z_i\, \psi(x, y, z_i) \wedge \phi(y))$ has finite Morley rank since ϕ has. From this and (3.7) it follows at once by Erimbetov's inequality (Lemma 9.3.5) that there is a uniform finite upper bound m_i on the rank of $\exists y(\forall z_i\, \psi(x, y, z_i) \wedge \phi(y))$, which is equivalent to $\theta_i(x, b_0, \ldots, c_{i-1})$. This proves the claim.

By the claim, $\theta_0(x)$ has finite Morley rank. But since $M \vDash \forall x\, \theta$, θ_0 is equivalent to $x = x$. $\qquad\square$

Complete types

Let X be a set of elements of M. The notion of a **complete n-type over X with respect to M** is defined in section 5.2. If $p(\bar{x})$ is a complete n-type over X with respect to M, then every formula $\phi(\bar{x})$ in p is a formula of L with parameters in M, and so it has a Morley rank (possibly ∞, but definitely not -1). We define the **Morley rank** of p, $\mathrm{RM}(p)$, to be the least value of $\mathrm{RM}(\phi)$ as ϕ ranges over the formulas in p.

Lemma 9.3.8. *Let $\Phi(\bar{x})$ be a set of formulas of L with parameters all lying in a set X of elements of M.*

(a) Let α be an ordinal such that for every finite subset Ψ of Φ, $\mathrm{RM}(\bigwedge \Psi) \geq \alpha$. Then there is a complete type $p(\bar{x})$ over X with respect to M, which has Morley rank $\geq \alpha$ and contains Φ.

(b) If Φ contains a formula of Morley rank α and Morley degree d, then

there are at most d complete types $p(\bar{x})$ over X which contain Φ and have Morley rank α.

Proof. (a) Let $\Phi'(\bar{x})$ be the set of all those formulas $\phi(\bar{x})$ of L with parameters from X, with the property that for some finite subset Ψ of Φ, $RM(\bigwedge\Psi \wedge \neg\phi) < \alpha$. Certainly every formula of Φ is in Φ', since contradictions have Morley rank -1.

We claim that for every finite subset Ψ' of Φ', $M \vDash \exists\bar{x}\bigwedge\Psi'$. If ϕ is in Ψ', then there is a finite set Ψ_ϕ in Φ such that $RM(\bigwedge\Psi_\phi \wedge \neg\phi) < \alpha$. Taking Ψ to be the union of the sets Ψ_ϕ with ϕ in Ψ', we have $RM(\bigwedge\Psi \wedge \neg\phi) < \alpha$ for each ψ in Ψ', by Lemma 9.3.2(a), and hence $RM(\bigwedge\Psi \wedge \neg\bigwedge\Psi') < \alpha$ by Lemma 9.3.2(b). But $RM(\bigwedge\Psi) \geqslant \alpha$, so $RM(\bigwedge\Psi \wedge \bigwedge\Psi') \geqslant \alpha$ by Lemma 9.3.2(b) again, and hence $RM(\bigwedge\Psi') \geqslant \alpha \geqslant 0$. The claim is proved.

So by Theorem 5.2.1, Ψ' can be extended to a complete type $p(\bar{x})$ over X with respect to M. This type p contains Ψ. If ϕ is a formula of Morley rank $< \alpha$ with parameters from X, then $\neg\phi$ lies in Φ' and hence in Ψ, so that ϕ is not in p (since types are consistent!). Thus p has Morley rank $\geqslant \alpha$.

(b) Suppose Φ contains the formula $\phi(\bar{x})$ of Morley rank α and Morley degree d. If $p_0, \ldots, p_d$ are distinct complete types over X which contain Φ and have Morley rank α, then there are formulas $\psi_0(\bar{x}), \ldots, \psi_d(\bar{x})$ in $p_0, \ldots, p_d$ respectively such that for all $i < j \leqslant d$, $\psi_i(M^n)$ is disjoint from $\psi_j(M^n)$. Then the formulas $\phi \wedge \psi_i$ define pairwise disjoint sets of Morley rank α, so that ϕ has Morley degree at least $d + 1$; contradiction. $\square$

Note that by Exercise 5.2.1, if A is an elementary substructure of M and X is a subset of A, then a complete type over X with respect to M is the same thing as a complete type over X with respect to A. As a special case of Lemma 9.3.8, suppose A is an L-structure and X, Y are sets of elements of A with $X \subseteq Y$. Then every complete type $p(\bar{x})$ over X can be extended to a complete type $q(\bar{x})$ over Y with the same Morley rank. We can ensure $RM(q) \geqslant RM(p)$ by the lemma, and $RM(q) \leqslant RM(p)$ since $p \subseteq q$.

Example 1: *The regular types of strongly minimal sets.* Let A be an L-structure and $\phi(x)$ a strongly minimal formula of L. By Theorem 9.3.4 and Lemma 9.3.8 there is a complete type $p(x)$ over $\varnothing$ which contains the formula $\phi(x)$ and has Morley rank 1. In fact p is uniquely determined: if $\psi(x)$ is any formula of L, then exactly one of $(\phi \wedge \psi)(A), (\phi \wedge \neg\psi)(A)$ is infinite. Say it is $\phi \wedge \psi$; then $\phi \wedge \neg\psi$ can't be in p, so ψ must be in p. For reasons that lie in stability theory, p is called the **regular type** of the strongly minimal set $\phi(A)$.

For example if A is an algebraically closed field, then its unique regular type is the type over $\varnothing$ of a transcendental element.

Suppose $p(\bar{x})$ is a complete type over a set X with respect to an L-structure A. If p has Morley rank α, its **Morley degree** is the least integer d such that p contains a formula $\phi(\bar{x})$ of Morley rank α and Morley degree d. The type p is the unique complete type over X which has rank α and contains ϕ. For suppose $q(\bar{x})$ is another such type, and $\psi(\bar{x})$ is a formula which is in p but not in q. Then both $\phi \wedge \psi$ and $\phi \wedge \neg \psi$ have Morley rank α, and so at least one of them must have lower Morley degree than ϕ, contradicting the choice of d. The Morley degree of a complete type may be any positive integer; see Exercise 6.

Clearly the regular type of a strongly minimal set has Morley degree 1.

Exercises for section 9.3

1. Let $\phi(\bar{x})$ be a formula of L with parameters in M. (a) Show that if $\bar{b}$ is any sequence of elements of M, and N is $(M, \bar{b})$, then ϕ has the same Morley rank calculated in N as it has in M. (You can assume that N is big, since if M is λ-big for some λ greater than the length of $\bar{b}$, then N is also λ-big.) (b) Give an example to show that if $L' \subseteq L$ and ϕ is a formula of L', then the Morley rank of ϕ calculated in $M|L'$ need not be the same as in M.

2. Write out a full proof of Lemma 9.3.2(a, c).

3. Let A be an L-structure and $\phi(\bar{x})$ a formula of L with parameters $\bar{a}$ in A. Show that there is a set T of sentences of L with parameters $\bar{a}$, such that the following are equivalent. (a) There is a set of formulas $\psi_s(\bar{x})$ with parameters from some elementary extension B of A, such that the properties (i)–(iv) of Lemma 9.3.1(b) hold with B for A. (b) A is a model of T. *This allows a more direct proof that if A is totally transcendental, then so is any structure elementarily equivalent to A.*

4. Let K and L be first-order languages, A a K-structure and B an L-structure. Suppose B is interpretable in A. Show that if A has Morley rank α, then the Morley rank of B is at most $(\alpha + 1)^n$ for some $n < \omega$. [Use Corollary 9.3.6(a).]

5. Let A be an L-structure, X a set of elements of A and $\bar{a}, \bar{b}$ tuples of elements of A. Show that if $\mathrm{tp}_A(\bar{a}/X)$ has Morley rank α and $\bar{b}$ is algebraic over X and $\bar{a}$, then $\mathrm{tp}_A(\bar{b}/X)$ has Morley rank $\leq \alpha$.

6. Give examples to show that for every positive integer d there are a structure A and a complete type $p(x)$ over $\varnothing$ with respect to A, such that p has Morley rank 1 and Morley rank d. [Equivalence relation with d classes, all infinite.]

7. Let $p(x)$ be a complete type over a set X with respect to the L-structure A. Suppose $\phi(x)$ and $\psi(x)$ are formulas of L with parameters in X, and suppose $\phi \in p$. Let $\theta(x, y)$ be a formula with parameters in X, which defines in A a bijection from $\phi(A)$ to $\psi(A)$. Show that there is a complete type $q(y)$ over X, such that for every formula $\sigma(y)$ of L with parameters in X, $\sigma \in q$ if and only if $\exists y(\theta(x, y) \wedge \sigma(y))$ is in p. Show that q has the same Morley rank as p.

8. Let G be a group which is totally transcendental. Prove: (a) there are no infinite strictly decreasing chains of definable subgroups of G. [Consider cosets and use Lemma 9.3.1.] (b) Every intersection of a family of definable subgroups of G is equal to the intersection of a finite subfamily.

9.4 Stability

Saharon Shelah introduced the distinction between stable and unstable theories in 1969, as an aid to counting the number of non-isomorphic models of a theory in a given cardinality. A year or so later he set his sights on the *classification problem*. Roughly speaking, this problem is to classify all complete first-order theories into two kinds: those where there is a good structure theory for the class of all models, and those where there isn't. After writing a book and about a hundred papers on the subject, Shelah finally gave a solution of the classification problem in the case of countable languages in 1982. On any reckoning this is one of the major achievements of mathematical logic since Aristotle.

However, it is not our topic here. Whatever Shelah's own interests, the results of stability theory (in Shelah's own words) 'give, and are intended to give, instances for fine structure investigation with considerable tools to start with'. This section is an introduction to those considerable tools.

Let T be a complete theory in a first-order language L. Then T is **unstable** if there are a formula $\phi(\bar{x}, \bar{y})$ of L and a model A of T containing tuples of elements $\bar{a}_i$ $(i < \omega)$ such that

(4.1)   for all $i, j < \omega$, $A \vDash \phi(\bar{a}_i, \bar{a}_j) \Leftrightarrow i < j$.

T is **stable** if it is not unstable. We say that a structure A is **stable** or **unstable** according as $\mathrm{Th}(A)$ is stable or unstable.

From the definition it's immediate that every finite structure is stable. *For the rest of this section we assume that T is a complete first-order theory in a first-order language L, and all the models of T are infinite.*

By the definition, any structure which contains an infinite set linearly ordered by some formula is unstable. Note that the set itself need not be definable, but there must be a first-order formula (possibly with parameters – see Exercise 1) which orders it.

Example 1: *Infinite boolean algebras.* Every infinite boolean algebra B contains an infinite set of elements which is linearly ordered by the algebra ordering '$x < y$'. So $\text{Th}(B)$ is unstable. In general the linearly ordered set will not be first-order definable.

Theorem 9.4.1. *Suppose A is stable.*

(a) *If $\bar{a}$ is a sequence of elements of A, then $(A, \bar{a})$ is stable.*

(b) *Suppose the structure B is interpretable in A; then B is stable.*

Proof. (a) is trivial. For (b), let Γ be an interpretation of the L-structure B in the K-structure A with coordinate map f, and suppose B is unstable. Then there are a formula $\phi(\bar{x}, \bar{y})$ of L and tuples $\bar{b}_i$ ($i < \omega$) in B such that for all $i, j < \omega$, $B \vDash \phi(\bar{b}_i, \bar{b}_j) \Leftrightarrow i < j$. For each i, choose $\bar{a}_i$ in A so that $f\bar{a}_i = \bar{b}_i$. Then by the reduction theorem (Theorem 4.2.1), $A \vDash \phi_\Gamma(\bar{a}_i, \bar{a}_j) \Leftrightarrow i < j$, and so A is unstable. $\qquad\square$

In particular every reduct of a stable structure is stable. It may be worth remarking here that stability is a robust property of theories – it is not affected by changing the language. Contrast this with the property of quantifier elimination, which comes and goes as we take definitional equivalents.

Sizes of Stone spaces

If M is a big model of T (cf. the beginnings of sections 9.2 and 9.3), then we can assume that a model A of T is an elementary substructure of M. So the set $S_n(X; A)$ of complete n-types over X with respect to A coincides with $S_n(X; M)$ (by Exercise 5.2.1). This allows us to assume that complete types are always with respect to M, and so we can simplify the notation by writing $S_n(X; A)$ as $S_n(X)$. Likewise we write $\text{tp}(\bar{a}/X)$ for $\text{tp}_M(\bar{a}/X)$.

To find examples of stable theories, we turn to the cardinalities of spaces of types (see section 5.2 above).

Let λ be an infinite cardinal. We say that T is λ-**stable** if for every model A of T and every set X of at most λ elements of A, $|S_1(X)| \leqslant \lambda$. We say that A is λ-**stable** if $\text{Th}(A)$ is λ-stable. A theory or structure which is not λ-stable is λ-**unstable**.

Theorem 9.4.2. *The following are equivalent:*

(a) *T is stable;*

(b) *For at least one infinite cardinal λ, T is λ-stable.*

The proof of (a) ⇒ (b) needs some preparation, and I postpone it to Theorem 9.4.14 below. The proof of (b) ⇒ (a) is not quite immediate either. It needs two lemmas. When P and Q are linear orderings and $P \subseteq Q$, we say that P is **dense in** Q if for any two elements b, c of Q with $b < c$ there is a in P such that $b < a < c$.

Lemma 9.4.3. *Let λ be any infinite cardinal. Then there are linear orderings P and Q with $P \subseteq Q$, such that $|P| \leq \lambda < |Q|$ and P is dense in Q.*

Proof. Let μ be the least cardinal such that $2^\mu > \lambda$; clearly $\mu \leq \lambda$. Consider the set $^\mu 2$ of all sequences of 0's and 1's of length μ; this set has cardinality $2^\mu > \lambda$. We can linearly order $^\mu 2$ as follows:

(4.2) $s < t \Leftrightarrow$ there is $i < \mu$ such that $s|i = t|i$ and $s(i) < t(i)$.

Let X be the set of all sequences in $^\mu 2$ which are 1 from some point onwards, and let Q be the linearly ordered set of all sequences in $^\mu 2$ which are not in X. Let P be the subset of Q consisting of those sequences which are 0 from some point onwards. By choice of μ, both X and P have cardinality $\Sigma_{\kappa < \mu} 2^\kappa \leq \lambda$, and so Q has cardinality $> \lambda$. Also one readily checks that P is dense in Q.   □ Lemma 9.4.3.

Lemma 9.4.4. *Let A be an L-structure, λ an infinite cardinal and X a set of at most λ elements of A. Suppose there is a positive integer n such that $|S_n(X)| > \lambda$. Then A is λ-unstable.*

Proof. We go by induction on n. When $n = 1$ there is nothing to prove. Suppose then that $n > 1$, and for each $r \in S_n(X)$ choose an $(n - 1)$-tuple $\bar{b}_r$ and an element c_r, both in the big model M, so that the n-tuple $\bar{b}_r \hat{\,} c_r$ realises r. Write q_r for the type $\mathrm{tp}(\bar{b}_r/X)$, which is a complete $(n - 1)$-type over X. There are two cases.

First, suppose the number of distinct types q_r is at most λ. Then since λ^+ is a regular cardinal, there must be a set $S \subseteq S_n(X)$ of cardinality λ^+, such that all the types q_r with $r \in S$ are equal. Choose some fixed $s \in S$, and for each $r \in S$ let q_r' be $\{\phi(\bar{b}_s, x) : \phi(x_0, \ldots, x_{n-1}) \in r\}$. Then it follows from Theorem 5.2.1 that the q_r' with $r \in S$ are λ^+ distinct 1-types over $X \cup \{\bar{b}_s\}$.

Second, suppose there are more than λ distinct types q_r. Then we have more than λ complete $(n - 1)$-types over X, and the conclusion holds by induction hypothesis.   □ Lemma 9.4.4

Now we prove (b) ⇒ (a) of the theorem. Assume T is unstable, so that there are a formula $\phi(\bar{x}, \bar{y})$ and a model of T as in (4.1). Replacing $\phi(\bar{x}, \bar{y})$ by $\phi(\bar{x}, \bar{y}) \wedge \neg \phi(\bar{y}, \bar{x})$ if necessary, we can suppose that ϕ defines an asymmetric relation on tuples in models of T. Let P and Q be as in Lemma 9.4.3. Take a new tuple of constants $\bar{c}_s$ for each $s \in P$, and consider the theory

$$(4.3) \qquad\qquad U = T \cup \{\phi(\bar{c}_s, \bar{c}_t): s < t \text{ in } P\}.$$

By the compactness theorem, U has a model B; we can choose B to be of cardinality $\leqslant \lambda$. Put $A = B|L$. For each element r of Q, consider the set of formulas $\Phi_r(\bar{x}) = \{\phi(\bar{c}_s, \bar{x}): s < r\} \cup \{\phi(\bar{x}, \bar{c}_t): r < t\}$. By Theorem 5.2.1, each set Φ_r is a type over $\text{dom}(A)$ with respect to A, and we can extend it to a complete type $p_r(\bar{x})$. If $r < r'$ in Q, then by density there is s in P such that $r < s < r'$, and it follows that $p_r \neq p_{r'}$. So there are more than λ complete types p_r with $r \in Q$.

We have proved that there are more than λ complete n-types over $\text{dom}(A)$, where n is the length of $\bar{x}$. By Lemma 9.4.4 this shows that T is λ-unstable. So (b) $\Rightarrow$ (a) of Theorem 9.4.2 is proved.   □ Theorem 9.4.2

Example 2: *Term algebras*. Let T be the complete theory of a term algebra of a language L. Then T is λ-stable whenever $\lambda^{|L|} = \lambda$, and hence T is stable. To prove this, we can suppose that T has infinite models. We use the language L_2 of Example 1 in section 2.6, and we assume the result of Theorem 2.7.5. By that theorem, if A is a model of T and $p(x)$ is a 1-type over $\text{dom}(A)$ with respect to A, then p is completely determined once we know which formulas of the following forms it contains: $\text{Is}_c(t(x))$, $\text{Is}_F(t(x))$, $s(x) = t(x)$, $t(x) = a$, where c, F are constant and function symbols of L, a is an element of A and s, t are terms built up from function symbols F_i of L_2. Note that for each $t(x)$, p contains at most one formula $t(x) = a$ with a in A. By this and a little cardinal arithmetic, the number of distinct types p is at most $|A|^{|L|}$. Let λ be any cardinal such that $\lambda^{|L|} = \lambda$ (for example any cardinal of the form $\mu^{|L|}$), and let A be any model of T of cardinality λ. Then there are at most $|A|$ complete 1-types over $\text{dom}(A)$. It follows that T is stable.

For totally transcendental theories T we can prove something stronger.

Theorem 9.4.5. *Let T be a complete first-order theory.*

(a) *If T is totally transcendental, then T is stable; in fact it is λ-stable for every $\lambda \geqslant |L|$.*

(b) *Conversely if T is ω-stable then T is totally transcendental.*

Proof. (a) Suppose T fails to be λ-stable for some $\lambda \geqslant |L|$. Then in the big model M there is a set X of λ elements such that $|S_1(X)| > \lambda$. Write $L(X)$ for L with the elements of X added as parameters. Let Ψ be the set of all those formulas $\psi(x)$ of $L(X)$ which lie in at least λ^+ of the types in $S_1(X)$, and let V be the set of all types in $S_1(X)$ which are subsets of Ψ. Then V contains all but at most λ of the types in $S_1(X)$. Hence every formula in Ψ lies in at least λ^+ types $\in V$. We claim

(4.4) if $\psi \in \Psi$ then there is χ such that $\psi \vee \chi$, $\psi \wedge \neg \chi$ are both in Ψ.

For let p, q be two distinct types $\in V$ which contain ψ. Since they are distinct, there is some formula χ such that $\psi \wedge \chi \in p$ and $\psi \wedge \neg \chi \in q$. This proves the claim.

Now it is straightforward to construct a tree of formulas as in (3.4) of Lemma 9.3.1, starting with the formula $x = x$. So by that lemma, T is not totally transcendental.

(b) Suppose T is not totally transcendental. Then by Lemma 9.3.1 there is a tree of formulas as in (3.4) of section 9.3 above (with $x = x$ as ϕ). There are countably many formulas in the tree, so between them they use at most countably many parameters from the big model M; let X be the set of these parameters. For each branch β of the tree, let p_β be some complete type over X which contains all the formulas in β; there is such a type of Theorem 5.2.1 and (i), (ii) of Lemma 9.3.1. By (iii) of that lemma, the types p_β are all distinct. But the tree has continuum many branches, so that T is not ω-stable. $\square$

Corollary 9.4.6. *If T is in a countable language L, and is μ-categorical for some uncountable cardinal μ, then T is totally transcendental and λ-stable for every $\lambda \geqslant \omega$.*

Proof. By Theorem 3.1.2 we can extend T to a Skolem theory T^Σ in a countable first-order language L^Σ extending L. By the Ehrenfeucht–Mostowski theorem (Theorem 9.1.6) there is an EF model B of T^Σ with spine μ. Let X be any countable set of elements of B. Then since μ is well-ordered, Theorem 9.1.7(b) told us that at most countably many complete 1-types over X are realised in B. Hence the same is true for the reduct $A = B|L$. It follows that T is ω-stable. For suppose not; let Y be a countable set of elements of the big model M such that $S_1(Y)$ is uncountable. By the downward Löwenheim–Skolem theorem there is an elementary substructure C of M with cardinality μ, which contains Y and elements realising uncountably many of the types in $S_1(Y)$. But μ-categoricity implies that C is isomorphic to A, and this contradicts what we have just proved about A. The rest is by Theorem 9.4.5. $\square$

Stable formulas

For the rest of this section we are headed for the proof that a stable theory is λ-stable for some λ. The proof is roundabout, and there are some interesting sights on the way. The first step is a slight adjustment of (4.1).

Let T be a complete theory in a first-order language L. Let $\phi(\bar{x}, \bar{y})$ be a formula of L, with the free variables divided into two groups $\bar{x}, \bar{y}$. An **n-ladder** for ϕ is a sequence $(\bar{a}_0, \ldots, \bar{a}_{n-1}, \bar{b}_0, \ldots, \bar{b}_{n-1})$ of tuples in some model A of T, such that

(4.5)   for all $i, j < n$, $A \vDash \phi(\bar{a}_i, \bar{b}_j) \Leftrightarrow i \leqslant j$.

We say that ϕ is a **stable** formula (for T, or for A) if there is some $n < \omega$ such that no n-ladder for ϕ exists; otherwise it is **unstable**. The least such n is the **ladder index** of $\phi(\bar{x}, \bar{y})$ (of course it may depend on the way we split the variables).

Lemma 9.4.7. *The theory T is unstable if and only if there is an unstable formula in L for T.*

Proof. Suppose first that T has an unstable formula $\phi(\bar{x}, \bar{y})$. Thus ϕ has an n-ladder for each $n < \omega$. Taking new tuples of constants $\bar{a}_i$, $\bar{b}_i$ $(i < \omega)$, consider the theory

(4.6) $T \cup \{\phi(\bar{a}_i, \bar{b}_j): i \leqslant j < \omega\} \cup \{\neg \phi(\bar{a}_i, \bar{b}_j): j < i < \omega\}$.

Each finite subset of (4.6) has a model – take an n-ladder for some large enough n. So by the compactness theorem there is a model A of T in which there are tuples $\bar{a}_i$, $\bar{b}_i$ $(i < \omega)$ obeying the conditions in (4.5). Put $\bar{c}_i = \bar{a}_i \bar{b}_i$ for each $i < \omega$, and let $\psi(\bar{x}\bar{x}', \bar{y}\bar{y}')$ be the formula $\neg \phi(\bar{y}, \bar{x}')$. Then

(4.7) $A \vDash \psi(\bar{c}_i, \bar{c}_j) \Leftrightarrow A \vDash \neg\phi(\bar{a}_j, \bar{b}_i) \Leftrightarrow i < j$,

so that ψ linearly orders $(\bar{c}_i: i < \omega)$. Thus T is unstable. The converse is an exercise.   □

Example 3: *Modules.* We show that every module is stable. Let A be a left R-module and L the language of left R-modules. By Lemma 9.4.7, the Baur–Monk quantifier elimination theorem (Exercise 2.7.12) and a small piece of combinatorics (see Exercise 7 below), it suffices to show that if $\phi(\bar{x}, \bar{y})$ is a p.p. formula of L then ϕ is stable. In fact we shall prove something stronger: ϕ has ladder index $\leqslant 2$. Since $\phi(\bar{x}, \bar{y})$ expresses that some finite set of linear equations in $\bar{x}$, $\bar{y}$ have a simultaneous solution, we can read it as

(4.8)   there is $\bar{z}$ such that $K\bar{x} + M\bar{y} + N\bar{z} = 0$,

where K, M, N are matrices over R, 0 is a column vector of zeroes, and $\bar{x}$, $\bar{y}$, $\bar{z}$ are read as column vectors.

Suppose $A \models \phi(\bar{a}_i, \bar{b}_j)$ whenever $i \leqslant j < 2$. Then by (4.8) there are tuples $\bar{c}_{00}, \bar{c}_{01}, \bar{c}_{11}$ in A such that

(4.9) $$K\bar{a}_0 + M\bar{b}_0 + N\bar{c}_{00} = 0,$$

(4.10) $$K\bar{a}_0 + M\bar{b}_1 + N\bar{c}_{01} = 0,$$

(4.11) $$K\bar{a}_1 + M\bar{b}_1 + N\bar{c}_{11} = 0.$$

Adding (4.9) to (4.11) and subtracting (4.10), we get

(4.12) $$K\bar{a}_1 + M\bar{b}_0 + N(\bar{c}_{00} - \bar{c}_{01} + \bar{c}_{11}) = 0,$$

proving that $A \models \phi(\bar{a}_1, \bar{a}_0)$.

Using Lemma 9.4.7 in the other direction, stability implies a rather strong chain condition on groups. We say that a structure A is **group-like** if some relativised reduct of A is a group; the relativised reduct that we have in mind is called the **group of** A. By convention a **stable group** is a stable group-like structure.

Lemma 9.4.8 *(Baldwin–Saxl lemma).* *Let L be a first-order language and A an L-structure which is a stable group; let G be the group of A. Let $\phi(x, \bar{y})$ be a formula of L. Write* **S** *for the set of all subsets of* $\mathrm{dom}(A)$ *which are subgroups of G of the form $\phi(A, \bar{b})$ with $\bar{b}$ in A; write* $\bigcap$**S** *for the set of all intersections of groups in* **S**.

(a) *There is $n < \omega$ such that every group in* $\bigcap$**S** *can be written as the intersection of at most n groups in* **S**.

(b) *There is $m < \omega$ such that no chain in* $\bigcap$**S** *(under inclusion) has length $> m$.*

Proof. Let n be the ladder index of ϕ. Suppose $n < k < \omega$ and there are subgroups $H_i = \phi(A, \bar{b}_i)$ $(i < k)$ such that no H_i contains all of $\bigcap_{j \neq i} H_j$. For each i choose $h_i \in \bigcap_{j \neq i} H_j \backslash H_i$, and write $a_0 = 1$, $a_{i+1} = h_0 \cdot \ldots \cdot h_i$. Then $A \models \phi(a_i, \bar{b}_j) \Leftrightarrow i \leqslant j$. This contradicts the choice of n. It follows that

(4.13) every intersection of finitely many groups in **S** is already an
        intersection of at most n of them.

Now let $\bigcap_f$**S** be the set of all intersections of finitely many groups in **S**. By (4.13) we can write each group in $\bigcap_f$**S** as a set of the form $\psi(A, \bar{c})$ where $\psi(x, \bar{z})$ is $\phi(x, \bar{y}_0) \wedge \ldots \wedge \phi(x, \bar{y}_{n-1})$. Since A is stable, ψ has a ladder index too; let it be m. Suppose there is a strictly ascending chain of subgroups of G,

(4.14) $$\psi(A, \bar{c}_0) \subset \ldots \subset \psi(A, \bar{c}_m).$$

Let a_0 be in $\psi(A, \bar{c}_0)$, and for each positive $i \leqslant m$ choose $a_i \in \psi(G, \bar{c}_i) \backslash \psi(G, \bar{c}_{i-1})$. Then $A \models \psi(a_i, \bar{c}_j) \Leftrightarrow i \leqslant j$, contradicting the choice of m. We have proved that every chain in $\bigcap_f$**S** has length at most m.

But now it follows that $\bigcap_f \mathbf{S} = \bigcap \mathbf{S}$. For otherwise we could find subgroups $\phi(A, \bar{d}_i)$ ($i < \omega$) such that the intersections $\bigcap_{j<i} \phi(A, \bar{d}_j)$ form a strictly decreasing sequence, contradicting the descending chain condition on $\bigcap_f \mathbf{S}$. Both (a) and (b) follow at once. $\square$

The **centraliser** $C_G(X)$ of a set of elements X in G is the subgroup of all elements g of G which commute with everything in X. Then $C_G(X) = \bigcap_{g \in X} C_G(g)$, and $C_G(g) = \phi(A, g)$ where $\phi(x, y)$ is the formula of L which expresses that $x \cdot y = y \cdot x$. So the Baldwin–Saxl lemma implies at once that *a stable group satisfies the descending chain condition on centralisers.*

Definability of types

Suppose A is a stable L-structure, X is a set of elements of A, and $\phi(y_0, \ldots, y_{n-1})$ is a formula of L with parameters from A. We write $\phi(X^n)$ for the set of all n-tuples $\bar{a}$ of elements of X such that $A \vDash \phi(\bar{a})$. Maybe ϕ has parameters lying outside X. But according to the next result, stability will allow us to define $\phi(X^n)$ using only parameters that come from X. This is a very counterintuitive fact, and it shows how strong the assumption of stability is.

Theorem 9.4.9. *Let A be a stable L-structure and $\phi(\bar{x}, \bar{y})$ a stable formula for A, with $\bar{y} = (y_0, \ldots, y_{n-1})$. Let X be a set of elements of A and $\bar{b}$ a tuple in A. Then there is a formula $\chi(\bar{y})$ of L with parameters in X, such that $\phi(\bar{b}, X^n) = \chi(X^n)$.*

Let us interpret this result for a moment. Write $p(\bar{x})$ for the type $\mathrm{tp}_A(\bar{b}/X)$. Since $\phi(\bar{b}, X^n)$ depends only on ϕ and $p(\bar{x})$, and not on the particular choice of tuple realising p, we can write the formula χ as $\mathrm{d}_p \phi(\bar{y})$. Then we can rephrase the equation at the end of the theorem as follows:

(4.15)   for every $\bar{c}$ in X, $\phi(\bar{x}, \bar{c}) \in p \Leftrightarrow A \vDash \mathrm{d}_p \phi(\bar{c})$.

A formula $\mathrm{d}_p \phi$ with the property of (4.15) is called a ϕ-**definition** of the type p.

The proof of the theorem depends on some combinatorial facts about ladders and trees.

We write $^n2$ for the set of sequences of length n whose terms are either 0 or 1, and $^{<n}2$ for $\bigcup_{j<n} {}^j2$. We write σ, τ for sequences; $\sigma | j$ is the initial segment of σ of length j. An n-**tree** for the formula $\phi(\bar{x}, \bar{y})$ is defined to consist of two families of tuples, $(\bar{b}_\sigma : \sigma \in {}^n2)$ and $(\bar{c}_\tau : \tau \in {}^{<n}2)$ such that for all $\sigma \in {}^n2$ and all $i < n$,

(4.16) $M \vDash \phi(\bar{b}_\sigma, \bar{c}_{\sigma|i}) \Leftrightarrow \sigma(i) = 0$.

The tuples $\bar{b}_\sigma$ are called the **branches** of the n-tree and the tuples $\bar{c}_\tau$ are called its **nodes**.

We say that a formula $\phi(\bar{x}, \bar{y})$ of L has **branching index** $\geq n$, in symbols $\mathrm{BI}(\phi) \geq n$, if there exists an n-tree for ϕ. This defines $\mathrm{BI}(\phi)$ uniquely as a natural number or ∞. The statement '$\mathrm{BI}(\phi) \geq n$' can be written as a sentence of L.

Lemma 9.4.10. *Let $\phi(\bar{x}, \bar{y})$ be a formula of L. If ϕ has branching index n, then ϕ has ladder index $< 2^{n+1}$. If ϕ has ladder index n, then ϕ has branching index $< 2^{n+2} - 2$.*

Proof. Typographical reasons force us to write $\bar{b}[i]$ for $\bar{b}_i$. For the first implication it is enough to note that if $\bar{b}[0], \ldots, \bar{b}[2^{n+1} - 1], \bar{c}[0], \ldots, \bar{c}[2^{n+1} - 1]$ form a 2^{n+1}-ladder for ϕ, then they can be turned into an $(n + 1)$-tree by taking the $\bar{b}[i]$'s as branches and the $\bar{c}[j]$'s as nodes, relabelling in an obvious way.

We prove the second implication by showing that if ϕ has branching index at least $2^{n+2} - 2$, then ϕ has ladder index at least $n + 1$; for ease of reading let us say rather that if ϕ has branching index at least $2^{n+1} - 2$, then ϕ has ladder index at least n.

8f1If H is an $(n + 1)$-tree for ϕ and $i = 0$ or 1, we write $H_{(i)}$ for the n-tree whose nodes and branches are the nodes $\bar{c}[\tau]$ and branches $\bar{b}[\sigma]$ of H such that $\tau(0) = \sigma(0) = i$. We shall say that a map $f: {}^{<n}2 \to {}^{<m}2$ is a **tree-map** if for any two sequences σ, τ in ${}^{<n}2$, $f(\sigma)$ is an end-extension of $f(\tau)$ precisely when σ is an end-extension of τ. If H is an m-tree and N is a set of nodes of H, we say that N **contains an n-tree** if there is a tree-map $f: {}^{<n}2 \to {}^{<m}2$ such that for each $\tau \in {}^{<n}2$, $\bar{c}[f(\tau)]$ is in N. Clearly this implies that there is an n-tree J for ϕ whose nodes all lie in N and whose branches are branches of H; we shall say also that N **contains the n-tree** J.

We claim the following. Consider $n, k \geq 0$ and let H be an $(n + k)$-tree for ϕ. If the nodes of H are partitioned into two sets N, P, then either N contains an n-tree or P contains a k-tree.

The claim is proved by induction on $n + k$. The case $n = k = 0$ is trivial. Suppose $n + k > 0$, and let the tuples $\bar{c}[\tau]$ be the nodes of H. Suppose $\bar{c}[\langle \rangle] \in N$. (The argument when $\bar{c}[\langle \rangle] \in P$ is parallel.) For $i = 0, 1$ let Z_i be the set of all nodes of $H_{(i)}$. By induction hypothesis, if $i = 0$ or 1 then either $N \cap Z_i$ contains an $(n - 1)$-tree or $P \cap Z_i$ contains a k-tree. If at least one of $P \cap Z_0$, $P \cap Z_1$ contains a k-tree, then so does P. If neither does, then both $N \cap Z_0$ and $N \cap Z_1$ contain $(n - 1)$-trees, and so, since $\bar{c}[\langle \rangle] \in N$, N contains an n-tree. This proves the claim.

To complete the proof of the lemma, assume that ϕ has branching index at least $2^{n+1} - 2$. We shall show, by induction on $n - r$, that for $1 \leqslant r \leqslant n$ the following situation S_r holds: there are

(4.17) $\bar{b}'[0], \bar{c}'[0], \ldots, \bar{b}'[q - 1], \bar{c}'[q - 1], H, \bar{b}'[q], \bar{c}'[q], \ldots,$
         $\bar{b}'[n - r - 1], \bar{c}'[n - r - 1]$

such that

(4.18) H is a $(2^{r+1} - 2)$-tree for ϕ,

(4.19) for all $i, j < n - r$, $A \vDash \phi(\bar{b}[i], \bar{c}[j]) \Leftrightarrow i \leqslant j$,

(4.20) if $\bar{c}$ is a node of H then $A \vDash \phi(\bar{b}[i], \bar{c}) \Leftrightarrow i < q$,

(4.21) if b is a branch of H then $A \vDash \phi(\bar{b}, \bar{c}[j]) \Leftrightarrow j \geqslant q$.

The initial case S_n states simply that there is a $(2^{n+1} - 2)$-tree for ϕ, which we have assumed. The final case S_1 implies that ϕ has ladder index at least n as follows. As H is a 2-tree, it has a node $\bar{c}$ and a branch $\bar{b}$ such that $M \vDash \phi(\bar{b}, \bar{c})$. Put $\bar{b}, \bar{c}$ in that order between $\bar{c}'[q - 1]$ and $\bar{b}'[q]$ in (4.17); then conditions (4.19)–(4.21) show that the resulting list of tuples yields an n-ladder for ϕ.

It remains to show that if $r > 1$ and S_r holds, then so does S_{r-1}. Assume S_r and put $h = 2^r - 2$. By (4.18), H is a $(2h + 2)$-tree. For each branch $\bar{b}$ of H, write $H(\bar{b})$ for the set of those nodes $\bar{c}$ of H such that $A \vDash \phi(\bar{b}, \bar{c})$. There are two cases.

Case 1: there is a branch $\bar{b}$ of H such that $H(\bar{b})$ contains an $(h + 1)$-tree. Then there are a node $\bar{c}$ in $H(\bar{b})$ and an h-tree H' for ϕ such that S_{r-1} holds when we replace H in (4.17) by $\bar{b}, \bar{c}, H'$ in that order.

Case 2: for every branch $\bar{b}$ of H, $H(\bar{b})$ contains no $(h + 1)$-tree. Then let $\bar{c}$ be the bottom node $\bar{c}[\langle \rangle]$ of H, let $\bar{b}$ be any branch of $H_{(0)}$ and let N be the set of all nodes of $H_{(0)}$. The case assumption tells us that $H(\bar{b}) \cap N$ contains no $(h + 1)$-tree. So by the claim applied to $H_{(0)}$, the set $N \backslash H(\bar{b})$ contains an h-tree H' for ϕ. Then S_{r-1} holds when we replace H in (4.17) by $H', \bar{b}, \bar{c}$ in that order.

So in either case S_{r-1} holds. This completes the induction.

$\square$ Lemma 9.4.10.

It follows at once that T is stable if and only if every formula $\phi(\bar{x}, \bar{y})$ of L has finite branching index.

Let $\phi(\bar{x}, \bar{y})$ be a formula of L and $\psi(\bar{x})$ a formula of L with parameters from some model A of T. We define the **relativised branching index** $\mathrm{BI}(\phi, \psi)$ to be $\geqslant n$ iff there is an n-tree for ϕ whose branches all satisfy ψ. Thus $\mathrm{BI}(\phi, \psi)$ is either a unique finite number or ∞. Clearly $\mathrm{BI}(\phi, \psi) \leqslant \mathrm{BI}(\phi)$, so that if ϕ is stable then $\mathrm{BI}(\phi, \psi)$ must be finite.

The statement '$\mathrm{BI}(\phi, \psi) \geqslant n$' can be written as a sentence of L with parameters from A. Since the variables $\bar{x}$ don't occur free in this sentence, it could be misleading to write it as '$\mathrm{BI}(\phi, \psi(\bar{x})) \geqslant n$'. So when it is necessary to mention the variables $\bar{x}$, I replace them with '$-$', thus: '$\mathrm{BI}(\phi, \psi(-)) \geqslant n$'.

The following lemma is crucial for proving Theorem 9.4.9.

Lemma 9.4.11. *If* $\mathrm{BI}(\phi, \psi)$ *is a finite number* n, *then for every tuple* $\bar{c}$ *of elements of M, either* $\mathrm{BI}(\phi, \psi \wedge \phi(-, \bar{c})) < n$ *or* $\mathrm{BI}(\phi, \psi \wedge \neg \phi(-, \bar{c})) < n$.

Proof. Suppose H_0 is an n-tree for ϕ whose branches satisfy $\psi(\bar{x}) \wedge \phi(\bar{x}, \bar{c})$, and H_1 is an n-tree for ϕ whose branches satisfy $\psi(\bar{x}) \wedge \neg \phi(\bar{x}, \bar{c})$. Then we can form an $(n + 1)$-tree H for ϕ whose branches satisfy $\psi(\bar{x})$, by putting $H_{(0)} = H_0$, $H_{(1)} = H_1$ (in the notation of the proof of Lemma 9.4.10), and taking $\bar{c}$ as the bottom node. $\qquad\qquad\square$

Suppose now that X is a set of elements of a model A of T, $p(\bar{x})$ a complete type over X, and $\phi(\bar{x}, \bar{y})$ a formula of L with finite branching index. Then the minimum value of $\mathrm{BI}(\phi, \psi)$, as $\psi(\bar{x})$ ranges over all formulas in p, is called the ϕ-**rank** of p.

Proof of Theorem 9.4.9. Let $p(\bar{x})$ be the type of $\bar{b}$ over X. Since ϕ is stable, the ϕ-rank of p must be some finite number n. Choose $\psi(\bar{x})$ in p so that $\mathrm{BI}(\phi, \psi) = n$, and let $\mathrm{d}_p \phi(\bar{y})$ be the formula '$\mathrm{BI}(\phi, \psi \wedge \phi(-, \bar{y})) \geqslant n$'.

We claim that (4.15) holds. From left to right, suppose $\phi(\bar{x}, \bar{c}) \in p$. Then $\psi(\bar{x}) \wedge \phi(\bar{x}, \bar{c}) \in p$ and hence $\mathrm{BI}(\phi, \psi \wedge \phi(-, \bar{c})) \geqslant n$, so that $A \vDash \mathrm{d}_p \phi(\bar{c})$. In the other direction, suppose $\phi(\bar{x}, \bar{c})$ is not in p. Then $\neg \phi(\bar{x}, \bar{c})$ is in p, and the same argument as before shows that $\mathrm{BI}(\phi, \psi \wedge \neg \phi(\bar{x}, \bar{c})) \geqslant n$. It follows by Lemma 9.4.11 that $\mathrm{BI}(\phi, \psi \wedge \phi(\bar{x}, \bar{c})) < n$, so that $A \vDash \neg \mathrm{d}_p \phi(\bar{c})$.

$\qquad\qquad\qquad\qquad\qquad\qquad\qquad\qquad\square$ Theorem 9.4.9

Let $p(\bar{x})$ be a complete type over a set X of elements of a model A of T. By a **definition schema of** p we mean a map d which takes each formula $\phi(\bar{x}, \bar{y})$ of L to a ϕ-definition $\mathrm{d}\phi(\bar{y})$ of p, where any parameters in $\mathrm{d}\phi$ come from X. A type is said to be **definable** if it has a definition schema. The next result says that *for stable theories, all complete types are definable.*

Corollary 9.4.12 *Let the theory* T *be stable. Let* A *be a model of* T, *let* X *be a set of elements of* A *and* p *a type over* X. *Then* p *is definable.*

Proof. Immediate from Theorem 9.4.9. $\qquad\qquad\qquad\qquad\qquad\qquad\square$

The moral of the next corollary is that if one is forming an heir–coheir amalgam of models of a stable theory, there is just one way to do it. The formulas that hold between the elements of the structures being amalgamated are completely determined by the definition schemas of the types of these tuples. In a slogan, *for stable theories, heir–coheir amalgams are unique.* The converse holds too, though we shall not prove it: if T is a complete theory and heir–coheir amalgams are unique for T, then T is stable.

Corollary 9.4.13. *Assuming T is stable, let A, B, C' and D be models of T which form an amalgam as in (3.3) of section 5.3 above. Let $\bar{b}$ be any tuple in B, let the type $p(\bar{x})$ be $\mathrm{tp}(\bar{b}/\mathrm{dom}\,A)$, and let d be a definition schema of p. Then d is also a definition schema of $\mathrm{tp}(\bar{b}/\mathrm{dom}\,C')$.*

Proof. We must show that

(4.22)   for every tuple $\bar{c}$ in C', $D \vDash \phi(\bar{b}, \bar{c})$ if and only if $C' \vDash \mathrm{d}\phi(\bar{c})$.

Let $\bar{c}$ be a counterexample. Then $D \vDash \phi(\bar{b}, \bar{c}) \leftrightarrow \neg\mathrm{d}\phi(\bar{c})$. By the definition of heir–coheir amalgams there is a tuple $\bar{a}$ in A such that $B \vDash \phi(\bar{b}, \bar{a}) \leftrightarrow \neg\mathrm{d}\phi(\bar{a})$. But this is impossible by the choice of d. $\square$

To complete the circle, we show (a) $\Rightarrow$ (b) in Theorem 9.4.1. In fact we show a little more.

Theorem 9.4.14. *Let T be a complete theory. The following are equivalent.*
(a) *T is stable.*
(b) *All types are definable for T.*
(c) *For every cardinal λ such that $\lambda = \lambda^{|L|}$, T is λ-stable.*

Proof. (a) $\Rightarrow$ (b) is by Corollary 9.4.12, and we proved (c) $\Rightarrow$ (a) at the beginning of the section. There remains (b) $\Rightarrow$ (c). Let A be a model of T and X a set of at most λ elements of A. By (b), each type p in $S_1(X)$ has a definition schema d_p, and if $p \neq q$ then $\mathrm{d}_p \neq \mathrm{d}_q$. So it suffices to count the number of possible definition schemas. Write $L(X)$ for L with the elements of X added as parameters. Then $|L(X)| \leq \lambda$, and each definition schema d_p is a map from a set of formulas of L to formulas of $L(X)$. So the number of schemas is at most $\lambda^{|L|} = \lambda$. $\square$

Exercises for section 9.4

1. Show that in the definition of 'unstable theory' it makes no difference if we allow parameters in the formula ϕ of (4.1).

2. Show that if $\mathbb{Q}$ and $\mathbb{R}$ are respectively the field of rational numbers and the field of reals, then both $\mathrm{Th}(\mathbb{Q})$ and $\mathrm{Th}(\mathbb{R})$ are unstable. [For $\mathbb{Q}$, find an infinite set of numbers which are linearly ordered by the relation '$y - x$ is a non-zero sum of four squares'. The same argument works for any formally real field.]

A formula $\phi(\bar{x}, \bar{y})$ is said to have the **strict order property** *(for a complete theory T) if in every model of T, ϕ defines a partial ordering relation on the set of all n-tuples, which contains arbitrarily long finite chains. T has the* **strict order property** *if some formula has the strict order property for T.*

3. Show that every complete theory with the strict order property is unstable.

A formula $\phi(\bar{x}, \bar{y})$ is said to have the **independence property** *(for a complete theory T) if in every model A of T there is, for each $n < \omega$, a family of tuples $\bar{b}_0, \ldots, \bar{b}_{n-1}$ such that for every subset X of n there is a tuple $\bar{a}$ in A for which $A \vDash \phi(\bar{a}, \bar{b}_i) \Leftrightarrow i \in X$. T has the* **independence property** *if some formula has the independence property for T.*

4. Show that every complete theory with the independence property is unstable.

5. Give examples of (a) a complete theory which has the strict order property but not the independence property, (b) a complete theory which has the independence property but not the strict order property. *Shelah shows that a complete theory T is unstable if and only if it has either the strict order property or the independence property.*

6. Show that if T is unstable then it has an unstable formula.

7. Let L be a first-order language and T a complete theory in L. (a) Show that if $\phi(\bar{x}, \bar{y})$ is a stable formula for T, then so is $\neg \phi(\bar{x}, \bar{y})$. (b) Show that if $\phi(\bar{x}, \bar{y})$ and $\psi(\bar{x}, \bar{y})$ are stable formulas for T, then so is $(\phi \wedge \psi)(\bar{x}, \bar{y})$. [Use the finite Ramsey theorem, Corollary 5.6.2.]

8. Suppose $\phi(\bar{x}, \bar{y})$ is a stable formula for T. (a) Show that if $\bar{y}'$ is a tuple of variables which includes all those in $\bar{y}$, and $\psi(\bar{x}, \bar{y}')$ is equivalent to $\phi(\bar{x}, \bar{y})$ modulo T, then ψ is also stable. (b) Show that if $\bar{x}'$ is $\bar{y}$ and $\bar{y}'$ is $\bar{x}$, and $\theta(\bar{x}', \bar{y}')$ is the formula $\phi(\bar{x}, \bar{y})$, then θ is stable.

9. Show that every stable integral domain is a field. [If t is any element $\neq 0$, t^n divides t^{n+1} for each positive n; so by stability there is some n such that t^{n+1} divides t^n too.]

The next exercise shows that unstable theories have the finite cover property in Keisler's sense; see Exercise 4.4.3.

10. Show that if T is an unstable complete theory in the first-order language L, then there is a formula $\phi(x, \bar{y})$ of L such that for arbitrarily large finite n, T implies that there are $\bar{a}_0, \ldots, \bar{a}_{n-1}$ for which $\neg \exists x \bigwedge_{i<n} \phi(x, \bar{a}_i)$ holds, but $\exists x \bigwedge_{i \in w} \phi(x, \bar{a}_i)$ holds for each proper subset w of n.

11. Let the L-structure A be a model of a stable theory T, let $p(\bar{x})$ be a complete type over X with respect to A, and let $\phi(\bar{x}, \bar{y})$ be a formula of L. Define the **strict ϕ-rank** of $p(\bar{x})$ to be the minimum value of $\mathrm{BI}(\phi, \psi)$ as $\psi(\bar{x})$ ranges over all formulas of p which are conjunctions of formulas of the form $\phi(\bar{x}, \bar{c})$ or $\neg \phi(\bar{x}, \bar{c})$. Prove Theorem 9.4.9 using strict ϕ-rank in place of ϕ-rank.

9.5 Morley's theorem

Throughout this section, L will be a countable first-order language and T a complete theory in L with infinite models. We shall prove:

Theorem 9.5.1 (Morley's theorem). *Suppose T is κ-categorical for some uncountable cardinal κ. Then T is λ-categorical for every uncountable cardinal λ.*

The *opening gambit* runs as follows. If T is categorical in cardinality κ, then there is no type which is realised in some models of cardinality κ and omitted in others. This suggests that all models of cardinality κ are saturated, and hence that there are not many types. But then there should not be many types realised in any model of T. This sketch is a pretty faithful summary of the first lemma below, except that we have to replace 'type' by the more sophisticated notion 'type over a set of elements'.

The proof of the first lemma uses Ehrenfeucht–Mostowski models; they are not needed anywhere else in the proof of Morley's theorem. For this reason some readers may prefer to take the lemma on trust.

Lemma 9.5.2. *T is totally transcendental, and hence λ-stable for all $\lambda \geq 0$.*

Proof. This was Corollary 9.4.6.   □

To paraphrase Lemma 9.5.2: Let A be a model of T and X a set of elements of A. Then the number of complete n-types over X, for any finite n, is at most $|X| + \omega$; in particular it is exactly $|X|$ if X is infinite.

The *body of the proof* revolves around the following idea. We would like to prove that every uncountable model of T is saturated. This would establish Morley's theorem by Theorem 8.1.8. But it requires us to show that every possible type must be realised in every uncountable model of T. (Here and henceforth, 'type' means 'complete type'.) How could we show this?

Algebraically closed fields provide a clue. Suppose we have an algebraically closed field A of uncountable cardinality, and a subfield B of smaller cardinality than A. Then we notice two things. First, every algebraic type over B must be realised in A, in order for A to be algebraically closed. Second, the unique non-algebraic 1-type over B must also be realised in A, since A has larger cardinality than the algebraic closure of B. We conjecture that the general case will run along the same lines as this. And so it will.

For the rest of the proof it is helpful to work inside a big model M of T, as in the previous three sections. From this point onwards, all 'elements' lie in M, and 'models' will always be elementary substructures of M. If X is any set of elements, we write $L(X)$ for the language L with the elements of X added as parameters (naming themselves unless we say otherwise). So a 'type over X' will be a complete type in the language $L(X)$ with respect to the structure M.

A type $p(\bar{x})$ over a set of elements X is said to be **principal** if there is a formula $\phi(\bar{x}) \in p$ such that p consists of exactly those formulas $\psi(\bar{x})$ of $L(X)$ such that

$$(5.1) \qquad M \vDash \forall \bar{x}(\phi \to \psi).$$

We call such a formula ϕ a **support** of p, and a **complete formula** over X. Note that $M \vDash \exists \bar{x} \phi$, since otherwise (5.1) holds for every formula ψ including the negations of formulas in p.

Prime models

If p is a principal type over X, and B is a model containing X, then p must be realised in B. For if ϕ is a support of p, then $M \vDash \exists \bar{x} \phi$, and so $B \vDash \exists \bar{x} \phi$ since B is an elementary substructure of A. So B contains a tuple $\bar{b}$ such that $B \vDash \phi(\bar{b})$ and hence $M \vDash \phi(\bar{b})$ too. But then $M \vDash \psi(\bar{b})$ for every formula ψ in p, and thus $\bar{b}$ realises p.

For totally transcendental theories a converse is true too: if X is any set of elements, then there is a model B which contains X and in which every tuple realises a principal type over X. The next two lemmas are devoted to proving this.

Before we launch into these lemmas, note that a totally transcendental theory has only countably many types over the empty set, and hence by Theorem 6.2.2 it has a prime model A_0, i.e. a model elementarily embeddable in every model. Also by Theorem 6.2.2, every tuple in this prime model is atomic, i.e. it realises a principal type over the empty set. Henceforth we fix a prime model A_0. Note that every model B of T is isomorphic to an elementary extension of A_0. For by the fact that A_0 is prime, there is an elementary embedding $f: A_0 \to B$, and since M is big, f extends to an automorphism g of M. Then $g^{-1}(B)$ is isomorphic to B and is an elementary extension of A_0.

Lemma 9.5.3. *Let M be a model of the totally transcendental theory T, X a set of elements of M and $\theta(\bar{x})$ a formula of $L(X)$ such that $M \vDash \exists \bar{x} \theta$. Then there is a complete formula $\phi(\bar{x})$ over X such that $M \vDash \exists \bar{x} \phi \wedge \forall \bar{x} (\phi \to \theta)$.*

Proof. Suppose not; then in particular θ itself is not complete, and so there are formulas $\theta_0(\bar{x})$, $\theta_1(\bar{x})$ of $L(X)$ such that

$$(5.2) \qquad M \vDash \exists \bar{x} \theta_1 \wedge \exists \bar{x} \theta_2,$$

(5.3) $M \vDash \forall \bar{x}(\theta_0 \to \theta) \wedge \forall \bar{x}(\theta_1 \to \theta)$

but

(5.4) $M \vDash \neg \exists \bar{x}(\theta_0 \wedge \theta_1).$

Likewise θ_0 is not a complete formula, and so there are formulas θ_{00}, θ_{01} of
$L(X)$ such that .... In this way we build up a tree as in Lemma 9.3.1, which
contradicts the assumption that T is totally transcendental. $\qquad\qquad\square$

Suppose now that X is a set of elements. By a **construction sequence** over
X (or length α, for some ordinal α) we mean a sequence $(a_i : i < \alpha)$ where
each element a_i realises a principal type over $X \cup \{a_j : j < i\}$.

Lemma 9.5.4. *Assuming M is totally transcendental, let μ be an infinite*
cardinal and X a set of at most μ elements. Then there is a construction
sequence $(a_i : i < \mu)$ over X such that the set of elements $\{a_i : i < \mu\}$ includes X
and forms a model. Moreover if A is any model constructed in this way, then
for every model B containing X, there is an elementary embedding $f : A \to B$
which is the identity on X.

Proof. First we construct A. By Tarski's criterion for elementary substructures (see Exercise 2.5.1), it suffices to ensure that if $i < \mu$ and $\theta(x)$ is any
formula of $L(X \cup \{a_j : j < i\})$ such that $M \vDash \exists x \theta$, then there is $k < \mu$ such
that $M \vDash \theta(a_k)$. One way to arrange this is to list as $\theta_i(\bar{y}, x)$ $(i < \mu)$ all the
formulas of $L(X)$ in which $\bar{y}$ is a tuple of distinct variables in the sequence
$(y_i : i < \mu)$ of distinct variables, in such a way that each formula θ_i contains as
free variables, besides the variable x, only variables y_j with $j < i$. This can
certainly be done, allowing formulas to occur more than once if necessary.
Now suppose we have constructed the sequence $\bar{a} = (a_j : j < i)$. Putting a_j for
y_j in θ_i, suppose $M \vDash \neg \exists x \ \theta_i(\bar{a}, x)$; then choose a_i arbitrarily. But if $M \vDash \exists x$
$\theta_i(\bar{a}, x)$, then use the previous lemma to find a complete formula $\phi(x)$ over
$X \cup \{a_j : j < i\}$ such that $M \vDash \forall x(\phi(x) \to \theta_i(\bar{a}, x))$, and choose a_i so that it
satisfies ϕ.

After μ steps, the set $A = \{a_i : i < \mu\}$ will contain X because the formulas
$c = x$ $(c \in X)$ will each appear as some θ_i. So by the Tarski criterion, A is a
model containing X.

Suppose B is also a model containing X; let $\bar{c}$ be a sequence listing the
elements of X. Then we can inductively choose elements b_i $(i < \mu)$ of B so
that if $\bar{a} = (a_j : j < \mu)$ and $\bar{b} = (b_j : j < \mu)$ then $(A, \bar{c}, \bar{a}|i) \equiv (B, \bar{c}, \bar{b}|i)$ for each
$i \leqslant \mu$. When $i = 0$ this says that $(A, \bar{c}) \equiv (B, \bar{c})$, which holds because both A
and B are elementary substructures of M containing X. When i is a limit
ordinal there is nothing to prove, since a first-order formula contains only
finitely many constants. Finally when $i = j + 1$, we chose a_j to realise a

principal type in $(A, \bar{c}, \bar{a} | j)$, say with support $\phi(\bar{c}, \bar{a} | j, x)$, and so now we can choose b_j to satisfy $\phi(\bar{c}, \bar{b} | j, x)$ in B. At the end of this construction, $(A, \bar{c}, \bar{a}) \equiv (B, \bar{c}, \bar{b})$, so that by the elementary diagram lemma (Lemma 2.5.3) there is an elementary embedding of A in B which is the identity on X. $\qquad\qquad\square$

If A is a model containing a set of elements X, and for every model B containing X there is an elementary embedding $f: A \to B$ which is the identity on X, then we say that A is **prime over** X. So we have shown that for every set X of elements, there is a prime model over X. WARNING: we have not yet shown that the prime model over X is determined up to isomorphism by X. Later we shall prove it in a special case (see Lemma 9.5.11 below).

The next lemma is needed for technical reasons.

Lemma 9.5.5. *Suppose X is a set of elements of M and $(a_i : i < \alpha)$ is a construction sequence over X. Then every element a_i in the sequence realises a principal type over X.*

Proof. Each element a_i realises a principal type p_i over $X \cup \{a_j : j < i\}$; since p_i is principal, it has a support $\psi_i(\bar{a}_i, x)$ where ψ_i is a formula of $L(X)$ and $\bar{a}_i$ is a tuple of elements of $\{a_j : j < i\}$. Choosing such a tuple $\bar{a}_i$, let $F(i)$ be a finite subset of i such that each element in $\bar{a}_i$ is a_j for some $j \in F(i)$.

Now consider a fixed $i < \alpha$. Since α is well-ordered, there is a finite set $W \subseteq i \cup \{i\}$ such that $i \in W$ and for every $j \in W$, $F(j) \subseteq W$. (Otherwise an application of König's tree lemma, Exercise 5.6.3, would find an infinite descending sequence in i.) To economise on subscripts, let us suppose for simplicity that i is finite and $W = i \cup \{i\}$. Adding redundant variables if necessary, each formula ψ_j $(j \leq i)$ can be written as $\psi_j(w_0, \ldots, w_{j-1}, x)$. Write $\chi(w_0, \ldots, w_i)$ for the formula

$$(5.5) \qquad \bigwedge_{j \leq i} \psi_j(w_0, \ldots, w_{j-1}, w_j).$$

Then $M \vDash \chi(a_0, \ldots, a_i)$. Moreover if $M \vDash \chi(b_0, \ldots, b_i)$ and $\bar{c}$ is a sequence listing the elements of X, then by induction on $j \leq i$ we can show that

$$(5.6) \qquad (M, \bar{c}, a_0, \ldots, a_j) \equiv (M, \bar{c}, b_0, \ldots, b_j).$$

It follows that if $M \vDash \exists w_0 \ldots w_{i-1} \chi(w_0, \ldots, w_{i-1}, b)$ then b realises the same type as a_i over X. So the formula $\exists w_0 \ldots w_{i-1} \chi(w_0, \ldots, w_{i-1}, x)$ is a support of the type of a_i over X, as required. $\qquad\qquad\square$

The strongly minimal formula

In an algebraically closed field, the prime model over a set X is precisely the set of elements which are algebraic over X. So we have captured and generalised one way in which types *must* be realised. We turn to the second, namely the unique non-algebraic type. The word 'unique' is the one to watch here: a crucial step in the proof of Steinitz' theorem is the observation that if F is an algebraically closed subfield of a field K, then all elements of K not in F have the same minimal polynomial over F.

We shall borrow from section 9.2 the notion of a strongly minimal formula. A formula $\phi(x)$, possibly with parameters from M, is said to be **strongly minimal** if $\phi(M)$ is infinite but for every formula $\psi(x)$ with parameters from M, at least one of $(\phi \wedge \psi)(M)$, $(\phi \wedge \neg \psi)(M)$ is finite. (Why does this definition define 'strongly minimal' rather than just 'minimal'? Answer: because the big model M is ω-saturated; cf. Fact 9.2.1.) As we showed in Theorem 9.2.3, every strongly minimal set carries notions of closed subset, independent subset, basis, dimension as in linear algebra.

Lemma 9.5.6. *If T is totally transcendental, then there is a strongly minimal formula $\phi(x)$.*

Proof. Suppose not. Since M is certainly infinite, there is a formula $\phi_{\langle\rangle}$ which is satisfied by infinitely many elements. If $\phi_{\langle\rangle}$ is not strongly minimal, then we can find another formula $\psi(x)$ such that

$$(5.7) \qquad \phi_{\langle\rangle}(M) \cap \psi(M), \quad \phi_{\langle\rangle}(M) \backslash \psi(M)$$

are both infinite. Putting $\phi_0 = \phi_{\langle\rangle} \wedge \psi$ and $\phi_1 = \phi_{\langle\rangle} \wedge \neg \psi$, and repeating the same argument with each of ϕ_0 and ϕ_1, we again build up a tree as in Lemma 9.3.1; again this contradicts the assumption that T is totally transcendental. $\square$

There is a problem here, that we have no control over where the parameters in ϕ come from, and so ϕ may be useless for analysing models which don't contain these parameters. Hence the importance of the next lemma.

Lemma 9.5.7. *Suppose that in T, infinity is definable. (This means: for every formula $\psi(\bar{x}, y)$ of L there is a formula $\theta_\psi(\bar{x})$ of L such that for all tuples $\bar{a}$, $M \vDash \theta_\psi(\bar{a})$ if and only if $\psi(\bar{a}, M)$ is infinite.) If T is totally transcendental, then the minimal formula ϕ in the previous lemma can be chosen so that its parameters lie in the prime model A_0.*

Proof. Carry out the proof of the previous lemma, but considering only formulas whose parameters come from A_0. This finds a formula $\phi(\bar{b}, x)$ such that $\phi(\bar{b}, A_0)$ is infinite but can't be split into two infinite pieces by any formula of $L(A_0)$. We claim that in fact $\phi(\bar{b}, x)$ is strongly minimal. Equivalently, we claim that if $\psi(\bar{c}, x)$ is any formula with parameters $\bar{c}$ from M, and we write $\phi_0(\bar{b}, \bar{c}, x)$ for $\phi(\bar{b}, x) \wedge \psi(\bar{c}, x)$ and $\phi_1(\bar{b}, \bar{c}, x)$ for $\phi(\bar{b}, x) \wedge \neg \psi(\bar{c}, x)$, then at least one of

(5.8) $$\phi_i(\bar{b}, \bar{c}, M) \ (i = 0, 1)$$

is finite. For if not, then

(5.9) $$M \vDash \exists \bar{z}(\theta_{\phi_0}(\bar{b}, \bar{z}) \wedge \theta_{\phi_1}(\bar{b}, \bar{z})).$$

Since A_0 is an elementary substructure of M, it follows that there is $\bar{d}$ already in A_0 such that $\phi_i(\bar{b}, \bar{d}, M) \ (i = 0, 1)$ are both infinite, contradicting the choice of ϕ. □

But we know from Corollary 8.5.9 that if T is λ-categorical for some uncountable λ, then T doesn't have the finite cover property, and so infinity is definable in T. The proof that the finite cover property fails in T seems to need ultraproducts. It is possible to show that infinity is definable in T without using ultraproducts, but one has to work harder – see the proof of Theorem 9.5.12 below.

Our strongly minimal formula ϕ is not necessarily complete over A_0. But it almost is: it pins down the type of a new element over a model, in the following sense.

Lemma 9.5.8. *Let $\phi(x)$ be a strongly minimal formula with parameters from A_0. Suppose A is a model containing A_0. Then there is a (complete) type $p_A(x)$ over A, such that if b is any element of M, then b realises p_A if and only if b is an element which satisfies ϕ but is not in A.*

Proof. Let $\psi(x)$ be any formula of $L(A)$. Then exactly one of $\phi(M) \cap \psi(M)$ and $\phi(M) \backslash \psi(M)$ is infinite. Suppose it is the former. Then for some natural number n, there are exactly n elements of A which satisfy $\phi \wedge \neg \psi$. Since we can write this fact as a sentence of $L(A)$, there are also exactly n elements of M which satisfy $\phi \wedge \neg \psi$, and these must be the same n elements. So any element b which is not in A but satisfies ϕ must satisfy $\phi \wedge \psi$. Likewise if $\phi(M) \backslash \psi(M)$ is infinite, then b satisfies $\phi \wedge \neg \psi$. Thus for every formula $\psi(x)$ of $L(A)$, exactly one of ψ and $\neg \psi$ finds an infinite subset of $\phi(A)$; we define p_A to consist of the formulas of $L(A)$ which do pick out an infinite subset of $\phi(A)$.

Our construction of p_A shows why every element b which satisfies ϕ but is not in A must realise p_A. The converse is easy, since p_A must contain ϕ and all the formulas $x \neq a$ with $a \in A$. □

In the terminology of Example 1 in section 9.3, p_A is the regular type of the strongly minimal set $\phi(A)$. We can define a type p_M over M in the same way. Then each type p_A is the restriction of p_M to the set of formulas of $L(A)$. (Types over the big model M are sometimes called **global types**.)

One-cardinal formulas

We want to show that every proper elementary extension of a model $A \supseteq A_0$ must contain an element realising p_A. This will generalise the fact that every algebraically closed field properly extending an algebraically closed field F must contain an element transcendental over F.

Put in another way, we want to show that it is impossible to find models A, B of T such that B is a proper elementary extension of A but $\phi(B) = \phi(A)$, where ϕ is our strongly minimal formula. In the language of section 8.4, we aim to show that ϕ is one-cardinal – meaning that $|\phi(A)| = |A|$ for any model A containing the parameters of ϕ.

If we happen to know that T is ω_1-categorical, then Vaught's two-cardinal theorem (Theorem 8.4.1, with $x = x$ for $\psi(x)$) already tells us that ϕ is one-cardinal. (This is because there is clearly a model A of T with $|A| = |\phi(A)| = \omega_1$, which can't be isomorphic to the model in part (a) of Vaught's theorem). So if we are only interested in showing that ω_1-categorical implies λ-categorical for all uncountable λ, we can skip straight away to Lemma 9.5.11 below.

Otherwise we need to call on some stability theory, recalling from Theorem 9.4.5 that our ω-stable theory T is stable. This will be the subtlest part of the proof. It is at heart an omitting types result for ω-stable theories.

Lemma 9.5.9. *Assume T is stable. Let $\phi(x)$ be any formula of L wth parameters. Let A be a model containing the parameters of ϕ. Suppose B and C are models which both contain the model A, and D is the prime model over $B \cup C$. Suppose also that*
(a) D is an heir-coheir amalgam of B and C over A, and
(b) $\phi(B) = \phi(A) = \phi(C)$.
Then $\phi(D) = \phi(A)$.

Proof. We consider any element d of $\phi(D)$; the aim is to show that it lies in A. Since D is prime over $B \cup C$, Lemma 9.5.5 finds us a complete formula over $B \cup C$,

(5.10) $\psi(\bar{b}, \bar{c}, x)$

with ψ in L, $\bar{b}$ in B and $\bar{c}$ in C, which supports the type of d over $B \cup C$. Let d be a definition of the type of $\bar{b}$ over A; there is such a definition by Corollary 9.4.12.

We claim that if $\chi(\bar{x}, \bar{y})$ is any formula of $L(A)$ and $\bar{e}$ is any tuple in $C \cup \{d\}$, then

(5.11) $$M \models \chi(\bar{b}, \bar{e}) \Leftrightarrow M \models d\chi(\bar{e}).$$

For suppose not; say we have

(5.12) $$M \models \neg \, (\chi(\bar{b}, \bar{c}', d) \leftrightarrow d\chi(\bar{c}', d)).$$

Then

(5.13) $$M \models \exists z(\phi(z) \wedge \neg \, (\chi(\bar{b}, \bar{c}', z) \leftrightarrow d\chi(\bar{c}'z))).$$

But B and C are heir–coheir over A, and so there is $\bar{c}''$ in A such that

(5.14) $$B \models \exists z(\phi(z) \wedge \neg \, (\chi(\bar{b}, \bar{c}'', z) \leftrightarrow d\chi(\bar{c}'', z))).$$

Then the element z can be found in B; but $\phi(B) = \phi(A)$, so it can be found in A. Let it be $d' \in A$. Then

(5.15) $$M \models \phi(d') \wedge \neg \, (\chi(\bar{b}, \bar{c}'', d') \leftrightarrow d\chi(\bar{c}'', d')).$$

The second conjunct contradicts the fact that d defines the type of $\bar{b}$ over A. The claim is proved.

Now apply the claim to formula (5.10). We infer that

(5.16) $$M \models d\psi(\bar{c}, d).$$

Hence $C \models \exists z \, d\psi(\bar{c}, z)$. Choose d'' in C so that $M \models d\psi(\bar{c}, d'')$ and hence $M \models \psi(\bar{b}, \bar{c}, d'')$ by the claim again. Then $M \models \phi(d'')$ since $\psi(\bar{b}, \bar{c}, x)$ supports the type of d over A. Since d'' lies in C, this implies that d'' is an element of A, and hence

(5.17) $$M \models \forall z \, (\psi(\bar{b}, \bar{c}, z) \rightarrow z = d'')$$

by choice of $\psi(\bar{b}, \bar{c}, z)$ as complete over $B \cup C$. We have proved that $d \in A$.

$\square$

Corollary 9.5.10. *Let T be κ-categorical for some uncountable κ. Let $\phi(x)$ be a formula of L with parameters, such that $\phi(M)$ is infinite. Then ϕ is one-cardinal.*

Proof. Suppose not. Then there are models $B_0 \subseteq B_1$ with $B_0 \neq B_1$ but $\phi(B_0) = \phi(B_1)$. Use the relativisation theorem (Theorem 4.2.1) as in the proof of Vaught's two-cardinal theorem (Theorem 8.4.1) to make B_0 and B_1 countable.

Now, using Lemma 6.4.3, choose an elementary map f and an extension B_2' of B_1 such that f is the identity on B_0 and B_2' is an heir-coheir amalgam of B_1 and $f(B_1)$ over B_0. Let $B_2 \subseteq B_2'$ be the prime model over $B_1 \cup f(B_1)$. Then by the lemma just proved, $\phi(B_2) = \phi(B_0)$. Moreover B_2 is a proper extension of B_1, since heir–coheir amalgams of models are always strong (see the remark after Theorem 5.3.3).

We can repeat this construction, using B_2 in place of B_1, to get a proper elementary extension B_3 of B_2 with $\phi(B_3) = \phi(B_0)$. If α is any ordinal, we can iterate the construction α times, taking unions of elementary chains (cf. Theorem 2.5.2) at limit ordinals. This gives us in particular a model B of cardinality κ with $\phi(B)$ countable, since $\phi(B) = \phi(B_0)$. But using compactness it is easy to find a model C of cardinality κ in which $\phi(C)$ is uncountable; this contradicts the assumption that T is κ-categorical. $\qquad \square$

Proof of Morley's theorem

Now we pull the threads together by showing that every model of T can be built up by first taking some elements which satisfy the strongly minimal formula ϕ, and then forming a prime model over these elements. This generalises the fact that every algebraically closed field can be built up by first taking a transcendence basis and then forming its algebraic closure.

Lemma 9.5.11. *Suppose T is κ-categorical for some uncountable cardinal κ, and let $\phi(x)$ be a strongly minimal formula with parameters from A_0. Then every model $A \supseteq A_0$ is the unique prime model over $\phi(A)$ (up to isomorphism over $\phi(A)$). In particular $|A| = |\phi(A)|$.*

Proof. Let B be any prime model over $\phi(A)$. Then there is an elementary map $f: B \to A$ which is the identity on $\phi(A)$. Putting $C = f(B)$, we have an elementary substructure C of A, and in particular $\phi(C) \subseteq \phi(A)$. Moreover if $a \in \phi(A)$ then $a \in \phi(B)$ and so $a = f(a) \in \phi(C)$. This proves that $\phi(A) \subseteq \phi(C)$ and hence $\phi(C) = \phi(A)$. But ϕ is one-cardinal, so that we can deduce $C = A$ and f is onto A. Thus f is an isomorphism from B to A which is the identity on $\phi(A)$.

We check the cardinalities. Put $|\phi(A)| = \lambda$. Then clearly $\lambda \leqslant |A|$. In the other direction, since $\phi(A)$ is infinite and the language is countable, the downward Löwenheim–Skolem theorem (Corollary 3.1.4) finds us a model D which contains $\phi(A)$ and has cardinality λ. Since A is prime over $\phi(A)$, there is an elementary embedding of A into D and hence $|A| \leqslant \lambda$. $\qquad \square$

The *final step* of the proof is to draw the conclusion, using facts that we proved earlier about strongly minimal formulas. Let A and B be models of T with the same uncountable cardinality λ; we must show that A and B are isomorphic. By the remarks before Lemma 9.5.3, we can assume that both A and B contain A_0. Then $\phi(A)$ and $\phi(B)$ both have cardinality λ by the last lemma. Now we recall what we know about strongly minimal sets. Choose in $\phi(A)$ a maximal independent set I, and in $\phi(B)$ a maximal independent set

J. Since every element of $\phi(A)$ is algebraic over I and the language is countable, I also has cardinality λ, and likewise J.

Let g be a bijection from I to J. Then by Lemma 9.2.6 (with X and Y empty), g is an elementary map. So by Lemma 9.2.5, g extends to an elementary map h from $\phi(A)$ onto $\phi(B)$ (since these sets lie inside the algebraic closures of I, J respectively). Since M is saturated, h can be extended to an automorphism f of M. Then $f(A)$ and B are two models with $f(A) \subseteq B$ and $\phi(f(A)) = \phi(B)$. So by the last lemma, $f(A)$ and B are isomorphic. Since A is isomorphic to $f(A)$, this proves Morley's theorem. $\square$

Tailpiece

We close by tying up some loose ends. If T is a complete theory, then an **extension of T by parameters** is a theory T' of the form $\text{Th}(A, \bar{a})$ where A is a model of T and $\bar{a}$ is a tuple of elements of A. We say that T' is **two-cardinal** if there is a formula $\phi(x)$ in the language of T such that in some model A of T', $|\phi(A)|$ is infinite but not equal to $|A|$. (This agrees with the definition in section 8.4, if we note that T' is a complete theory.)

Theorem 9.5.12. *Let T be a complete theory in a countable first-order language L, and suppose that T has infinite models. Then the following are equivalent:*

(a) T is uncountably categorical.

(b) T is totally transcendental, and no extension of T by parameters is two-cardinal.

Moreover if T has a strongly minimal one-cardinal formula in L, then (a) and (b) hold.

Proof. (a) $\Rightarrow$ (b): (a) implies that T is totally transcendental by Lemma 9.5.2; it implies the rest of (b) by Corollary 9.5.10.

(b) $\Rightarrow$ (a): Assume (b). Since T is totally transcendental, it has a prime model A_0. Also by Lemma 9.5.6 there is a strongly minimal formula $\phi(x)$; by Lemma 9.5.7 this formula can be chosen so that its parameters come from A_0, provided that infinity is definable in T.

We show that infinity is definable in T. For this we refer back to Theorem 8.4.2. Let $\phi(x, \bar{y})$ be a formula of L, $\bar{c}$ a tuple of new constants and L' the language got by adding $\bar{c}$ to L. Let U be the theory consisting of the sentences of L' of the form $\neg \forall x \theta$ where $\theta(x)$ is a layering by $\phi(x, \bar{c})$. We claim that the following theory has no models:

$$(5.18) \qquad U \cup T \cup \{\exists_{\geqslant n} x \phi(x, \bar{c}): n < \omega\}.$$

For if A is a model of (5.18), we can regard A as an elementary substructure of the big model M in which some tuple $\bar{a}$ interprets the constants $\bar{c}$. Let T'

be $\mathrm{Th}(M, \bar{a})$, so that T' is an extension of T by parameters. Since $\phi(M, \bar{a})$ is infinite but not two-cardinal, we have $(x = x) \leqslant \phi(x, \bar{c})$ in the notation of Theorem 8.4.2. So by that theorem, there is a layering $\theta(x)$ by $\phi(x, \bar{c})$ such that $M \vDash \forall x \theta$. This contradicts the assumption that A, and hence $(M, \bar{a})$, is a model of U. The claim is proved.

So by compactness there are some conjunction $\chi(\bar{c})$ of finitely many sentences in U, and some $n < \omega$, such that

$$(5.19) \qquad T \vdash \forall \bar{y}(\exists_{\geqslant n} x \phi(x, \bar{y}) \rightarrow \neg \chi(\bar{y})).$$

By Theorem 8.4.2, if $\neg\chi(\bar{b})$ holds in a model A of T then $|\phi(A, \bar{b})| = |A|$. So for any tuple $\bar{b}$ in any model A of T, the cardinality of $\phi(A, \bar{b})$ is either infinite or $< n$. This shows that infinity is definable in T.

From this point onwards, the proof that T is λ-categorical for every uncountable λ proceeds as before.

Finally when there is a strongly minimal one-cardinal formula in L, Corollary 9.3.7 implies that some extension of T by parameters is totally transcendental, and hence so is T. The rest of the argument is as for (b) $\Rightarrow$ (c), noting that the only reason we needed infinity to be definable was to ensure that the parameters of the strongly minimal formula lie in the prime model – which is trivially true when ϕ lies in L. $\qquad\qquad\square$

Further reading

The paper which gave us Morley's Theorem is a jewel of the logical literature. Almost every line in it contains a new idea which later model theorists have worked out in the more general setting of stability theory; and yet it is beautifully readable:

Morley, M. Categoricity in power. *Transactions of American Mathematical Society* **114** (1965), pp. 514–538.

Shelah developed Morley's insights and added his own. His account of his main findings is the Mount Everest of model theory:

Shelah, S. *Classification theory*, revised edition. Amsterdam: North-Holland, 1990.

Most readers will be thankful for a more user-friendly account. There are now several, and two which can be highly recommended are:

Lascar, D. *Stability in model theory*. Harlow: Longman Scientific & Technical, 1987.

Buechler, S. *Essential stability theory*. Perspectives in Mathematical Logic. Berlin: Springer-Verlag 1996.

The next two books are more advanced, at least in the sense that they discuss the more recent 'geometric' stability theory. This powerful theory, created by Boris Zil'ber and developed above all by Ehud Hrushovski, has been at centre stage since the early 1980s.

Pillay, A. *Geometric stability theory*, Oxford University Press 1996.

Zilber, B. *Uncountably categorical theories*. Providence RI: American Mathematical Society 1993.

INDEX TO SYMBOLS

$\equiv$ (elementarily equivalent) 39

$\equiv_L$ 39

$\equiv_{\infty\omega}$, $\equiv_{\kappa\omega}$ 39

$\equiv_0$ 75

$\Rightarrow_1$, $\Rightarrow_1^+$ 141

$\Rightarrow_2$ 144

$\preccurlyeq$ 50

$\preccurlyeq_n$ 146

$EF_n(A, B)$ (game) 75

$EF[A, B]$ (unnested game) 82

$\sim_\gamma$ 75

$\approx$ 82

Aut(A) (automorphism group) 94

G_X, $G_{\{X\}}$ (setwise stabiliser) 94

$G_{(X)}$ (pointwise stabiliser) 94

Set-theoretic

η_α (saturated dense linear orderings) 217

$\lambda \to (\mu)_\nu^k$ (partition relation) 153

See also the Note On Notation, p. ix.

Definable sets and classes

$\phi(A^n)$ (n-ary relation defined by formula
   ϕ) 23f

Mod(T) (class of models of theory T) 30

EC (first-order definable) 31

EC$_\Delta$ (first-order axiomatisable) 31

PC (pseudo-elementary) 104

PC$_\Delta$, PC$_\Delta'$ 104f

acl(X), acl$_L(X)$ (algebraic closure of set)
   138, 259

RM$(-)$, RM$_A(-)$ (Morley rank) 265

I_θ (class of elements of A^{eq} defined by θ)
   114

$\bar{a}/\phi$ (equivalence class of $\bar{a}$) 113

G° (connected component of group) 30

Permutations and automorphisms

1_A (identity automorphism) 6

Sym(Ω) (symmetric group) 94

INDEX

Some brackets have been added to determine the alphabetic listing. Thus 'ω-categorical' (without brackets) is listed as if it was 'omega-categorical', and '(λ)-categorical' (with brackets) is listed as if it was 'categorical'.